Spatial Data Quality

Spatial Data Quality

Edited by

Wenzhong Shi
The Hong Kong Polytechnic University

Peter F. Fisher
University of Leicester

Michael F. Goodchild
University of California

CRC Press
Taylor & Francis Group
Boca Raton London New York

CRC Press is an imprint of the
Taylor & Francis Group, an **informa** business
A TAYLOR & FRANCIS BOOK

CRC Press
Taylor & Francis Group
6000 Broken Sound Parkway NW, Suite 300
Boca Raton, FL 33487-2742

First issued in paperback 2019

© 2002 by Taylor & Francis Group, LLC
CRC Press is an imprint of Taylor & Francis Group, an Informa business

No claim to original U.S. Government works

ISBN-13: 978-0-367-39585-8

Publisher's Note
This book has been produced from camera-ready copy provided by the editors

British Library Cataloguing in Publication Data
A catalogue record for this book is available from the British Library

Library of Congress Cataloging in Publication Data
A catalogue record has been requested

**Visit the Taylor & Francis Web site at
http://www.taylorandfrancis.com**

**and the CRC Press Web site at
http://www.crcpress.com**

Contents

Contributors

The editors

Wenzhong Shi
Director, Advanced Research Centre for Spatial Information Technology, and Associate Professor, Department of Land Surveying and Geo-Informatics, The Hong Kong Polytechnic University. Research interests include GIS, spatial data quality, three dimensional and dynamic data modelling in GIS, design and development of GIS, integration of GIS and remote sensing, and feature extraction from remotely sensed images.

Department of Land Surveying and Geo-Informatics
The Hong Kong Polytechnic University
Hung Hom, Kowloon
Hong Kong
E-mail   lswzshi@polyu.edu.hk

Peter F. Fisher
Professor of Geographical Information. Research interests include GIS in general, visualization and uncertainty in particular.

Department of Geography
University of Leicester
Leicester
LE1 7RH
United Kingdom
E-mail   pff1@le.ac.uk

Michael F. Goodchild
Chair of the Executive Committee of the National Center for Geographic Information and Analysis (NCGIA) and Professor of Geography at the University of California, Santa Barbara. Research interests include GIS, environmental modelling, geographical data modelling, spatial analysis, location theory, accuracy of spatial databases and statistical geometry.

Department of Geography
University of California
Santa Barbara
CA 93106-4060
USA
E-mail   good@ncgia.ucsb.edu

The contributors

Henri J. G. L. Aalders
Professor in Topography. Research interests include GIS, GIS standardization, GIS quality, spatial access techniques, GIS application fields in large scale topography and cadastres.

Department of Geodesy, GIS Technology
Faculty of Civil Engineering and Geo Sciences
TU Delft
P.O.B. 5030
NL-5030 GA
The Netherlands Delft
E-mail   h.j.g.l.aalders@geo.tudelft.nl

Tinghua Ai
Associate Professor of the School of Resource & Environment Science, Wuhan University, Wuhan, China. Research interests include GIS, cartography, map generalization, spatial data modeling, spatial analysis, quality of spatial data.

School of Resource & Environment Science
Wuhan University
129 Luoyu Rd
Wuhan
430079
P. R. China
E-mail   aith@wuhan.cngb.com

Kate Beard
Professor and Chair, Department of Spatial Information Science and Engineering. Research interests include uncertainty in spatial databases, metadata and digital libraries.

University of Maine
348 Boardman Hall
Orono
Maine
04469-0110
USA
E-mail   beard@spatial.maine.edu

Thomas Bittner
Research interests include qualitative spatial reasoning, ontology, vagueness, granularity and uncertainty.

1890 Maple Ave. #300
Evanston
IL
60201

USA
E-mail   bittner@cs.northwestern.edu

Sébastien Brunet
Fellow researcher in the Scientific and Public Involvement in Risk Allocation Laboratory (SPIRAL) – University of Liège. Research interests include public decision-making, deliberative democracy, governance and risk.

University of Liège
B.31
bte 29
Sart Tilman
Belgium
E-mail   Sebastien.Brunet@ulg.ac.be

Peter A. Burrough
Director of the Utrecht Centre for Environment and Landscape Dynamics (UCEL) and Professor of Physical Geography at Utrecht University, the Netherlands. Research interests include GIS and environmental modelling, including geostatistics, applications of fuzzy logic and the development of dynamic modelling languages for environmental modelling; land degradation and erosion, soil and water quality.

Faculty of Geographical Science
Utrecht University
Post box 80 115
3508 TC Utrecht
The Netherlands
E-mail   p.burrough@geog.uu.nl

Zhongliang Cai
Research Assistant of the School of Resource & Environment Science, Wuhan University. Research interests include GIS, cartography, map generalization, spatial data modeling, multimedia electronic atlas and quality of spatial data.

School of Resource & Environment Science
Wuhan University
129 Luoyu Rd.
Wuhan
430079
P. R. China
E-mail   caisoft@public.wh.hb.cn

Wilhelm Caspary
Professor of Geodetic Engineering at the University of the Bundeswehr Munich. Research interests include GIS-data acquisition, data quality, location based mobile services and Geodesy-reference systems, positioning, mobile mapping, spatial statistics, data analysis.

Institute of Geodesy
University of the Bundeswehr Munich
D-85577 Neubiberg
Germany
E-mail   wilhelm.caspary@unibw-muenchen.de

Bernard Cornélis
Lecturer in Geomatics at the Université de Liège. Research interests include GIS,
spatial decision-making, climatology, environmental modelling and vulgarization
of scientific concepts.

Département de Géographie
Université de Liège
Rue Richard Orban 35
B-4257 Berloz
Belgique/Belgium
E-mail   Bernard.Cornelis@ulg.ac.be

Laure Dassonville
Manager of the cartographic business department. Ex-chair of the CERCO working
group on quality. Research interests include GIS and quality.

COGIT Laboratory
Institut Géographique National
2 avenue Pasteur
94160 Saint-Mandé
France
E-mail   Laure.Dassonville@ign.fr

Igor Drecki
Cartographer and Researcher. Research interests include geo-visualization,
uncertainty communication, spatial analysis, cartographic design and education.

Environment B·O·P
Quay Street
Whakatane
New Zealand
E-mail   igor@envbop.govt.nz

Matt Duckham
Doctor. Research interests include GIS and object-orientation, uncertainty in spatial
information, qualitative spatial reasoning, social context of GIS.

Department of Computer Science
University of Keele
Keele
Staffordshire
ST7 8AA

UK
E-mail   matt@cs.keele.ac.uk

Ernesto Gan
Technical Officer in Lands Department Survey and Mapping Office of The Government of the HKSAR. Research interests include GIS, accuracy of spatial databases, metadata, cartography, system development, marketing strategy, internet marketing, e-commerce and marketing communications with GIS.

Rm 2331
23/F
North Point Government Offices
North Point
Hong Kong
E-mail   egan@hutchcity.com

Guo Wei
Associate Professor of Fiberhome Telecommunication Technologies Limited. Research interests include GIS, spatial data modelling, telecommunication network management and spatial databases.

Software R & D Section
No. 88 Youkeyuan Road
Hongshan District
Wuhan
P.R.China
E-mail   guowei@fiberhome.com.cn

Gary J. Hunter
Associate Professor and Reader. Deputy-Head of Department. Research interests include GIS, managing spatial data quality and applying risk management techniques to understanding uncertainty in spatial decision making.

Department of Geomatics
University of Melbourne
Victoria 3010
Australia
E-mail   garyh@unimelb.edu.au

Antti Jakobsson
Chief Engineer in the National Land Survey of Finland, a post-graduate student in Helsinki University of Technology. Research interests include data quality, quality management and topographic data management.

National Land Survey of Finland
P.O.BOX 84
FIN-00521 Helsinki

Finland
E-mail   antti.jakobsson@nls.fi

Gerhard Joos
Senior scientist at the GIS laboratory (AGIS) of the University of the Bundeswehr. Research interests include GIS, location-based services, interoperability, quality, standardization and geodata fusion.

AGIS
Universität der Bundeswehr München
D-85577 Neubiberg
Germany
E-mail   Gerhard.Joos@unibw-muenchen.de

Ho Chung Law
Doctor. Research interests include geographic information science and operational research.

Research Development & Statistics Directorate
Home Office
50 Queen Anne's Gate
London
SW1H 9AT
UK
E-mail   Ho.law@homeoffice.gsi.gov.uk

Claude Luzet
EuroGeographics Executive Director. Member of the Steering Committee of GSDI (Global Spatial Data Infrastructure) and of ISCGM (International Steering Committee for Global Mapping). Research interests include quality of spatial reference data and data quality management. With particular focus on issues raised by cross-sector, cross-border, cross-culture, cross-language interoperability and harmonization.

6-8 avenue Blaise Pascal
77455 Marne-la-Vallée CEDEX 2
France
E-mail   Claude.luzet@eurogeographics.org

Ewan Masters
Senior Lecturer, School of Geomatic Engineering, The University of New South Wales. Research interests include spatial information, accuracy of spatial databases, data quality management and global positioning system.

School of Geomatic Engineering
The University of New South Wales
Sydney
NSW 2052

Australia
E-mail   e.masters@unsw.edu.au

John E. McCreadie
Sales & Marketing Director. Research interests include various general areas within
the field of geo-spatial engineering

SDS
3 Hope Street
Bo'ness
West Lothian
EH51 0AA
Scotland
E-mail   jem@sds.co.uk

Roger Merritt
Research interests include upgrading of Cadastre/parcel maps using survey accurate
measurements, parametric least squares adjustments, and measurement databases.
Upgrading of layers which are spatially dependent on the Cadastre, such as utilities
and annotation layers.

2 Cherana Pl
Kareela
Sydney
New South Wales
Australia, 2232
E-mail   Roger@SpatialWeb.com.au

Jibo Qiu
PhD student. Research interests include GIS, modelling and updating spatial data
quality information.

Department of Geomatics
University of Melbourne
Victoria 3010
Australia
E-mail   jqi@sunrise.sli.unimelb.edu.au

Karin Reinke
PhD student. Research interests include GIS and communicating spatial data
quality.

Department of Geomatics
University of Melbourne
Victoria 3010
Australia
E-mail   karjr@sunburn.sli.unimelb.edu.au

François Vauglin
Research interests include spatial data quality, spatial accuracy, resolution, multiple representations, data fusion and multiscale.

COGIT Laboratory
Institut Géographique National
2 avenue Pasteur
94160 Saint-Mandé
France
E-mail   Francois.Vauglin@ign.fr

Michael J. C. Weir
Lecturer and Programme Director of Natural Resources Management, International Institute for Aerospace Survey & Earth Sciences (ITC). Research interests include spatial data acquisition for natural resource management, GIS data quality and remote sensing.

International Institute for Aerospace Survey & Earth Sciences (ITC)
P.O. Box 6
7500 AA Enschede
The Netherlands
E-mail   weir@itc.nl

Stephan Winter
Assistant Professor. Research interests include GI science, topology, computational geometry, uncertainty, GI services and navigation.

Institute for Geoinformation
Technical University Vienna
Gusshausstrasse 27-29
1040 Vienna
Austria
E-mail   winter@geoinfo.tuwien.ac.at

Thomas K. Windholz
Postdoctoral Researcher at the University of Maine. Research interests include GIS, uncertainty modelling and spatial analysis.

Department of Spatial Information Science and Engineering
University of Maine
Orono
ME 04469
USA
E-mail   twindhol@spatial.maine.edu

Preface

Quality is an issue for any kind of information. One wants to know that bus timetables are accurate, that aircraft positions are accurately measured by radar, that weather maps reflect accurate measurement and prediction of temperature, that people's names are spelled correctly in telephone directories, and so on. Information is increasingly the basis of modern society, of its economic well-being, and of the institutions that govern our actions. In all of these respects the quality of information is a major determinant of efficiency and efficacy. If modern society cannot depend on information to be accurate, then many of its activities quickly become impossible. Unfortunately even the highest quality information may still have deficiencies. Some information may never be completely accurate. In this case people do not require any assurance that the information is accurate, but they need to be able to assess the consequences of the inaccuracy and uncertainty for their use of the information.

This book is about the quality of a particular type of information. The term *spatial* is used in the title as synonymous with *geographical* (and also with the modern coinage *geospatial*), and in the sense of this book spatial data consist of information about the surface and near-surface of the Earth. Such information is used today for many purposes, including the management of natural resources, war-fighting, navigation, the maintenance of utility networks, agriculture, the monitoring of disease outbreaks, and a host of other everyday activities. In all of these cases it is obvious that data must be of high quality, but it is not at all clear why the fact that the data are spatial prompts any kind of special concern. Yet over the past two decades a large research literature has emerged on the subject of spatial data quality. Why, then, are spatial data subject to particular problems regarding quality?

First, spatial data must of necessity be approximate, since it is impossible to capture all of the infinite complexity of the Earth's surface in any form, whether as a map, a globe, a digital database, or a narrative. This issue is not unique to spatial data, of course, but its validity is especially obvious when one considers the role of scale in spatial data. Scale is defined for a map as the ratio of distance on the map to distance on the ground, and it determines the level of detail captured by the map. Its equivalent for a digital database is termed *spatial resolution*, and relates directly to the size of the smallest features recorded in the database. Thus a highly detailed map is clearly a better approximation to the Earth's surface than a less detailed one, and scale and spatial resolution are important components of quality. For example, an airline reservation system can and does maintain records of every reservation, and in principle can be of perfect quality; but a spatial database cannot record every feature on the Earth's surface.

Second, the creation of a map or a spatial database is a long and complex process, involving many stages of acquisition, interpretation, analysis, and generalization. Unlike a simple scientific measurement of a property like temperature, the process by which an elevation appears on a map may involve many measurements, and some degree of subjectivity. Inaccurate processes, such as the misregistration of aerial photographs, produce errors that appear simultaneously in many parts of the map. Other processes, particularly ones involving human interpretation, may be undocumented and unreplicable. In short, the lineage of a spatial database is rarely fully known, and conventional methods of analyzing and modeling error are far too simple to be applicable to spatial data quality.

Third, many types of spatial data involve human judgment. A soil scientist must use judgment in creating a map of soils, and a landscape ecologist must similarly use judgment in drawing boundaries on maps between different classes of vegetation. Two scientists may not agree on where to place the boundaries, or how to assign the classes, even though they may be highly respected and experienced. Thus the concept of accuracy, which implies the existence of a truth that can be approached by better and more precise measurement, simply does not apply to many types of spatial data. Instead, truth must sometimes be regarded as a vague concept that is in principle unattainable.

Finally, spatial databases record the locations and attributes of objects on the Earth's surface. Many of these objects will be well defined, such that one person can successfully and precisely locate the object based on another's description of its characteristics and location. This would be true, for example, of buildings, survey monuments, streets, and many other human modifications to the natural landscape. But other features, such as mountains, coastlines, or valleys, are artifacts of human interpretation. Two people might well not agree on the precise boundaries of a mountain, or the existence of a particular swamp, or the number of lakes in a specific state. In such instances, the objects that appear so precisely and crisply in a spatial database simply do not exist on the ground with the degree of reliability of a building, or a survey monument.

Many efforts have been made over the past decades to define spatial data quality, and to identify its core elements. The U.S. Spatial Data Transfer Standard (1994) defines five components of spatial data quality—positional accuracy, attribute accuracy, lineage, logical consistency, and currency—and its approach has been adopted in many areas. A compendium edited by Guptill and Morrison (1995) provides an excellent discussion of the various components, suggests some extensions, and describes the ways in which each element can be characterized. Users of the U.S. Federal Geographic Data Committee's Content Standard for Digital Geospatial Metadata are expected to describe the quality of a data set in terms of the five components, and this approach is being mirrored in new ISO (1998) standards that are under development by Technical Committee 211.

Much effort has also gone into developing reliable methods of accuracy assessment, especially in areas such as resources management where it is important to know the accuracy of forest stand maps, maps of vegetation cover, and other information essential to the management of the forest resource. Several books have appeared on this topic (Leung, 1988; Goodchild and Gopal, 1989; Hunter, 1991; Burrough and Frank, 1996; Goodchild and Jeansoulin, 1998; Heuvelink, 1998; Lowell and Jaton, 1999; Mowrer and Congalton, 2000).

Many other research threads related to data quality are explored in this book. The first section looks at efforts to advance the theoretical framework for spatial data quality, in the form of models of quality, links to existing areas such as geostatistics that can provide ready-made techniques and approaches, and designs for spatial databases that incorporate knowledge of data quality and directly address problems of its analysis and impacts. The second section looks in detail at a selection of methods for handling spatial data quality, with discussions that range from new techniques for visualization, to the perspectives of policy-makers. The final section has two themes: case studies of data quality as metadata, and assessments of the consequences of data quality on specific applications. Each section is introduced by one of the editors, and the book ends with an epilog.

As a collection of edited research papers and introductions, the book attempts to create a sense of the current state of the art in spatial data quality, and the directions currently being pursued by the research community. As such it is directed at two communities of readers: first, researchers, graduate students, and advanced undergraduates interested in spatial data quality issues and in gaining an understanding of current and potential research directions; and second, practitioners and software developers interested in implementing the state of the art in practice, or in providing the tools that will enable such implementation.

The chapters grew out of an International Symposium on Spatial Data Quality held in Hong Kong in the summer of 1999, and each has been extensively modified and edited to suit the purposes of the book. The editors wish to acknowledge the efforts of the authors, and also of the many other researchers whose work has contributed to the advancement of our understanding of spatial data quality over the past decades. The list is far too long to reproduce here, since it includes many mentors, colleagues, former students, and collaborators, and in most cases their contributions are evident in the citations at the end of each chapter.

Above all, however, we wish to thank all those people who made the Symposium a success, Geoffrey Shea, Tracy Cheung, Matthew Pang, Mannars Chan, all the members of Program Committee and Organizing Committee, and the student helpers. The appreciation also goes to the sponsors and supporters of the Symposium. And Edith Chan, Chun Liu and Ahmed Shaker who have assisted in the task of bringing together the contributions to this book.

Michael F. Goodchild
Peter F. Fisher
Wenzhong Shi

REFERENCES

Burrough, P.A. and Frank, A., 1996, *Geographic Objects with Indeterminate Boundaries*, (London: Taylor & Francis).

Goodchild, M. F and Gopal, S., 1989, *Accuracy of Spatial Databases*, (London: Taylor & Francis).

Goodchild, M. F. and Jeansoulin, R., 1998, *Data Quality in Geographic Information, From Error to Uncertainty*, (Paris: Hermes).

Guptill, S.C. and Morrison, J.L., 1995, *Elements of Spatial Data Quality*, (Oxford: Elsevier), p.203.

Heuvelink, G., 1998, *Error Propagation in Environmental Modelling with GIS*, (London: Taylor & Francis).

Hunter, G.J., 1991, *Proceedings of a Symposium on Spatial Database Accuracy*, Department of Surveying and Land Information, University of Melbourne.

ISO / TC211 1998, URL: http://www.statkart.no/isotc211/

Jeansoulin, R. and Goodchild, M.F., 1998, *Data Quality in Geographic Information: From Error to Uncertainty,* (Paris: Hermes).

Leung, Y., 1988, *Spatial Analysis and Planning under Imprecision,*(New York: Elsevier).

Lowell, K. and Jaton, A., 1999, *Spatial Accuracy Assessment: Land Information Uncertainty in Natural Resources*, (Michigan: Ann Arbour Press).

Mowrer, H.T. and Congalton, R.G., 2000, *Quantifying Spatial Uncertainty in Natural Resources,* (Chelsea, Michigan: Ann Arbor Press).

SDTS National Institute of Standards and Technology, 1994, *Federal Information Processing Standard Publication 173.* (Spatial Data Transfer Standard Part 1.Version 1.1) U.S Department of Commerce.

Acknowledgements

Peter A. Burrough would like to thank the Survey Department of the Netherlands Directorate for Public Works and Water Management for providing Laser altimetry data for Figure 2.7. The data are part of the Actueel Hoogtebestand Nederland (AHN). (See http:// www.ahn.nl or http://www.geo-loket.nl)

Bernard Cornélis and Sébastien Brunet would like to thank the Université de Liège Research Council for its support and encouragement to the transdisciplinary investigations.

Laure Dassonville, François Vauglin, Antti Jakobsson and Claude Luzet wish to acknowledge the experts of the CERCO Working Group on Quality and especially François Salgé who launched the working group on quality as the first chairman, Pascal Fasquel who has been its reporter, Jean Poulit chairman of the CERCO and John Leonard, the CERCO General Secretary, who gave their support for writing their chapter. They also acknowledge reporters of the sub-working groups, Antonio Rodriguez and Dave Sharman.

Matt Duckham and John E. McCreadie acknowledge that their work was supported by NERC and SDS U.K. under a CASE award at the University of Glasgow. Gothic software and support was generously supplied under a development license by Laser-Scan, U.K. Finally, the help, interest and openness of the employees of Kingston Communications and Informed Solutions is acknowledged with thanks. The author is currently funded by the EPSRC at the University of Keele under grant GR/M 56685 "Managing vagueness, uncertainty and granularity in spatial information systems".

Antti Jakobsson wishes to acknowledge the help given by the experts in ISO TC 211 working group that developed the forthcoming standards, especially editors Ms. Leslie Godwin (quality principles), Mr. Frederick Broome and Dr. Ryosuke Shibasaki (quality evaluation procedures). The author is also grateful for the information he gained from visits to the IGN and from the working group on quality in EuroGeographics/CERCO (Comité Européen des Responsables de la Cartographie Officielle), which is an association of European National Mapping Agencies.

Ho Chungh Law and Peter F. Fisher would like to state that the Safer Cities Programme Evaluation was one of the projects within the research programme of the Home Office Research Development & Statistic Directorate between 1988-1995. They wish to acknowledge Dr Paul Ekblom of the Home Office for his support and encouragement throughout their project.

Jibo Qiu and Gary J. Hunter wish to acknowledge the support received for their research through Australian Research Council Large Grant No. A49601183 titled "Modelling Uncertainty in Spatial Databases".

Karin Reinke and Gary J. Hunter wish to acknowledge the support received for their research through Australian Research Council Large Grant No. A49601183 titled "Modelling Uncertainty in Spatial Databases".

Wenzhong Shi and Wei Guo acknowledge the support from the Research Grants Council of the Hong Kong SAR (Project No. PolyU 5050) and The Hong Kong Polytechnic University.

Thomas K. Windholz, Kate M. Beard and Michael F. Goodchild wish to thank Gerard Heuvelink for his valuable inputs to focus on the essential parts of the problem. Furthermore, he provided some interesting guidelines for future work on the model in their chapter. They gratefully acknowledge NIMA for support of their research under grant NMA202-97-1-1021.

Introduction to Part I: Theoretical Models for Uncertain GIS

Michael F. Goodchild

Statistics provides one of the most general theoretical frameworks for discussion of uncertainty, and interest in an approach based on simple models of probability has a long history. For example, uncertainty about the outcome of a single toss of a six-sided dice can be formalized within the framework of discrete statistics, using a simple model in which each of the six possible outcomes is assigned an equal probability of 1/6. Uncertainty about a single measurement on a continuous scale can be addressed in the framework of continuous statistical distributions, notably the Gaussian or normal distribution. The theory of measurement errors conceptualizes error as a form of uncertainty (for example, a single measurement of temperature made with a thermometer with an accuracy of 1 degree Celsius would leave the user of the measurement uncertain by the same amount about the actual or true temperature).

But although they are useful as foundations, such simple models prove to be far from comprehensive with respect to uncertainties in maps, images, and geographical databases. First, these forms of uncertainty are far too complex to be conceptualized in terms of discrete outcomes, or measurement error. For many types of geographical data there is no clear concept of *truth*, so models that address the differences between measurements and the truth are clearly inappropriate. Instead, many geographical terms and concepts are poorly defined, such that uncertainty exists in the concepts themselves. It is unlikely, for example, that two professional soil scientists operating independently would produce the same, correct map of an area, however extensive their training. Second, geographical data is often derivative, in the sense that many stages of processing, interpretation, compilation, and transformation occurred between the initial stage of measurement and the final product. These stages are generally not documented and not replicable. As a result, uncertainties and errors in the final product tend to have common sources, and to be strongly correlated. Thus the theory of error, which generally assumes that measurement errors are independent, is invalid.

In fact the errors present in geographical data are of such magnitude that if correlations were not present the data would be far less useful. For example, slope is often estimated by comparing adjacent elevation estimates in digital elevation models that are subject to substantial measurement errors. If the errors in adjacent elevation estimates were independent, the uncertainties associated with the slope estimates would be unacceptable. But because adjacent elevation estimates tend to be subject to strongly (and positively) correlated errors, uncertainties in slope estimates are surprisingly small.

The six chapters in this part all address theoretical models, and in total they provide a comprehensive cross-section of the models that researchers have found useful in this context. As befits any contribution to geographical information science, they run the full gamut from models of specific types of uncertainty, to models of entire geographical information systems, and models of the process by which people communicate their knowledge of the planet's surface using digital technology. The next section provides a brief introduction to the six chapters and the relationships between them, and this introduction ends with a brief discussion of the sufficiency of the set, and the possibility of other theoretical models.

PART OVERVIEW

Early work on errors in maps dates from well before the advent of geographic information systems (GIS). By 1980, for example, the work of Maling (1989) and others had provided a solid foundation for theorizing about the accuracy of measurements made from maps, that went well beyond the limitations of standard statistics. But the theory of regionalized variables initially developed by Matheron (1971) and others, and now known as *geostatistics*, has proven to be a very rich source of ideas and theoretical concepts with which to address spatial data quality. The GIS community became seriously engaged with geostatistics in the 1990s, partly through a key editorial by Journel (1996) in the field's most influential journal, and partly through the work of Peter Burrough, whose GIS text (Burrough, 1986) contained a substantive and readily understood introduction.

It is highly appropriate, therefore, that Chapter 2 of this book provides a current, comprehensive, and insightful overview of the importance of geostatistics and its applications. In it, Burrough introduces all of the main concepts of geostatistics, interprets them in the context of spatial data quality, and provides a series of compelling examples of geostatistical applications to practical problems.

Geostatistics addresses uncertainties in representations of phenomena conceptualized as *fields*, that is, variables that are functions of geographical location. Fields are the more appropriate of the two geographical conceptualizations for phenomena that are continuous, where each location in space is associated with a single value of the variable. They are used to conceptualize geographical variation in atmospheric temperature and pressure; land ownership, cover class and use; population density; and a host of other phenomena. Some phenomena are represented as fields of variables measured on continuous scales, and subject to measurement error; others are represented as fields of classifications, where the variable's value must equal one of a number of defined classes; and others are represented as fields of names, as in the case of land ownership or the field of *county* or *nation*.

Other phenomena, however, are more appropriately conceptualized using the other framework of *discrete objects* embedded in an otherwise empty space. Discrete objects are countable, and maintain their integrity if they move (or are edited in geographical information systems). Animals clearly fit this model, as do vehicles, and other easily identified and manipulable objects, such as street signs, or trees. But other phenomena fall between the two alternatives, exhibiting behaviors and properties that are better conceptualized as field-like for some purposes, but as discrete-object-like for others. Many loosely defined geographical features have these characteristics, including mountains (countable for some purposes, but better addressed as parts of a continuous elevation surface for others), lakes, and many natural geographical features. Road and river networks are particularly problematic, since to be considered as collections of discrete objects somewhat arbitrary decisions must be made about the limits of each object. Often this is done by breaking the network at intersections, but this approach makes it difficult to describe properties that vary continuously over the network, such as stream flow rate, or road pavement quality.

Geostatistics is clearly inappropriate as a model for uncertainty in discrete objects. Instead, a variety of approaches have been proposed based on concepts of fuzzy sets, rough sets, and logical inference. In Chapter 3, Stephan Winter and Thomas Bittner make a substantial contribution to this literature by examining formal methods for reasoning about uncertain regions. The differences between the two approaches are obvious, and grounded in entirely different theoretical traditions.

One of the advantages of adopting a probabilistic framework is that it opens a vast resource of mathematical structures, including long-discovered distributions. Although the standard models of statistics rarely describe geographical uncertainty precisely, they can be readily adapted and extended for this purpose. Surveyors have long used models derived from the Gaussian distribution to describe uncertainties in position, and in Chapter 4 Wenzhong Shi and Wei Guo extend this tradition by examining the modeling of uncertainty in topological relationships. They build a suitable model of underlying geometric uncertainty, and are able to obtain useful results concerning qualitative relationships between objects. Much more of this kind of work is needed if a full suite of methods for describing uncertainty in geometric and topological operations in GIS is to be constructed.

This part contains two chapters that attempt to integrate understanding of uncertainty into new designs for geographical information systems. In Chapter 1, I ask whether the original GIS design of the 1960s was appropriate given contemporary concerns for spatial data quality and the research results of the past decade. The answer, not surprisingly, is no, and I lay a foundation of ideas for what I term a *measurement-based* GIS, to distinguish it from the traditional *coordinate-based* GIS. This concept has a long history in surveying, but has rarely been implemented in GIS.

In Chapter 5, Matt Duckham and John McCreadie describe another, more comprehensive approach to the same problem. They make use of modern ideas of object orientation, which emerged as a strong theme in computer science in the 1980s, and apply them to the design of an error-sensitive GIS.

Finally, in Chapter 6 Karin Reinke and Gary J. Hunter take an even more holistic view of the uncertainty problem, by presenting a framework for addressing uncertainty within a communication model of GIS. In recent years, and particularly following the advent of the World Wide Web Internet application, it has become clear that the emphasis in GIS has shifted from one in which the technology performed a specific role for a user, to one in which it acts as a communication channel between users. GIS began in the 1960s as a tool for performing tasks which individual users found difficult to execute by hand, but this paradigm has been substantially replaced by another in which the technology acts as a means whereby one person supplies another with information about the Earth's surface. The authors apply this same concept to the issue of spatial data quality, and construct a multi-stage framework in which they address questions of how information on uncertainty can be transmitted effectively. This is perhaps the most compelling of the outstanding problems associated with spatial data quality, because it holds the key to making the rapidly increasing population of GIS users, including the general public and children, aware of the uncertainties that inevitably exist in the data input to a GIS and the results that are obtained from it.

SUMMARY

This part of six chapters provides a cross-section of theoretical frameworks, and an introduction to the methodological and applications-oriented chapters of the later sections. It seems important to ask whether the cross-section is complete, or whether any substantially useful framework has been omitted.

Some of the later chapters rely on theoretical frameworks that are not discussed in this part. For example, Chapter 10 addresses visualization, and Chapter 12 examines decision-making, but neither of these areas are addressed in this part as of theoretical frameworks, because both are far more general than the context of spatial data quality. Chapter 19 addresses metadata, but similarly the theory behind generalization and abstraction is not sufficiently specific to spatial data quality to be the subject of a chapter in this part.

In recent years geographical information scientists have discovered several important literatures that have obvious relevance to spatial data quality. Rough sets and fuzzy sets are compelling examples, and although these approaches are mentioned in this part we have not included extensive discussion, partly because to date there have been no substantial theoretical extensions in the geographical context. However, it seems likely that such extensions will be needed, and that progress will be made in the not too distant future.

REFERENCES

Burrough, P.A., 1986, *Principles of Geographical Information Systems for Land Resources Assessment* (Oxford: Oxford University Press).

Journel, A.G., 1996, Modeling uncertainty and spatial dependence: stochastic imaging. *International Journal of Geographical Information Systems*, **10(5)**, pp. 517–522.

Maling, D.H., 1989, *Measurements from Maps: Principles and Methods of Cartometry* (New York: Pergamon).

Matheron, G., 1971, *The Theory of Regionalized Variables and Its Applications* (Paris: Ecole National Superieure des Mines).

CHAPTER ONE

Measurement-based GIS

Michael F. Goodchild

ABSTRACT

GIS designs have evolved over the past 30 years, and once adopted a design tends to persist as a legacy despite progress in fundamental research. A feature of most GIS designs is the representation of position by derived coordinates, rather than by original measurements. In such coordinate-based GIS it is impossible to apply traditional error analysis, or to estimate uncertainties in derived products, or to make partial adjustments to the observations affected by a given measurement error, and to propagate their effects correctly. Thus concern for accuracy issues forces a rethinking of fundamental GIS design. Measurement-based GIS is defined as retaining details of measurements, such that error analysis is possible, and such that corrections to positions can be appropriately propagated through the database. It is shown that measurement-based designs have major economic advantages, in addition to supporting a more comprehensive approach to uncertainty.

1.1 INTRODUCTION

The earliest geographical information system (GIS) was designed and developed in the 1960s, and since then much progress has been made in developing tools for working with digital geographical information, and in bringing these tools to a wider audience through commercialisation (for reviews of the history of GIS see Coppock and Rhind, 1991; Foresman, 1998). Today, GIS has become a widely available approach to the solution of a large number of problems, from Earth science to local decision-making.

The literature on accuracy and uncertainty in GIS is much more recent. Although Maling and others were writing about analytic cartography from a statistical perspective in the 1960s and 1970s (Maling, 1989), and although the field of geometric probability traces its roots back to the work of Buffon and others in the 18th Century, the first reviews of these issues and their importance for GIS date only from the 1980s (see Burrough, 1986; Goodchild and Gopal, 1989). GIS designers have many options to choose from, since there are many ways of representing the same geographical phenomena in digital form. So if such design decisions were being made as much as two decades before the first discussions of accuracy, an interesting question arises: were the early design decisions appropriate, or should they be re-examined in the light of newer concerns for uncertainty, and if so, what design decisions are affected?

Burrough and Frank (1996) have already identified one area that casts light on this issue: the representation of objects with uncertain boundaries. It is clearly difficult to characterise such objects entirely in the vector domain, since there are no simple vector-based models of uncertainty in the position of a boundary or of heterogeneity of the object's contained area, although use has been made of simple descriptive statistics such as the epsilon band (Mark and Csillag, 1989). Instead, it is necessary to switch into a raster representation (or more generally, a field-based view), so that every pixel can be assigned some measure of belonging, either to a general class or to a specific object. Goodchild (1989) provides an early discussion of the field/object dichotomy from this perspective.

The purpose of this chapter is to discuss and explore another instance of the same general problem. Although maps and geographical databases are ultimately constructed from measurements and observations, it is common for such data to be processed, interpreted and interpolated in creating a final product. Moreover, such early data are commonly not retained—a typical soil map, for example, contains no representation of the original observations from which the map was compiled. In this chapter I explore the consequences of this fundamental design decision, within the context of uncertainty. I show that the decision was flawed, because it severely constrains the value of spatial databases to their users, particularly when the impacts of uncertainty must be assessed, or when the opportunity exists for update or the reduction of uncertainty. I propose an alternative design, termed *measurement-based* GIS to distinguish it from the traditional *coordinate-based* GIS. The chapter shows how it avoids some of these problems, and explores some of its details and implications.

This central idea is not entirely new. The field of adjustment in surveying is a well-developed area of theory connecting uncertainty in measurements to uncertainty in compiled maps. Some years ago Kjerne and Dueker (1988) showed how object-oriented ideas could be exploited to code these connections. Buyong and Frank (1989), Buyong *et al.* (1991), and Buyong and Kuhn (1992) have also written about the inheritance of measurement uncertainty, and Duckham's chapter 5 also addresses this issue in the context of modern object-oriented design. In this chapter I have attempted to generalise some of these ideas, to place them in a broader framework, and to explore some of their wider implications.

1.2 MEASUREMENTS AND COORDINATES

1.2.1 Geographical information defined

The fundamental atom of geographical information is the tuple $<x,z>$, which links a spatiotemporal location x (time may be present or absent, depending on whether the information is time-dependent) to a set of attributes z, drawn from the things that can be known about a location—its temperature, soil type, county, owner, the name of a geographical feature type known to be present at the location, *etc.* Because space and time are continuous, the creation of a spatially continuous map or image of an area would require an infinite number of tuples, even if it were limited to representing a single instant in time. In practice, we resort to numerous schemes for reducing or compressing what would otherwise be an infinite set of tuples, by ignoring areas that are outside the area of geographical coverage, or areas that are empty with respect to the topic of interest, or through various forms of generalisation and abstraction. For example, we identify a set of locations with a region, represent the region as a polygon formed by a finite number of vertices, and assign attributes to the region. Goodchild *et al.* (1999) review many of these methods.

In this chapter I focus on **x**, and issues of positional uncertainty, though much of the discussion also applies to **z**. I ignore also the issue of separability: whether it is possible to distinguish uncertainty in **x** from uncertainty in **z** (can one distinguish between correct attributes of the wrong location and wrong attributes of the correct location?).

The representation of geographical location (the spatial dimensions of **x**) is almost always absolute in traditional GIS—that is, location is with respect to the absolute Earth frame, through geographical coordinates (latitude and longitude) or some convenient planar coordinate system, such as UTM (Universal Transverse Mercator). Thus a GIS is able to associate attributes with geographical locations, and service queries of the form "Where is **z**?", or "What is at **x**?". Without absolute location, it would be impossible to integrate different databases by location, a function that is often claimed to be one of GIS's greatest strengths (note, however, that this does not imply that all locations in the data structure are in absolute form, only that absolute location can be determined as a service of the GIS).

In this chapter I distinguish two bases for determination of **x**—those in which **x** is *measured* directly, using global positioning systems (GPS) or geometric techniques, and those in which **x** is *interpolated* between measured locations. The latter occurs, for example, when the position of some feature recognisable on an aerial photograph is established with respect to registered tics or control points. It also occurs when a surveyor establishes the location of a boundary by linking two surveyed monuments with a mathematically straight line.

Let the set of measurements required to establish a measured location be denoted by **m**, and let the function linking these measurements to the location be denoted by f, that is, $\mathbf{x} = f(\mathbf{m})$. The inverse of f is denoted by f^{-1}, that is, the function that allows measurements to be determined from locations. Note that I do not wish to imply that the transformation of measurements into positions is always a simple mathematical function; rather, f denotes what in practice may be a complex series of transformations involving several stages of human intervention. In what follows, this expression is also used to describe the derivation of an array of locations from a complex set of measurements.

1.2.2 The theory of measurement error

Errors in scientific measurements are commonly analysed using a simple theory based on the Gaussian distribution. In this section I review the basic elements of this theory, although it will be well known to many readers, because it provides an essential basis for what follows.

Suppose that some scalar measurement, such as a measurement of temperature using a thermometer, is distorted by an error generated by the measuring instrument. The apparent value of temperature x' can be represented as the sum of a true value x and a distortion δx. If some manipulation of x is required, the theory of measurement error provides a simple basis for estimating how error in x will propagate through the manipulation, and thus for estimating error in the products of manipulation (Taylor, 1982; and see Heuvelink, 1998, and Heuvelink *et al.*, 1989, for discussions of this in the context of GIS). Suppose that the manipulation is a simple squaring, $y = x^2$, and write δy as the distortion that results. Then:

$$y + \delta y = (x + \delta x)^2$$

$$y + \delta y = x^2 + 2x\delta x + \text{terms of order } \delta x^2$$

Ignoring higher-order terms, we have:

$$\delta y = 2x\delta x$$

More generally, given a measure of uncertainty in x such as its standard error σ_x, the uncertainty in some $y = f(x)$, denoted by σ_y, is given by:

$$\sigma_y = df/dx \, \sigma_x$$

The analysis can be readily extended to the multivariate case and the associated partial derivatives.

1.2.3 Errors in position

Suppose that position has been distorted by error, such that the observed location x' is distorted by a vector $\varepsilon(\mathbf{x})$ that is a function of location. Kiiveri (1997) and Hunter and Goodchild (1996) have discussed this model, and the conditions that must be imposed on $\varepsilon(\mathbf{x})$ to ensure that normal conditions are not violated—that the space is not torn or folded, ensuring that its basic topological properties are preserved. We also typically assume that $\varepsilon(\mathbf{x})$ varies smoothly in space, with continuity and strong spatial autocorrelation, in order to permit locations to be interpolated with reasonable accuracy, and to allow the use of rubber-sheeting methods in registration. That is, we assume:

$$\varepsilon(\mathbf{x} + \delta\mathbf{x}) - \varepsilon(\mathbf{x}) \text{ tends to } 0 \text{ as } \delta\mathbf{x} \text{ tends to } 0$$

and that strong covariances exist among ε at different locations.

In practice a single database may contain objects with many different lineages. If two objects occupy the same location, it does not follow that ε is the same for both objects. Instead, it may be necessary to model many different error fields, and to associate each object or even parts of objects with distinct fields. The implications of this are discussed in the next section.

1.2.4 Relative and absolute accuracy

In practice, it is common to distinguish two forms of positional error, though only informally. In this section I attempt to formalise their definitions, in the interests of clarity, and in order to understand their relevance to the central issue.

Consider two locations x_1 and x_2, and suppose that distance must be measured between them. The error in the distance will be determined by the variance–covariance matrix of their positional errors. If $\varepsilon(x_1)=\varepsilon(x_2)$, in other words perfect correlation exists between the two errors, then covariances will equal the products of the square roots of the respective variances, and the error in distance will be 0. But if correlation is zero (errors are independent), then covariances will be zero, and the error in distance will show the combined effects of both positional errors. *Absolute* error is defined for a single location as $\varepsilon(x)$. *Relative* error is defined only for pairs of points, and describes error in the determination of distance. Moreover, a continuum of levels of relative error exist depending on the degree of correlation between the two positional errors. In principle it is possible for negative correlations to exist, such that relative error can exceed the errors inherent in the independent case, but in practice we suspect that correlations are almost always non-negative. Since relative and absolute error are not commensurate, one being a function of two locations and the other of one, and since one term, *relative*, describes an entire continuum, the dichotomy does not seem to provide much basis for formal treatment.

Conceptually, however, the distinction may help. Consider the objects shown in Figure 1.1. Suppose the building's form is described by four vertices, generated by interpretation of an aerial photograph and the use of a standard template that enforces parallel edges and rectangular corners. Assume that the building might have been located in various positions, described by an error model. The error

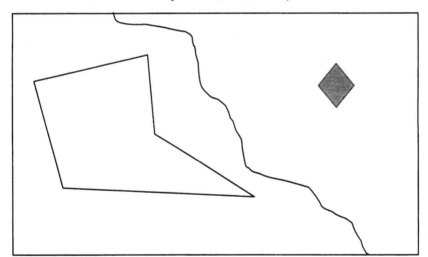

Figure 1.1 A parcel of land, a river, and a building. Each object and all of their parts are subject to distortion due to errors in positioning, but complex covariances exist between these errors. Because it was created using a template, the building behaves under positional distortion as a rigid frame able to move and rotate, but not to change shape.

model might be used to simulate equally likely observed locations (Openshaw, 1989). Because of correlations among the errors, it is easiest to think of the entire ensemble as a single sample from a population of equally likely realisations of the entire ensemble, rather than as a collection of error models for each individual object or vertex.

In practice, because a template was used there will be very strong correlations between the errors distorting the positions of the four vertices of the building. Thus the model would have only three degrees of freedom—for example, distortions of one vertex in two spatial dimensions, plus distortion by rotation of the building about that vertex as a rigid body. It is possible to think of positional error in terms of the movements of objects and their parts that result from resampling of the error distribution. Some objects will change shape under resampling, if the larger errors occurred during independent determination of the locations of their vertices, while others will remain rigid. Some objects will move independently of each other, if their positions were determined by independent processes, whereas others that were produced by the same process and share substantial lineage will move in concert. Such objects may retain some degree of correctness in their relative positions while their absolute positions change. Thus a display of many distinct realisations of the ensemble error model in rapid succession will convey a fairly accurate impression of the error correlation structure.

1.2.5 Adjustment and update

In principle, the variance–covariance matrix of positional errors in an ensemble of locations can be derived from knowledge of the characteristics of measurement errors, through extensions of the theoretical framework outlined above. Moreover, if the actual error present in a point's location can be determined, it should be possible to correct the associated measurements, and then to adjust the locations of other points appropriately. Of course, if all covariances are zero then the point's location can be corrected independently. But in the normal case of strong covariances, especially within objects and between objects that share aspects of their lineage, correction of one location without simultaneous correction of locations with correlated errors will not be helpful. For example, correction of one vertex of the building without simultaneous correction of the remaining three vertices, based on knowledge of the variance–covariance matrix of errors, will change the building's shape.

1.2.6 The geodetic model

These issues are to some extent resolved by use of what will be termed here the *geodetic model*. In this model locations are arranged in a hierarchy, as shown in Figure 1.2. At the top are a small number of locations termed control points or *monuments* that are established with great accuracy by geodetic survey. From these a much larger number of locations are established by measurement, through a process of *densification*. Since these measurements are not as accurate as those used to establish the monuments, the second tier of locations is also less accurately known. Further measurements using even less accurate instruments are used to register aerial photographs, lay out boundary lines, and determine the contents of geographical databases.

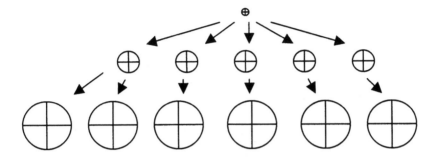

Figure 1.2 Inheritance hierarchy of the geodetic model. A single monument, located with high accuracy, is used to determine the positions of a denser network of less-accurate points, which are in turn used to determine the positions of denser and even-less-accurate points. Further locations may be interpolated between these points.

Following earlier arguments, there will be strong correlations in errors between any locations whose lineages share part or all of the tree. All points inherit the errors present in the monuments, but distances between points that share the same monument are not affected by errors in the location of the monument itself. Thus it is possible to achieve comparatively high accuracy in the results of simple GIS operations like area measurement despite inaccuracies in positioning, if those errors occur in common parents. However, if the structure of the hierarchy is not known, it is not possible to know how much shared lineage exists between pairs of objects that are the subject of analysis, *even though such objects may be in the same layers, or may be parts of the same complex object*. The internal structures used by the organisation of the spatial database may mask the hierarchical structure of the geodetic model.

1.3 TOWARDS MEASUREMENT-BASED GIS

1.3.1 Definitions

I define a *coordinate-based GIS* as one that provides access to the locations of measured objects x, but not to the measurements m from which those locations were derived, or to the function f used to derive x from m. The GIS may or may not provide access to the rules used to determine the locations of interpolated objects from measured objects (for example, the rule defining the edge of a parcel as mathematically straight may be implicit in the data structure, but the rule defining a tree relative to the control points of an aerial photograph may not).

A *measurement-based GIS* is defined as one that provides access to the measurements **m** used to determine the locations of objects, to the function *f*, and to the rules used to determine interpolated positions. It also provides access to the locations, which may either be stored, or derived on the fly from measurements.

In the following, it is assumed that the spatial database is of sufficient complexity that multiple object types exist, with complex lineage. More specifically, it is assumed that covariances between errors in the positions of pairs of measured locations are positive. It follows from the nature of interpolation that covariances are also frequently positive between pairs of interpolated locations.

In a coordinate-based GIS it is not possible to correct positions for part of the database, since the knowledge of error covariances needed to adjust other positions is not available. Partial correction may improve the absolute positions of corrected points, but will affect the relative positions of corrected and uncorrected points in unknown ways. These impacts include changes of shape and other geometric inconsistencies, such as non-existent bends or offsets in linear features, and violations of topological constraints.

In the annals of GIS there are many anecdotes about the costs of attempting partial correction of coordinate-based databases. For example, Goodchild and Kemp (1990, Unit 64) describe the costs to a utility company when partial update moved a large proportion of features across land ownership boundaries. In such situations many agencies have resorted to recompilation, abandoning old, low-accuracy data completely because of the problems of partial correction.

By contrast, no such problems exist in measurement-based GIS. If a location is observed to be distorted, the means exist to determine the offending measurements, correct them, and propagate the effects of correction to all other dependent positions, because **m** and *f* are known.

In addition, it is possible in measurement-based GIS to calibrate error models fully, allowing determination of the impacts of propagating positional errors through GIS operations. The properties of the error field ε could be determined, allowing interoperation between two distorted maps of the same area (for a practical motivation see Church *et al.*, 1998). In the attribute domain, knowledge of measurements could allow the spatial dependence parameters identified by Goodchild, Sun, and Yang (1992), Heuvelink (1998), and Hunter and Goodchild (1997) to be defined and utilised in Monte Carlo simulations.

1.3.2 Hierarchy

A measurement-based GIS is structured as a hierarchy, as outlined in the discussion of the geodetic model above. Let $x^{(i)}$ denote a location at level *i* in the hierarchy. Then locations at level *i*+1 are derived from level *i* locations through equations of the form:

$$x^{(i+1)} = f(m, x^{(i)})$$

At the top (or root) of the tree are locations $x^{(0)}$ which *anchor* the tree. At each level the measurements **m** and function *f* are stored, and the locations x are either stored or derived as needed.

Consider, for example, a utility database in which locations of underground pipes are stored. In such examples the locations of pipes are typically recorded by measurement from other features of known location, such as property lines, or street kerbs. A pipe might be recorded as 3 feet (1 foot = 0.3048m) from a given property line, offset to the left looking in the direction in which the property line is recorded. In this case **m** would be recorded as {3.0,L} or in some other suitable notation. If the pipe is resurveyed, or moved, its position can be re-established by correcting the measurement, or by changing other aspects of the measurement data. But since the dependence is explicit, there will be no need to worry about corrupting the relative positions of pipe and property line, as there would in a coordinate-based GIS.

1.3.3 Beyond the geodetic model

Situations often arise in GIS where the root of the tree is not determined with great accuracy. Suppose, for example, that a national database of major highways is built, and anchored to no better than 100m accuracy (according to the U.S. National Map Accuracy Standards such a database could be described as having a *scale* of 1:200,000, but see Goodchild and Proctor, 1997). It follows that all other locations in the database are absolutely located to no better than 100m. However, it is likely that relative accuracies are higher, since independent distortion of as much as 100m in the elements of such a database would be unacceptable because of the geometric and topological distortions it would produce (there would be sharp bends in what should be straight lines or smooth curves). Again, the metaphor of a semi-rigid frame floating in space is helpful in conceptualising situations like this.

Suppose now that a local agency wishes to link its own database of streets to the national database. This database is likely to be much more accurate, perhaps anchored to 1m to the geodetic frame. This database could be conceptualised as a second tree, but in this case the positional standard error associated with $x^{(0)}$ would be only 1m. In essence, the example can be characterised as two trees, with no common root, and with one tree having a standard error that is much larger than that typical of the geodetic model, in which there is only one tree and a highly accurate anchor (see Figure 1.3).

To link the two databases together, the highest level of the more accurate database is established as the common root. Suppose that its anchor consists of a well-defined point resolvable to better than 1m, such as a survey monument or a photography control point. Suppose also that this location corresponds to that of one of the points in the highest level of the less accurate tree, although the resolution of this anchor point in the less accurate tree is probably much lower (for example, this anchor point might be described as an intersection between two streets, and the intersection might contain the monument or control point anchor of the less accurate tree, see Figure 1.4). Figure 1.3 shows the link that is now built between the new common anchor and the anchor of the less accurate tree, integrating the two trees into one. This link appears as a pseudo-measurement, with a displacement of zero and a standard error equal to 100m. Since the two trees were established and anchored independently, it is reasonable to assume zero covariance between the errors in the measurements in the two subtrees.

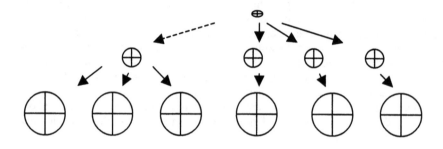

Figure 1.3 Merging of two data sets with distinct inheritance hierarchies. The three least-accurate points on the left depend on a single parent node of moderate accuracy, while the three least-accurate points on the right depend on a more-accurate set, and on a single high-accuracy monument. The dashed arrow represents the pseudo-measurement that is inserted to merge the two trees.

1.4 DISCUSSION AND CONCLUSIONS

If the measurements used to establish position in a spatial database are not retained, but instead all positions are defined only by coordinates with respect to the Earth frame, then it is impossible to correct or update parts of the database without creating geometric and topological distortions that are frequently unacceptable. The almost universal adoption of this design by the GIS software industry is based on the perception that it is possible to know location exactly, and is reflected in the frequent use of precision that greatly exceeds accuracy in the internal representation of coordinates. But in practice exact location is not knowable, and all measurements on which locations are based are subject to some level of error.

By retaining measurements and the functions needed to derive coordinates, it is possible to support incremental update and correction, and to provide much more informed estimates of the impacts of uncertainty in GIS operations. Thus measurement-based GIS designs offer the potential for dramatic reductions in the cost of database maintenance, support for transaction-based operations, and much greater usefulness. Measurement-based principles can be implemented to support database integration, even when the databases being integrated have very different levels of positional accuracy. But such integration is much more problematic using traditional coordinate-based designs.

To return to a point made at the outset, this distinction between coordinate-based and measurement-based GIS strikes at a fundamental issue: are the designs that were developed early in the history of GIS, and remain influential as legacy systems, still optimal given current concepts? Clearly the answer based on this example is *no*.

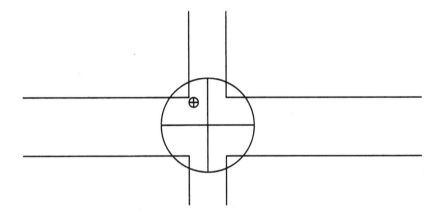

Figure 1.4 Example of two inheritance hierarchies (see Figure 1.3). One hierarchy is anchored to the high-accuracy monument shown as the small circle, located at the kerb. The other is anchored with low accuracy to the entire intersection.

Two distinct strategies are available to support measurement-based GIS: one can design such a GIS from ground up, or one can adopt a framework that incorporates the necessary elements. Hierarchical databases have fallen out of fashion in the past two decades, but the relational model that has largely replaced them has no inherent concept of hierarchy. On the other hand, object-oriented database designs include concepts of inheritance, and provide some of the necessary forms of support. Smallworld's GIS has support for measurement-based designs, through appropriate use of its inheritance features. But full adoption of a measurement-based paradigm is more problematic, since it involves respecification of many functions to include explicit propagation of error characteristics, and explicit storage of error variance–covariance matrices. Thus implementation of a prototype measurement-based GIS, with its associated database and functionality, remains as a challenge to the GIS research community.

1.5 REFERENCES

Burrough, P.A., 1986, *Principles of Geographical Information Systems for Land Resources Assessment* (Oxford: Oxford University Press).

Burrough, P.A. and Frank, A.U., editors, 1996, *Geographic Objects with Indeterminate Boundaries* (London: Taylor and Francis).

Buyong, T. and Frank, A.U., 1989, Measurement-based multipurpose cadastre. In *Proceedings, ACSM/ASPRS Annual Convention, Baltimore, MD, April 2–7*.

Buyong, T., Kuhn, W. *et al*, 1991, A conceptual model of measurement-based multipurpose cadastral systems. *Journal of the Urban and Regional Information Systems Association*, **3(2)**, pp. 35–49.

Buyong, T. and Kuhn, W., 1992, Local adjustment for measurement-based cadastral systems. *Surveying and Land Information Systems*, **52(1)**, pp. 25–33.

Church, R.L., Curtin, K.M., Fohl, P., Funk, C., Goodchild, M.F., Noronha, V.T. and Kyriakidis, P., 1998, Positional distortion in geographic data sets as a barrier to interoperation. In *Technical Papers, ACSM Annual Conference* (Bethesda, Maryland: American Congress on Surveying and Mapping).

Coppock, J.T. and Rhind, D.W., 1991, The history of GIS. In *Geographical Information Systems: Principles and applications,* edited by Maguire, D.J., Goodchild, M.F. and Rhind, D.W. (Harlow, UK: Longman Scientific and Technical), pp. 21–43.

Foresman, T.W., editor, 1998, *The History of Geographic Information Systems: Perspectives from the Pioneers* (Upper Saddle River, NJ: Prentice Hall PTR).

Goodchild, M.F., 1989, Modeling error in objects and fields. In *Accuracy of Spatial Databases*, edited by Goodchild, M.F. and Gopal, S. (Basingstoke, UK: Taylor & Francis), pp. 107–114.

Goodchild, M.F. and Gopal, S., editors, 1989, *Accuracy of Spatial Databases* (London: Taylor & Francis).

Goodchild, M.F. and Kemp, K.K., 1990, *NCGIA Core Curriculum. Volume 3: Application Issues in GIS.* (Santa Barbara, CA: National Center for Geographic Information and Analysis).

Goodchild, M.F. and Proctor, J., 1997, Scale in a digital geographic world. *Geographical and Environmental Modelling*, **1(1)**, pp. 5–23.

Goodchild, M.F., Sun, G. and Yang, S., 1992, Development and test of an error model for categorical data. *International Journal of Geographical Information Systems*, **6(2)**, pp. 87–104.

Goodchild, M.F., Egenhofer, M.J., Kemp, K.K., Mark, D.M. and Sheppard, E.S., 1999, Introduction to the Varenius project. *International Journal of Geographical Information Science*, **13(8)**, pp. 731–746.

Heuvelink, G.B.M., 1998, *Error Propagation in Environmental Modelling with GIS* (London: Taylor and Francis).

Heuvelink, G.B.M., Burrough, P.A. and Stein, A., 1989, Propagation of errors in spatial modelling with GIS. *International Journal of Geographical Information Systems*, **3**, pp. 303–322.

Hunter, G.J. and Goodchild, M.F., 1996, A new model for handling vector data uncertainty in geographic information systems. *Journal of the Urban and Regional Information Systems Association*, **8(1)**, pp. 51–57.

Hunter, G.J. and Goodchild M.F., 1997, Modeling the uncertainty in slope and aspect estimates derived from spatial databases. *Geographical Analysis*, **29(1)**, pp. 35–49.

Kiiveri, H.T., 1997, Assessing, representing and transmitting positional uncertainty in maps. *International Journal of Geographical Information Science*, **11(1)**, pp. 33–52.

Kjerne, D. and Dueker, K.J., 1988, Modeling cadastral spatial relationships using Smalltalk-80. In *Proceedings, GIS/LIS 88, San Antonio, TX* (Falls Church, VA: ASPRS/ACSM/AAG/URISA), pp. 373–385.

Maling, D.H., 1989, *Measurement from Maps: Principles and Methods of Cartometry* (Oxford: Pergamon).

Mark, D.M. and Csillag, F., 1989, The nature of boundaries on 'area-class' maps. *Cartographica* **26(1)**, pp. 65–78.

Openshaw, S., 1989, Learning to live with errors in spatial databases. In *Accuracy of Spatial Databases*, edited by Goodchild, M.F. and Gopal, S. (Basingstoke, UK: Taylor and Francis), pp. 263–276.

Taylor, J.R., 1982, *An Introduction to Error Analysis: The Study of Uncertainties in Physical Measurements* (Mill Valley, CA: University Science Books).

CHAPTER TWO

GIS and Geostatistics for Environmental Modelling

ABSTRACT

To date, the use of geographical information systems (GIS) in environmental modelling has largely focused on two main issues: the digitising and automation of large amounts of spatial data for a wide range of applications, and the use of these data in spatial analysis and numerical models. Frequently, the spatial data are considered to be near perfect representations of the real world, and only rarely are they considered as having an associated degree of uncertainty in either their locational, topological or non-geographic attributes. Therefore, when these data are used as inputs to numerical models of environmental processes the user has no idea of how uncertainties in data and parameters propagate through the models.

In contrast, the discipline of geostatistics developed from the need to make spatial predictions of known reliability from geographically sparse, often expensive, sampled data. Geostatistics uses probability theory to extend the information gained from sampled data to unsampled sites. These interpolations are accompanied by estimates of their associated error bounds, which provide powerful means of studying the propagation of uncertainties through both linear and non-linear numerical spatial models.

This chapter explores some of the ways that geostatistics can be used to aid spatial analysis of point data. The value of geostatistics for GIS lies in the provision of reliable interpolation methods with known errors, in methods of upscaling and generalisation, and for generating multiple realisations of spatial patterns. These stochastic methods improve understanding of how uncertainties in the output of models of spatial processes accrue from errors in data and imperfections in the structure of the models. In addition, GIS can support Geostatistics by aiding the geo-registration of data, and providing a spatial context for interpolation and conditional simulation, as well as providing easy-to-use and effective tools for data display and visualisation. The conclusion is that in spite of differences in history and approach, the marriage of GIS and geostatistics provides a powerful, and complementary suite of tools for spatial analysis in the agricultural, earth and environmental sciences.

2.1 THE NEED FOR GEOSTATISTICS IN GIS

Geographical information systems or GIS (Burrough and McDonnell, 1998) have been used since the late 1960s (Coppock and Rhind, 1991) for automating map making and using digital spatial data. Today we see that in many cases, particularly in environmental studies, the paper map has been replaced by the much more flexible electronic database. Digital GIS provide powerful spatial databases with many applications (see Longley *et al.*, 2001) ranging from land registration, marketing and in-car navigation systems to the dynamic modelling of environmental processes such as soil erosion and land degradation (de Roo and Jetten, 1999) or nutrient transport of over hydrological catchments (de Wit, 1999) .

Though GIS were initially sets of computer tools for the storage, retrieval, analysis and display of spatial data, today they may also include links to standard packages for statistical and spatial analysis, or may include intrinsic or external links to numerical models of environmental processes. Recent articles have illustrated the linkage and applications of several packages for spatial statistics in GIS (e.g. Bao *et al.*, 2000, Bivand and Gebhart, 2000, Wise *et al.*, 2001). Interestingly enough, these papers make little reference to Geostatistics in spite of the increasing availability of geostatistical software in both shared and commercial software (Gstat – Pebesma and Wesseling, 1998, Journel, 1996) so the first task is to explain which application in spatial analysis require a geostatistical rather than a conventional statistical approach.

Geostatistics are useful for upscaling the data on attributes that have been collected at points to provide complete areal coverage (Bierkens *et al.*, 2000). For reasons of cost, accessibility or measurement technique, the essential data may only have been sampled at a limited number of locations and need to be extended to the rest of the study area. Extending limited data from "hard" locations to "soft" unsampled areas is generally known as "interpolation" and can be carried out by a wide range of mathematical techniques (c.f. Burrough and MacDonnell, 1998). Of these, only geostatistics provide methods for associating a quantitative estimate of uncertainty with the interpolated values. This information may be important when the interpolated surfaces are to be used as input variables or control parameters for numerical models of environmental processes.

For example, suppose that we have a possibly non-linear, empirical relation between the spatial variations of the soil moisture capacity as a function of the slope of the land and the thickness of the soil layer. Using the commonly available "Map Algebra" approach (Burrough and MacDonnell, 1998, Tomlin, 1990) we would want to compute a new gridded overlay (called SOILM) for soil moisture as a function of slope and soil depth at all grid cells:

SOILM = f(SLOPE, SOILDEPTH)

In raster-GIS it is easy to generate maps of slope directly from gridded digital elevation models so that at the level of resolution of the DEM the slope is known for every cell – there is 100 per cent coverage. Measurements of soil depth, however, are usually made by augering, which is both time-consuming and expensive; moreover, if the soil depth were to be sampled everywhere, the whole soil would be disturbed! Clearly, it is essential to be able to extend the data from the limited coverage of the sampled sites to all other locations in order to compute the soil moisture in terms of slope and soil depth.

In addition to computing the soil moisture for each cell, we may also be interested in the uncertainties of the result, i.e. what are the uncertainties in SOILM, and how do they depend on the uncertainties in SLOPE and SOILDEPTH? These uncertainties include a) those due to the formulation of the model, and b) those due to uncertainties in the values of the input variables, in this case slope and soil depth. The uncertainties in slope accrue only from the elevation data of the DEM and the slope algorithm used, but the errors in soil depth accrue ·from both the measurements and the interpolation. Together these errors may have an important effect on the outcome of any form of spatial analysis that is carried out using the interpolated data. Clearly, the smaller the uncertainties in the inputs, the better the output, so it behoves us to reduce the input uncertainties as much as possible. Note that in this chapter I do not explore how the propagation of errors depends on the model, a topic that is ably covered by Heuvelink, 1999.

For the purposes of this chapter, the role of geostatistics in GIS, particularly for applications in the environmental sciences, is the following:

- Provide optimal methods of spatial interpolation that include estimates of reliability at unsampled locations;
- Provide means of estimating the effects of uncertainty in the sampled data on the outcomes of spatial analysis and numerical models, particularly of environmental processes.

The rest of this chapter explains and illustrates how these roles are carried out.

2.2 A QUICK OVERVIEW OF THE PRINCIPLES OF GEOSTATISTICS

Information on data quality is of paramount importance. It affects the reliability of data analysis and presentation, as well as impacting on the credibility attached to decision-making with GIS (Goodchild et al., 1994a).

First, here are some definitions of important terms (Bierkens *et al.*, 2000).

The extent is the total area of interest – it could include an arbitrary area of terrain, a single mapping unit (e.g. all river valleys), or a single polygon (a given floodplain).

The support is the largest area for which z, a property of interest, can be thought of as homogeneous – frequently this is the size of an individual observation (e.g. a soil core), the area from which several sub-samples have been collected and bulked, or a single pixel of remotely sensed data. The discretisation of space (i.e. the level of resolution) in GIS means that all point data must be linked to a support, but frequently the support size is not recorded unless we are dealing with remotely sensed data. It is essential that all data entered into a numerical model relate to the same support otherwise patterns and processes at different levels of spatial resolution may be confounded.

A sample is a subset of n support units (observations) that have been selected from a maximum possible number N of non-overlapping support units within the extents of the area in question.

The coverage is the proportion of the extent that is covered by the sample. Clearly, the larger the coverage, the less the need for interpolation.

So, the main task of geostatistics is, given a sample size n from an area of known extent, is to provide an optimal extension of the coverage of z from the sampled sites to the whole area. Unlike simple interpolation techniques such as inverse distance weighting or splines that impose an ad hoc mathematical function (see Goovaerts, 1999 or Burrough and MacDonnell, 1998 for details), geostatistics starts by attempting to characterise the spatial autocovariance structure of the sampled data. Put simply, this means that the approach is first to determine whether the spatial variation of the property of interest is merely "noise" – i.e. the variation has no relation with location, or "signal" – i.e. the variation shows a strong relation with location (high spots are close to other high spots and vice versa), or some mixture of the two. These situations are illustrated by the three surfaces in the upper part of Figure 2.1. All three surfaces are stochastic, in the sense that they represent random outcomes, but the difference is that each surface has a different correlation structure – that is, the distance over which the data values are correlated.

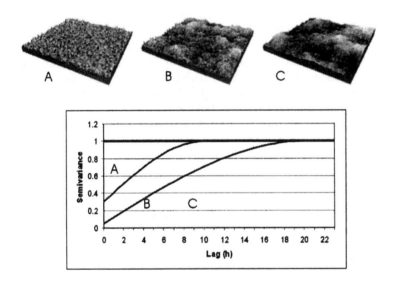

Figure 2.1 Three random surfaces with different autocorrelation structures and their associated variograms. A: no spatial correlation, B: weak spatial correlation up to a lag of 8 distance units, C: strong spatial correlation up to a lag of 18 distance units.

Given that sampled data include stochastic spatial variation, the challenge is to parametrise the spatial variation so that it can be dealt with systematically. Standard geostatistics does this by means of spatial autocovariance structures, often represented by the (semi)variogram, or its cousin the autocovariogram. The semivariance indicates the degree of similarity of values of a regionalised variable Z over a given sample spacing or lag, h.

The value of a random variable Z at x is given by

$$Z(x) = m(x) + \varepsilon'(x) + \varepsilon'' \tag{2.1}$$

where m(x) is a deterministic function describing the 'structural' component of Z at x, $\varepsilon'(x)$ is the term denoting the stochastic, locally varying but spatially dependent residuals from m(x)—the regionalised variable—, and ε'' is a residual, spatially independent Gaussian noise term having zero mean and variance σ^2. Note the use of the capital letter to indicate that Z is a random function and not a measured attribute z.

In the simplest case, where no trend or drift is present, m(x) equals the mean value in the sampling area and the average or expected difference between any two places x and x + h separated by a distance vector h, will be zero:

$$E[Z(x) - Z(x+h)] = 0 \qquad (2.2)$$

where Z (x), Z (x + h) are the values of random variable Z at locations x, x + h. Also, it is assumed that the variance of differences depends only on the distance between sites, h, so that

$$E[\{Z(x) - Z(x+h)\}^2] = E[\{\varepsilon'(x) - \varepsilon'(x+h)\}^2] = 2\gamma(h) \qquad (2.3)$$

where $\gamma(h)$ is known as the semivariance. The two conditions, stationarity of difference and variance of differences, define the requirements for the intrinsic hypothesis of regionalised variable theory. This means that once structural effects have been accounted for, the remaining variation is homogeneous in its variation so that differences between sites are merely a function of the distance between them. So, for a given distance h, the variance of the random component of Z (x) is described by the semivariance:

$$\text{var}[\varepsilon'(x) - \varepsilon'(x+h)] = 2\gamma(h) \qquad (2.4)$$

If the conditions specified by the intrinsic hypothesis are fulfilled, the semivariance can be estimated from sample data. (Semi)variograms (shown in the lower part of Figure 2.1) are graphs of the semivariance $\gamma(h)$ against sample spacing or lag, h: they are estimated by:

$$\gamma(h) = \frac{1}{2n} \sum_i^n \{z(x_i) - z(x_{i+h})\}^2 \qquad (2.5)$$

where n is the number of samples, and $z(x_i)$, $z(x_{i+h})$ are measurements separated by a distance h. Figure 2.1 shows how the form of the semivariogram model reflects the roughness of the sampled surface.

Because experimentally derived semivariances do not always follow a smooth increase with sample spacing, a suitable mathematical model is fitted to the data (Goovaerts, 1997; Burrough and McDonnell, 1998; Deutsch and Journel, 1998). Variogram fitting is perhaps the most difficult practical aspect of using geostatistical interpolation in GIS because it requires a certain amount of experience and insight into the modelling of spatial autocovariation. Practical experience suggests that at least 50 and preferably 100 observations are needed to yield coherent experimental semivariograms. Much practical geostatistics is concerned with the estimation and fitting of variograms to experimental data (Pannatier, 1996; Pebesma and Wesseling, 1998).

Once a variogram model has been fitted, it is used to compute the interpolation weights at any given location. The value $z(x_0)$ (i.e. the unknown value of z at location x_0) is given by:

$$z(x_0) = \sum_{i=1}^{n} \lambda_i \cdot z(x_i) \qquad (2.6)$$

with $\sum_{i=1}^{n}\lambda_i = 1$. The weights λ_i are chosen so that the estimate $z(x_0)$ is unbiased, and that the estimation variance σ_e^2 is less than for any other linear combination of the observed values.

The minimum variance of $[z(x_0) - z(x_0)]$, the prediction error, or 'kriging variance' is given by:

$$\hat{\sigma}_e^2 = \sum_{i=1}^{n} \lambda_i \gamma(x_i,x_0) + \phi \qquad (2.7)$$

and is obtained when

$$\sum_{i=1}^{n} \lambda_i \gamma(x_i,x_j) + \phi = \gamma(x_j,x_0) \qquad (2.8)$$

for all j .

The quantity $\gamma(x_i,x_j)$ is the semivariance of z between the sampling points x_i and x_j; $\gamma(x_i,x_0)$ is the semivariance between the sampling point x_i and the unvisited point x_0. Both these quantities are obtained from the fitted variogram. The quantity ϕ is a Lagrange multiplier required for solving the equations. The method is known as ordinary point kriging (OPK): it is an exact interpolator in the sense that when the equations given above are used, the interpolated values, or best local average, will coincide with the values at the data points. The interpolated values relate to areas of land the same size as the original support. Predictions can also be computed for units of land (blocks) that are larger than those sampled: this is called block kriging. This smoothes out local variations and enables the resolution of the interpolated data to match that of other data (e.g. the pixel size of remotely sensed imagery).

For GIS modelling applications, the data will usually be interpolated to a regular grid. Similarly, the estimation error σ_e^2, known as the kriging variance, can also be mapped to the same grid to give valuable information about the reliability of the interpolated values over the area of interest. Often the kriging variance is mapped as the kriging standard deviation (or kriging error), because this has the same units as the predictions.

2.2.1 An example

The above theory is illustrated with the data set of heavy metal pollution of floodplain soils on the eastern side of the Maas floodplain in the south of the Netherlands (see Burrough and MacDonnell, 1998 Appendix 3), but here extended to a larger area with 157 observations. As the data are only used here to illustrate methods, we shall call the attribute to be mapped "an index of pollution". Figure 2.2 gives an example of an ordinary point kriging interpolation of this attribute and its accompanying error surface. Note the smaller values of interpolation error (lighter shades of grey) where data are abundant.

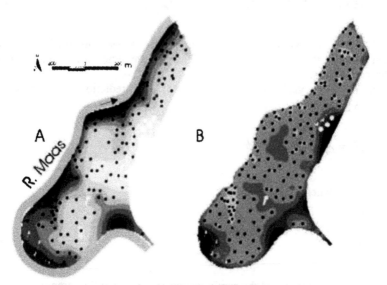

Figure 2.2 A, surface of interpolated pollution index; B, associated kriging errors. Black dots are sample locations. Darker tones indicate greater degree of pollution in A and larger kriging variance in B.

2.3 ERROR SURFACES AND NUMERICAL MODELLING

Now that we have not only an interpolated surface that covers the extent completely but also an estimate of the associated errors it is easy in GIS to compute new surfaces that can function as probability envelopes for the map algebra model, for example the lower and upper bounds of the predicted values ± 2 standard errors. In this situation we have interpolated first, and then computed the model. This is appropriate when the model is linear and the errors are normally distributed (Bierkens et al, 2000). When the model is non-linear or it is expensive to compute or there are insufficient data to compute the variogram then it is more appropriate to compute the model at the known data points and then interpolate the model results. In both situations we may be able to improve the results by using ancillary data from the GIS.

2.3.1 Using ancillary information to aid interpolation

The simple use of geostatistical interpolation assumes that there is but one spatial correlation structure (i.e. level of variability) over the extent of the area in question. In practice there may be more information that we can use to improve the errors associated with interpolation. For example, we may have a suitable stratification of the area into different units, each of which may have a different pattern of spatial variation, and hence a different semivariogram. If there are sufficient data points to compute the semivariogram then it may be worthwhile carrying out the variogram modelling and interpolation for the different components separately. This is known as stratified kriging.

Stratified kriging is easy to implement in a GIS as follows. First, you need to have a polygon map showing the different map units. Second, it is sensible to carry out a straightforward ANOVA to see if the units differ in mean and variance, though they might still have different spatial correlation as in Figure 2.1. Finally, there must be sufficient data points in all units to support variogram estimation.

Figure 2.3A shows the stratification of the example area into three hydrological units which are related to the flooding frequency of the area: unit 1 is flooded annually, unit 2 every 2-5 years and unit 3 less than once every 5 years. The respective means and standard deviations of the pollution index units are given in Table 2.1. An F test ($F = 26.595$, degrees of freedom 2,154) shows the division is highly significant, but a post-hoc Scheffé test indicates that the significance holds only for unit 1 versus units 2+3 combined. Given the limited number of samples in unit 3 it seems sensible to combine these and to limit the stratification to 2 strata, which increases F to 52.810 (degrees of freedom 1,155). Note that by obtaining the per cent coverage of the area from a GIS, we see that unit 1 only covers a fifth of the area, yet is the most important stratum.

Table 2.1 Summary statistics of the pollution index of the hydrological units.

Unit	Coverage per cent	N samples	Mean	Standard deviation
1	22.66	85	6.2164	0.6864
2	44.70	48	5.4522	0.5371
3	32.64	24	5.5666	0.5497
2+3	77.34	72	5.4904	0.5402
Total	100.00	157	5.8834	0.7199

Figure 2.4 shows the experimental variograms for the whole area, and the two strata separately. Clearly, the variogram for the whole area is a sort of average of the other two. Figure 2.3B, C show the results of the interpolation using the strata and gives a good impression of the effect of the fine surface drainage structure of the annual flooding zones on the results. The degree of improvement is evident from the maps of kriging error which are in most parts of the area are approximately at least a half of those obtained without strata.

Other methods for incorporating ancillary data in geostatistical interpolation are co-kriging (the use of a supplementary cheap-to-measure attribute that is correlated with the attribute of interest), multivariate kriging (data reduction from many attributes to one or a few such as with Principal Components or fuzzy k-means), KT kriging (the use of spatial regression to model continuous trends) and indicator kriging (for working with binary data). More details and examples can be found in Goovaerts (1997, 1999) and Burrough and MacDonnell (1998). In all cases the aim is to reduce the errors of interpolation so that when the resulting surfaces are used as inputs to numerical models the errors associated with the model outputs will be as small as possible.

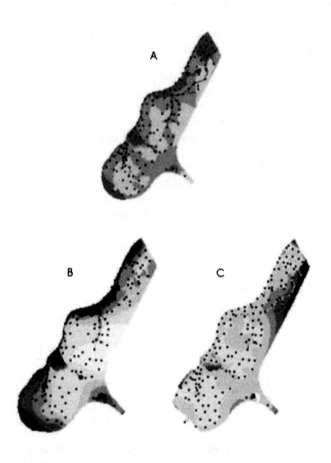

Figure 2.3 Hydrological strata (A) and stratified kriging (B, estimates; C, kriging variance). Dark tones in B, C indicate higher pollution levels.

2.4 STOCHASTIC INPUTS TO THE MODELLING OF SPATIAL PROCESSES: ERROR PROPAGATION IN SPATIAL MODELS

The estimates of errors and error surfaces provided by geostatistics can be used in several ways to understand the propagation of errors through spatial models. The simplest way is to run the model with data for the three situations of the original interpolation and the two ± 2 standard deviation surfaces shown in Figure 2.5. This approach assumes that all the errors are spatially correlated, however, and that adjacent grid cells are not able to take up any value that is allowed by the probability distribution for each cell separately.

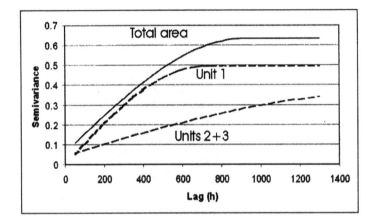

Figure 2.4 Spherical models of variograms fitting the pollution index for the whole area, and the two strata separately.

Heuvelink (1999) gives an alternative approach in which the propagated errors are computed using a Taylor series approximation with information from the means, variances and covariances of the data in question. The main advantages of the Taylor series approach are that error propagation is computed for each cell in terms of its probability distribution, and the computations are easily carried out. The disadvantage is that the method requires a rigorous approach to theory with important assumptions, and increasingly complicated computations as model complexity increases – see Heuvelink (1999) or Burrough and McDonnell (1998).

As computers have become more powerful, however, Monte Carlo methods have become more attractive for estimating error propagation. In these methods, multiple realisations of the input data are used to drive many runs of the model so that the probability distribution of the model outputs can be determined and model sensitivity to key variables can be assessed. A major advantage is that this approach is not affected by the complexity or non-linearity of the model.

2.5 CONDITIONAL SIMULATION – USING THE VARIOGRAM TO GENERATE EQUIPROBABLE SURFACES

Conditional simulation is an alternative method for estimating error that uses the information from the sampled data and the variogram model to create a series of equally probable realisations of the surface of interest that are all different, all pass through the known data points and yet all have the same mean, variance and autocovariance structure (Goméz-Hernández and Journel, 1992). For each realisation one first sets up the simple kriging equations with an overall mean for each stratum.   An unsampled point is selected at random and its value and its kriging error are calculated from the sample data.   The point is added to the data and the process repeated for another random point until all points have been visited, when the whole realisation is saved as a map.

A

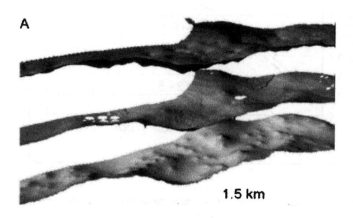

1.5 km

B

Figure 2.5 Close up view of interpolated surfaces and upper and lower surfaces showing interpolation
± 2 kriging standard errors. A, ordinary point kriging for whole area;
B, stratified kriging for 2 hydrological units.
Vertical differences indicating variations in pollution index are exaggerated.

Figure 2.6 Four realisations of conditionally simulated surfaces using stratified kriging with two units.

Figure 2.6 displays four different realisations of the index of pollution computed using the variograms for stratified kriging in the 2 hydrological units (strata). Each realisation is an equally likely picture of the variation of the pollution index over the area of interest. Computing time has reduced enormously in recent years, with each realisation in Figure 2.6 taking a mere 0.76 seconds on a 600 MHz laptop (compared with 2.2 minutes given in Burrough and MacDonnell, 1998).

The error in model output due to the input data is determined by running the model for as many times as there are combinations of simulated error surfaces, though the absolute number may be reduced using Latin hypercube sampling (Pebesma and Wesseling, 1998). The combined results of all the runs define the probability distribution of the model output for each cell. Examples of conditional simulation for error propagation in numerical models linked to GIS - e.g. the 3D groundwater model 'MODFLOW' - are given by Bierkens (1994) and Goméz-Hernández and Journel (1992): Kyriakidis et al., (1999) provide an example of using conditional simulation to the assessment of accuracy of USGS one-degree digital elevation models (DEMs).

2.6 GEOSTATISTICS FOR OPTIMISING SAMPLING

In very many cases point data are interpolated using information from an existing GIS database so the spatial analyst will have had no control of the design and layout of the spatial sampling. The result may be excessively large interpolation errors in zones that have been undersampled. McBratney *et al.*, (1981) demonstrated that spatial sampling for interpolation was best carried out on an equilateral triangular grid, or failing that, a square grid when spatial variation was isotropic. They also showed that the kriging errors depended on sample spacing and the size of the block being interpolated. McBratney and Webster (1981) published a computer program called OSSFIM which given a variogram, computed combinations of sample spacing and block size. This meant that once the spatial correlation structure is known, one can optimise sampling to meet given standards.

Van Groenigen and Stein, 1998 and van Groenigen *et al.*, 1999 have extended this work, developing a general approach to optimising the design of spatial sampling schemes using methods of simulated annealing. They demonstrate the value of designing sampling schemes to match the aims of a study rather than making do with what is available. The method is available as a prototype interactive program (see http://agronomy.ucdavis.edu/groenigen). Methods of optimising sampling to minimise interpolation errors are an important accessory to the spatial analyst's tool kit when data must be collected for a specific purpose, and not just downloaded from any convenient database.

2.7 THE VALUE OF GIS FOR GEOSTATISTICS

Besides acting as a spatial database, GIS provides several benefits to geostatisticians that are largely concerned with the correct geometric registration of sample data, prior data analysis, the linking of hard and soft data, and the presentation of results.

2.7.1 Geo-registration

As with all spatial data, spatial analysis must be carried out on data that have been collected with reference to a properly defined coordinate system. GIS can provide the means to register the locations of samples directly (via GPS or other methods), or to convert local coordinates to standard coordinates. The use of standard coordinates ensures that data collected at different times can be properly combined and printed on conventional maps. The use of standard coordinate systems is particularly important when international databases are created, such as occurs in Europe, for example.

2.7.2 Exploratory data analysis

The presence of spatial outliers, or other irregularities in the data may have important consequences for the fitting of variogram models, or for determining whether data should be transformed to logarithms. GIS often provide search engines that can be linked to statistical packages to determine whether any given data set contains anomalies or unexpected structure. The underlying reasons for such anomalies may sometimes be easily seen when these data are displayed on a map together with other information. Not all users of EDA in GIS use conventional geostatistics, however, and other measures of spatial autocorrelation such as Moran's I statistic are often used (Griffith and Layne, 1999).

2.7.3 Spatial context and the use of external information

Increasingly, the suite of geostatistical methods currently available allow the user to incorporate external information that can be used to modify, and possibly improve, the predictions or simulations required. Geostatisticians term the external information "secondary", because they believe that the "hard data" measured at the sample locations is most important. But GIS practitioners might prefer to call the "primary data" that which separates a landscape into its main components - different soils, or rock types, or land cover classes, regarding the sampled data as merely filling in the details that were not apparent at the smaller map scale. For example, Figure 2.7 shows the difference between the interpolated elevation map created by OPK from the 157 data points and the detail apparent on a modern 5x5m DEM obtained by laser altimetry! Clearly, the interpolation misses much essential detail, a fact that should not be forgotten when interpreting the interpolated maps of soil pollution.

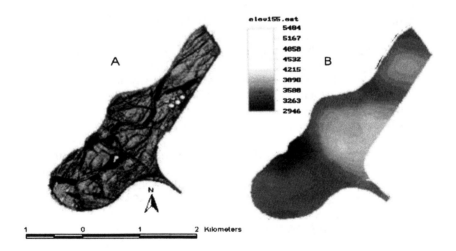

Figure 2.7 An illustration to show how much detail is missed by interpolation. A, shaded relief of elevation surface as recorded with continuous laser altimetry with a pixel of 5x5m; B, elevation interpolated by OPK from the 157 observations to a 5x5m grid. (Elevations in mm).

2.7.4 Display and visualisation – 2D, 3D, plus time

Who is the recipient of a geostatistical interpolation? If a geostatistician, then simple maps and tables of numbers may suffice, but environmental managers need to see how the results relate to other aspects of the terrain. Today it is easy to import the results of a kriging interpolation into a GIS and display the results in conjunction with a scanned topographic map, or to display them in 3D over a digital elevation model (DEM) of the landscape from which the samples were taken. Such presentation invites visual interpretation, the re-evaluation of results and the discovery of more information, and therefore is an essential part of the spatial analysis process.

2.8 CONCLUSIONS.

This chapter shows that GIS and geostatistics have much to give to each other, particularly when GIS users need to model environmental processes using data that have associated spatial variation and error. Geostatistics benefit from having standard methods of geographical registration, data storage, retrieval and display, while GIS benefits by being able to incorporate proven methods for handling and understanding errors in data and their effects on the outcomes of models used for environmental management

REFERENCES

Bao, S., Anselin, L., Martin, D. and Stralberg, D., 2000, Seamless integration of spatial statistics and GIS: The S-PLUS for Arc View and the S+Grassland links. *Journal of Geographical Systems* 2: 287-306.

Bierkens, M.F.P., 1994, *Complex confining layers: a stochastic analysis of hydraulic properties at various scales.* Royal Dutch Geographical Association (KNAW) /Faculty of Geographical Sciences, University of Utrecht, Utrecht, NL.

Bierkens, M.F.P., Finke, P. A. and de Willigen, P., 2000, Upscaling and Downscaling methods for Environmental Research. *Developments in Plant and Soil Sciences,* Vol 88, Kluwer Academic Publishers, Dordrecht, 190pp.

Bivand, R. and Gebhart, A., 2000, Implementing functions for spatial statistical analysis using the R language. *Journal of Geographical Systems* 2: 307-317.

Burrough, P. A., 1996, Opportunities and limitations of GIS-based modeling of solute transport at the regional scale. In: *Application of GIS to the Modeling of Non-Point Source Pollutants in the Vadose Zone,* SSSA Special Publication 48, Soil Science Society of America, Madison, 19-37.

Burrough, P.A., and McDonnell, R.A., 1998, *Principles of Geographical Information Systems.* Oxford University Press, Oxford, 330pp.

Coppock, J.T. and Rhind, D.W., 1991, The History of GIS. In D. J. Maguire, M.F. Goodchild and D. W. Rhind (eds), *Geographical Information System,* Vol. 1:Principles. Longman Scientific Technical , New York , pp. 21-43.

Deutsch, C. and Journel, A.G., 1998, *GSLIB Geostatistical Handbook.* Oxford University Press, New York.

Griffith, D.A. and Layne, L.J., 1999, *A Case Book for Spatial Data Analysis: a compilation of analyses of different thematic data sets*. Oxford University Press, New York, 506pp.

Goméz-Hernández, J.J, Journel A.G., 1992, Joint sequential simulation of multigaussian fields. In: A. Soares (ed), *Proc. Fourth Geostatistics Congress*, Troia, Portugal. Quantitative Geology and Geostatistics (5) 85-94, Kluwer Academic Publishers, Dordrecht.

Goovaerts, P., 1997, *Geostatistics for Natural Resources Evaluation*. Oxford University Press, New York, 483pp.

Goovaerts, P., 1999, Using elevation to aid the geostatistical mapping of rainfall erosivity. *CATENA* 34: 227-242.

Heuvelink, G.B.M., 1998, *Error Propagation in Environmental Modelling*. Taylor & Francis, London, 127pp.

Heuvelink, G.B.M. and Burrough, P.A., 1993, Error propagation in cartographic modelling using Boolean logic and continuous classification. *International Journal of Geographical Information Systems* 7: 231-246.

Journel, A.G., 1996, Modelling uncertainty and spatial dependence: stochastic imaging. *International Journal of Goegraphical Information Systems* 10: 517-522.

Kyriakidis, P.C., Shortridge A.M. and Goodchild, M.F., 1999, Geostatistics for conflation and accuracy assessment of digital elevation models. *International Journal of Geographical Information Science* 13: 677-707.

Longley, P.A., Goodchild, M.F., Maguire, D.J. and Rhind D.W., 2001, *Geographic Information Systems and Science*. J. Wiley, Chichester.

McBratney A.B. and Webster, R., 1981, The design of optimal sampling schemes for local estimation and mapping of regionalised variables. II. Program and examples. *Computers & Geosciences* 7: 335–365.

McBratney, A.B., Webster, R. and Burgess, T.M.,1981, The design of optimal sampling schemes for local estimation and mapping of regionalised variables. I. Theory and method. *Computers & Geosciences* 7: 331–334.

Pannatier, Y., 1996, *Variowin. Software for Spatial Data Analysis in 2D*. Statistics and Computing, Springer Verlag, Berlin, 91pp.

Pebesma, E. and Wesseling, C.G., 1998, GSTAT: a program for geostatistical modelling, prediction and simulation. *Computers and Geosciences* 24: 17-31.

Roo, A.P.J. de and Jettcn, V. (Eds), 1999, Modelling of soil erosion by water on a catchment scale. GCTE Focus 3 workshop, 14-18 April, 1997, Utrecht University. *CATENA* 37 Special Issue, 546pp.

Tomlin, C.D., 1990, *Geographic Information Systems and Cartographic Modeling*. Prentice Hall, Englewood Cliffs, NJ, 249 pp.

Van Groenigen, J.W. and Stein, A., 1998, Constrained optimisation of spatial sampling using continuous simulated annealing. *Journal of Environmental Quality* 27:1078-1086.

Van Groenigen, J.W., Siderius, W. and Stein, A., 1999, Constrained optimisation of soil sampling for minimisation of the kriging variance. *Geoderma* 87: 239-259.

Wise, S., Haining, R. and Ma, J., 2001, Providing spatial statistical data analysis functionality for the GIS user: the SAGE project. *International Journal of Geographical Information Science* 15: 239-254.

Wit, M.J.M., de, 1999, *Nutrient fluxes in the Rhine and Elbe basins*. Royal Dutch Geographical Society/Faculty of Geographical Sciences, Netherlands Geographical Studies 259, Universiteit Utrecht, Utrecht, 176 pp.

ACKNOWLEDGEMENTS

Laser altimetry data for Figure 2.7 were provided by the Survey Department of the Netherlands' Directorate for Public Works and Water Management. The data are part of the Actueel Hoogtebestand Nederland (AHN). (http://www.minvenw.nl/rws/mdi/home/product/index4.htm)

CHAPTER THREE

Hierarchical Topological Reasoning with Vague Regions

Stephan Winter and Thomas Bittner

ABSTRACT

Topological relations can be determined efficiently in raster hierarchies (image pyramids, quadtrees), by applying a recursive procedure for an adapted intersection model. At least at the higher levels of a hierarchy, the procedure has to deal with mixed pixels. That requires a more flexible use and interpretation of intersection schemata. It is shown that the found methodology can be transferred to deal with regions with spatial uncertainty, too.

3.1 INTRODUCTION

Topological relations between regions can be determined in hierarchical representations efficiently. This was shown for raster hierarchies (Winter, 1999a), like image pyramids or quadtrees (Samet, 1990). In a first approach to hierarchical reasoning, regions are considered to be crisp and each represented by a binary image each. However, a hierarchy on binary images introduces a third value on higher levels, where elements occur that are totally inside (i), totally outside (e), or of mixed content (x). A reasoning procedure that starts at the top level of a hierarchy with a breadth-first strategy leaves the results of mixed pixels open, but refines the results from level to level. This allows stopping whenever the reasoning result is sufficiently concrete to answer a given query. Thus the result can consist of a set of possible relations. Furthermore, the procedure can be formulated incrementally, searching in depth only at positions that can contribute new information (Winter, 1999a).

The assumption that regions are crisp is too strict for regions representing physical objects in the geographical domain. Physical objects are observable, and observation induces definition uncertainty in abstracting the continuous world to discrete entities (Burrough and Frank, 1996) and (at least) random noise in measurement (Baarda, 1967). Therefore, a more sophisticated representation of the location of physical objects is the concept of vague regions (Cohn *et al.*, 1997) or rough location (Worboys, 1998; Bittner, 1999). A vague region distinguishes parts that belong to the region for sure, parts that may belong to the region, and parts that are definitely outside (Figure 3.1). On vague regions also a hierarchy can be built, which requires three values at all levels. The meaning of the third value changes in this case from *mixed* to *maybe*.

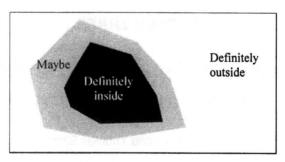

Figure 3.1 A vague region partitions the space into parts that belong to the region for sure, parts that do not belong to the region, and parts that may belong to the region.

In this chapter we apply hierarchical topological reasoning to vague regions, using a raster representation for the regions. Hence the four-intersection model (Egenhofer and Franzosa, 1991) is adapted with regard to raster representation, which in binary images provides only the topological sets of *interior* and *exterior*, and with regard to three-valued images with the third topological 'set' of *maybe*. We also investigate the differences to the hierarchical reasoning with crisp regions. Of central interest is the different meaning of *mixed* and *maybe* pixels, corresponding to the conceptual difference of generalisation and uncertainty. The difference in meaning causes some changes in decision rules. But in principle the properties of the hierarchical procedure are preserved.

The chapter starts with the four-intersection for crisp raster regions as our reasoning tool (Section 3.2). Then we introduce vague raster regions (Section 3.3). We investigate the determination of topological relations between vague regions, comparing them with existing methods (Section 3.4). We will derive the same results, but in contrast to the existing methods we can apply the hierarchical reasoning procedure directly (Sections 3.5 and 3.6).

3.2 RELATIONS BETWEEN CRISP RASTER REGIONS

In this section we apply the point-set based *four-intersection model* (Egenhofer, 1989; Egenhofer and Franzosa, 1991) for regions given their interior and exterior sets. In total, five relations can be distinguished. We will make use of this intersection model in the chapter. In the logic-based *region connection calculus* (RCC), primitives are regions and a connectedness axiom, instead of point sets and their intersection sets (Randell *et al.*, 1992; Cohn *et al.*, 1997). The boundary insensitive RCC-5 is equivalent to our approach.

The intersection model is based on the intersection sets of point-sets. The considered sets are the *interior*, the *boundary*, and the *exterior* of two regions. The nine intersection sets in total are binarized to be empty or non-empty and written into a 3×3-array, the nine-intersection. Single arrays refer to specific topological relations.

In contrast to point sets (with a concept of boundaries), we are dealing with raster images, i.e. regular partitions without explicit boundaries. The partitions form a set of two-dimensional jointly exhaustive and pair-wise disjoint elements. In regular partitions the elements are called *pixels* or sometimes *resels* (Tobler, 1984). Raster representations only allow the interior X° and the exterior X^c of a region X to be distinguished. Instead of artificial two-dimensional concepts of boundaries (Rosenfeld, 1979; Kong and Rosenfeld, 1989), we adapt the intersection model for the given two topological sets. Then the topological relation between two (independent) regions X and Y is characterized by four intersection sets Z_k with:

$$Z_1 = X^\circ \cap Y^\circ, \quad Z_2 = X^\circ \cap Y^c, \quad Z_3 = X^c \cap Y^\circ, \quad Z_4 = X^c \cap Y^c$$

Analogue to Egenhofer's model, these intersection sets can be classified as empty (labelled by F) or not (labelled by T) by a function m:

$$m = \begin{cases} F & \text{if } Z_k = \varnothing \\ T & \text{else} \end{cases}$$

and ordered in a 2×2-array, the four-intersection *I4* for region representations:

$$I4 = \begin{pmatrix} m(Z_1) & m(Z_2) \\ m(Z_3) & m(Z_4) \end{pmatrix} \tag{3.1}$$

Using this schema, five relations can be distinguished between two regions (Table 3.1) (Winter, 2000).

Table 3.1 The five distinct four-intersections for regions in region representations, and the names of the corresponding relations from the vector representations.

$\begin{pmatrix} F & T \\ T & T \end{pmatrix}$	$\begin{pmatrix} T & T \\ T & T \end{pmatrix}$	$\begin{pmatrix} T & F \\ F & T \end{pmatrix}$	$\begin{pmatrix} T & T \\ F & T \end{pmatrix}$	$\begin{pmatrix} T & F \\ T & T \end{pmatrix}$
Disjunt, meet	Overlap	Equal	Contains, covers	ContBy, covBy

Some of Egenhofer's relations for boundary representations are fused in this case to more general ones. We conclude that region representations without a concept of a boundary cannot express topological relations with the granularity of boundary representations.

The conceptual neighborhood graph in the boundary-oriented four-intersection model (Egenhofer and Al-Taha, 1992), motivated by possible changes of relations in continuous movement or deformation, is simplified in this case to Figure 3.2. The graph is equivalent to the continuity network in the RCC-5 model (Cohn *et al.*, 1997).

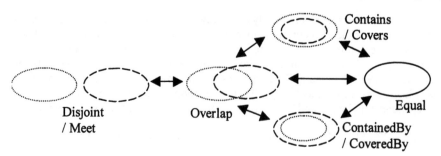

Figure 3.2 The conceptual neighborhood graph of the five topological relations in Table 3.1, equivalent to the continuity network of Cohn *et al.*

3.3 CONTINUOUS AND DISCRETE VAGUE REGIONS

We assume that every spatial object is exactly located at a single region of space in each moment of time (Casati and Varzi, 1995). This region may be a simple region of three-dimensional space, think of your body and the region of space it carves out of the air. However, the exact region may be a complex region also, consisting of multiple regions of three-dimensional space, as in the case of the Hawaiian Islands, or consisting of multiple regions of two-dimensional space, as in the case of the representation of the Hawaiian Islands on a paper map.

3.3.1 Definitorial vagueness and indeterminacy of location

Spatial objects can be distinguished into *bona fide* and *fiat* objects (Smith, 1995). The boundaries of bona fide objects correspond to discontinuities in physical reality. Bona fide boundaries are observable in physical reality. Examples for bona fide objects are highways or buildings. Fiat objects are the product of human conceptualization. The boundaries of fiat objects do not necessarily correspond to discontinuities of the underlying physical reality. Examples are objects like seas or the federal states of the USA.

In the scientific realm fiat objects are created by definitions. Vagueness is a property of definitions of fiat objects. Consider fiat objects like valleys and mountains. The definitions of those objects are too vague in order to fix the boundaries of a particular instance. Consider for example Mont Blanc. The boundary between rock and air is bona fide, determinate, and observable but "where is the boundary of Mont Blanc among its foothills" (Smith and Mark, 1998). Within certain limits there are arbitrary choices of boundary location possible.

Often, the vague definition might be strong enough to identify regions of space at which observably parts of the object are located and to identify regions of space where no parts of the object are located. This often allows us to draw boundaries around a *certain core* and a *certain exterior*. This results in a concentric partition of space consisting of the certain core, a broad boundary, and a certain exterior. Given our assumption above that every spatial object is located at a single region of space in every moment of time, every region of space that satisfies:

- the interior contains the core region,
- the boundary is located in the broad boundary part between core and exterior,

is an equally good candidate to be the exact region of the object in question. This reflects the definitorial indeterminacy caused by the vagueness of the definition of a fiat object. In the remainder we use the notion *vague region* in order to refer to concentric regional partitions of space consisting of a core region, a broad boundary region and an exterior region. We introduce a label *maybe* for the broad boundary region, i.e., the real boundary is located somewhere in this region. Several formalisms are proposed for the representation of vague regions, e.g., Clementini and Di Felice (1996) and Cohn and Gotts (1996).

The special aspect of relations between vague regions is the following: A single vague region corresponds to a whole class of exact regions. Consequently, every single relation between two vague regions corresponds to the set of all relations that can possibly hold between the exact regions subsumed by the vague regions. This will be shown in detail in Section 3.4.2.

3.3.2 Finite resolution and location indeterminacy

Finite resolution in observation leads to approximations and finally to vague regions again. Spatial analysis of discrete observations, like remote sensing, or point sampling, results in approximations of spatial objects with respect to elements of regional partitions, e.g., regular raster partitions, created by the underlying measurement or observation process. Consider Figure 3.3. There are raster cells that are parts of the object's exact region, raster cells that overlap the object's exact region but are not part of it, and raster cells that are neither part of raster cells, nor overlap them. We call the set of raster cells which are part of the object's exact region the *lower approximation set* and the set of all raster cells which are part of or overlap the object's exact region the *upper approximation set* (Pawlak, 1982; Worboys, 1998).

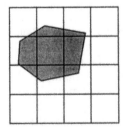 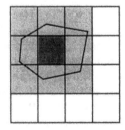

Figure 3.3 Left: A regular partition of any resolution will never fit to continuous regions. Right: Instead, a certain interior (lower approximation, *dark*) and a certain exterior (*white*) can be defined. The broad boundary area (*bright*) in union with the lower approximation is the upper approximation.

Spatial analysis yields classifications of raster pixels that are assumed to coincide with lower and upper approximations of spatial objects observed in physical reality, with respect to the regional partition created by the underlying measurement process (Bittner and Winter, 1999). The underlying assumption is that the lower approximation region coincides with a core part of the observed object, and that the boundary is somewhere located in the region formed corresponding to the set-theoretic difference of lower and upper approximation sets. For example, consider a spatial object in physical reality, observed by means of remote sensing. Spatial analysis essentially yields *rough set* representations of the observed object in the underlying raster partition created by remote sensing imaging: object interior pixels, mixed pixels, and exterior pixels. Given the assumptions above, each region of space whose interior covers the lower approximation set and which boundary is located in the region formed by the set of mixed pixels is an equally good candidate for the exact region of the object. Consequently, we have the same situation we had in the case of vague regions above.

In the remainder of this chapter the representation of relations between vague regions representing the indeterminacy caused by definitorial vagueness is applied to relations between vague regions derived from lower and upper approximation sets representing the outcome of spatial analysis applied to (e.g., remote sensed) images. We exploit the discrete raster structure of the underlying raster to define relations between vague regions consisting of finite sums of raster cells. We show that those relations correspond to relations between vague regions in continuous domains. We further exploit the hierarchical organization of the underlying raster partition in order to provide effective means of computation of relations between vague rasterized regions.

3.4 RELATIONS BETWEEN VAGUE REGIONS

Topological relations between vague regions are already investigated in the nine-intersection, extended for two-dimensional boundaries (Clementini and Di Felice, 1996), as well as in the RCC theory in the form of the *egg-yolk model* (Cohn and Gotts, 1996). In contrast to the nine-intersection approach, using a two-valued logic and an explicit notion of boundary, we apply a three-valued logic and do not use the explicit notion of boundary. The interpretation of the broad boundary region as *boundary* is misleading in our view. The three-valued logic will require calculating the four-intersection locally. This will be advantageous later, when we determine the relations in hierarchical representations of the considered regions; in particular an incremental refinement takes local computability for granted.

However, reasoning between vague regions is by no means the only way to handle spatial uncertainty in topological reasoning. For example, Winter takes the resolution as an indication for the size of the uncertainty. Then he uses a stochastic model of the observation process to classify the relation to the most likely one (Winter, 1999b). But the choice of vague regions allows a more general point of view. There are geometric assumptions in Winter's approach that can be neglected here.

3.4.1 Nine-intersection and region connection calculus for vague regions

A vague region represents a real-world object with boundaries anywhere in between its lower approximation and its upper approximation, i.e., we can draw any number of valid crisp realizations (crispings) inside these two limits. For two vague regions, one topological relation exists between any pair of crispings. There are a number of relations possible between the members of the sets of crisp regions on which the vague regions stand. The relation between two vague regions is therefore a set of (possible) topological relations, with a size depending on how many relations can occur. The set is limited by the relation of the two lower approximations on one side, and the relation of the two upper approximations on the other side. The conceptual neighbourhood graph (Figure 3.2) is needed to prove that the remaining possible relations are in the graph in between these two relations, and that there cannot be more (Clementini and Di Felice, 1996; Cohn and Gotts, 1996).

Consider, for example, Figure 3.4. The two lower approximations (dark) are related with *disjoint/meet*. The two upper approximations are related by *Overlap*. Other pairs of crispings will have one of these relations, because there is no other relation in between *disjoint/meet* and *overlap* in the conceptual neighbourhood graph. However, the example is not the only configuration of vague regions with this set of possible topological relations. There are three other relations between pairs of vague regions that have crispings in the relation *disjoint/meet* or *overlap* only. That means, in the context of topological relations it is not necessary to find all configurations of two vague regions, but it is necessary to find all sets of topological relations that occur between two vague regions.

Figure 3.4 One of 46 topological relations in the egg-yolk model.

We count thirteen sets in tota, according with Cohn and Gotts (1996). These are:

- four sets of relations containing only one relation between all pairs of crispings of the vague regions:
  {disjoint/meet}, {overlap}, {contains/covers}, {containedBy/coveredBy};
- five sets of relations containing two relations between pairs of crispings:
  {disjoint/meet, overlap}, {overlap, contains/covers}, {overlap, containedBy/coveredBy}, {contains/covers, equals}, {containedBy/coveredBy, equals};
- two sets of relations containing three relations between pairs of crispings:
  {disjoint/meet, overlap, contains/covers}, {disjoint/meet, overlap, containedBy/coveredBy};
- one set of relations containing four relations between pairs of crispings:
  {overlap, contains/covers, containedBy/coveredBy, equals};

- one set of relations containing all five relations, indicating indeterminacy of the real relation:
  {disjoint/meet, overlap, contains/covers, containedBy/coveredBy, equals}.

Within this frame we present now a representation of the sets of possible relations by the four-intersection introduced in Section 3.2. For that reason we have to extend the four-intersection for handling properly the third type of region, the difference between upper and lower approximation, which contains the crispings of the region boundaries.

3.4.2 The interior/exterior intersection model for vague regions

We present here the four-intersection for the interior and exterior sets (Equation 3.1) extended for vague regions. We develop a pixel-wise calculation of the extended four-intersection matrix, and also a convolution of the local results for the global four-intersection between two rasterized vague regions. At the end, we find one characteristic four-intersection matrix that relates to the set of possible relations between two vague regions.

Consider the overlay of two images for the intersection operation. Pixel-wise, each corresponding pair of pixels of the two images contributes to specific intersection sets. Intersecting pixels that belong to the interior or exterior contribute exactly to one intersection set. With three pixel labels (*interior*, *maybe*, and *exterior*, see Section 3.3.1) the intersection of a pair of pixels can contribute to more than one intersection set, for which reason we determine a complete four-intersection matrix for each pixel pair. We call this matrix a *local four-intersection*. Furthermore, the Boolean logic introduced with binary intersection sets ($\varnothing$ or $\neg\varnothing$, or F and T, respectively) has to be extended to a three-valued logic: F, T, and in case of indeterminacy, *Maybe* (*M*). Note that this capitalized *Maybe* refers to the certainty of having an empty intersection set, in contrast to the pixel label describing the certainty of the location of a region.

There are nine rules needed to intersect two corresponding pixels. We present six, and the remaining three follow from symmetry. For shortness, we denote i for a pixel belonging to the lower approximation (interior for sure), e for a pixel that does not belong to the region for sure, and m for a pixel from the broad boundary. The following intersections shall be ordered; the left pixel is from **A**, and the right pixel from **B**.

- $i \cap i$: if an interior pixel of image **A** intersects with an interior pixel of image **B**, then the local four-intersection is not empty in exactly one intersection set:

$$I4_{\text{local}} = \begin{pmatrix} T & F \\ F & F \end{pmatrix}$$

- similar for $i \cap e$ and for $e \cap e$.

More complex are the remaining three cases:

- $i \cap m$: if an interior pixel of image **A** intersects with a maybe pixel of image **B**, then the local four-intersection is:

$$I4_{\text{local}} = \begin{pmatrix} M & M \\ F & F \end{pmatrix}$$

- $m \cap m$ results in a local four-intersection of:

$$I4_{\text{local}} = \begin{pmatrix} M & M \\ M & M \end{pmatrix}$$

- $e \cap m$ results in a local four-intersection of:

$$I4_{\text{local}} = \begin{pmatrix} F & F \\ M & M \end{pmatrix}$$

The global four-intersection matrix for two raster images is calculated by accumulating all local four-intersections with a logical *or*-operation. This logical *or*-operation has to be defined on three values, applying the three-valued logic of Lukasiewicz (Sinowjew, 1968). The definition is according to Table 3.2. It realizes simply that any evidence that an intersection set is not empty overrules (weaker) statements of being empty.

Table 3.2 Truth table of a three-valued OR.

	T	M	F
T	T	T	T
M	T	M	M
F	T	M	F

Additionally we assume *a-priori* that the intersection sets of the two exteriors are never empty, i.e. we deal with limited regions only, which overrules other possible outcomes of a bounded image overlay.

The Maybe in a local or global four-intersection can be interpreted as a wildcard, to be replaced by T as well as by F. With that, a four-intersection containing an M no longer refers to one topological relation of the set defined in Section 3.2, but to a set of relations. Consider again the example in Figure 3.4: the four-intersection characterizing both relations, *disjoint/meet* as well as *overlap*, is:

$$I4 = \begin{pmatrix} M & T \\ T & T \end{pmatrix}$$

For each of the sets of relations in Section 3.4.1 a corresponding three-valued four-intersection can be found. Combinatorially more than 3 four-intersection arrays can be designed. But as it is known from the egg-yolk model, other four-intersections have no corresponding geometric interpretation. That means other than the 3 four-intersections do not occur from the overlay of three-valued images (here in the order of Section 3.4.1):

- four sets of relations containing one relation between pairs of crispings of the vague regions:

$$\begin{pmatrix} F & T \\ T & T \end{pmatrix}, \begin{pmatrix} T & T \\ T & T \end{pmatrix}, \begin{pmatrix} T & T \\ F & T \end{pmatrix}, \begin{pmatrix} T & F \\ T & T \end{pmatrix};$$

- five sets of relations containing two relations between the pairs of crispings:

$$\begin{pmatrix} M & T \\ T & T \end{pmatrix}, \begin{pmatrix} T & T \\ M & T \end{pmatrix}, \begin{pmatrix} T & M \\ T & T \end{pmatrix}, \begin{pmatrix} T & M \\ F & T \end{pmatrix}\begin{pmatrix} T & F \\ M & T \end{pmatrix};$$

- two sets of relations containing three relations between the pairs of crispings:

$$\begin{pmatrix} M & T \\ M & T \end{pmatrix}, \begin{pmatrix} M & M \\ T & T \end{pmatrix};$$

- one set of relations containing four relations between the pairs of crispings:

$$\begin{pmatrix} T & M \\ M & T \end{pmatrix};$$

- one set of relations containing all possible relations between the pairs of crispings:

$$\begin{pmatrix} M & M \\ M & T \end{pmatrix}.$$

Note that with increasing indeterminacy the number of *Maybe*'s increased, too. So far, two images containing a vague region each can be overlaid, and the resulting four-intersection will be one of the set of 3 four-intersections corresponding to the sets in Section 3.4.

3.5 HIERARCHICAL DETERMINATION OF TOPOLOGICAL RELATION BETWEEN CRISP REGIONS

Overlaying two images in the described manner grows in complexity with the size of the images. One idea to reduce the effort is by exploiting hierarchical representations of the images. Traversal through the hierarchy levels is necessary only at locations that could contribute new information to the four-intersection of a coarse level. We summarize this idea first for crisp regions (Winter, 1999a). In the next section the comparison is made for hierarchical reasoning with vague regions.

The hierarchical approach is based on a regular hierarchical partition of space, let us say an image pyramid. At the root level, knowledge of the exact location of a region in an image is uncertain. The only safe statement at the root level is that the region is a proper part of the covered image space. Therefore, we label the root node by a predicate *miXed* (x): it contains *certainly* in parts the interior and in parts the exterior of the region. At the next levels, we label homogeneous nodes by *interior* or *exterior*, and inhomogeneous nodes still by *miXed*. A pyramid of a binary image has at least at the bottom level only binary values, and *miXed* values occur at higher levels only. Let us presume that the two image pyramids cover the same space and have the same tree depth. To intersect two regions in two hierarchies requires local intersection rules and a definition for an *or*-operation.

Assume a breadth-first progress through the pyramids. At root level, the *miXed* root of **A** intersects with the *miXed* root of **B**, $x \cap x$. In this situation there is no evidence for any intersection set given, so we decide to label each intersection set with *Maybe*:

$$I4_{root} = \begin{pmatrix} M & M \\ M & M \end{pmatrix}$$

The fourth intersection set will be overruled by our *a-priori* assumption that the intersection set between the two exteriors is never empty ($Z_4 = T$). The meaning of this four-intersection is that all relations of the range of relations are possible, i.e. total indeterminacy. To specify the intersection sets at positions of wildcards one has to evaluate the next lower level of the pyramid. The second level consists of four nodes, where also other labels than *miXed* can occur. One finds:

- $i \cap i$: if an interior node of pyramid **A** intersects with an interior node of pyramid **B**, then the local four-intersection is not empty in exactly one intersection set:

$$\begin{pmatrix} T & F \\ F & F \end{pmatrix};$$

and similarly for $i \cap e$ and $e \cap e$. Again, the interesting cases are:

- $i \cap x$: if an interior node of pyramid **A** intersects with a *miXed* node of pyramid **B**, then the local four-intersection is:

$$\begin{pmatrix} T & T \\ F & F \end{pmatrix};$$

- $x \cap x$ results in a local four-intersection of

$$\begin{pmatrix} M & M \\ M & M \end{pmatrix};$$

- $e \cap x$ results in a local four-intersection of

$$\begin{pmatrix} F & F \\ T & T \end{pmatrix}.$$

The four-intersections of one level need to be accumulated by an *or*-operation again. Here the same three-valued logic holds as above, so the operation remains unchanged (Table 3.2). Finer levels need to be visited only where still a wildcard in the four-intersection exists. This incremental in-depth progress reduces the costs enormously. Reasoning with crisp regions guarantees that always a completely determined four-intersection exists, which is found at the latest at the bottom level, but sometimes earlier. The hierarchical procedure allows a determination of the relation from coarse to fine, where coarse means the full range of relations possible at root level, and refining the set step by step. The procedure could stop at any time, when the resolution of the result is fine enough to answer a given query; this reduces the number of calculations again.

3.6 HIERARCHICAL DETERMINATION OF TOPOLOGICAL RELATIONS BETWEEN VAGUE REGIONS

The reasoning for vague regions is compared here with the hierarchical reasoning for crisp regions. We show that the principle is the same. The difference is in the detail: partition elements are considered as *maybe* in the context of vague regions, compared to *miXed* in the context of crisp representations. This difference leads to slightly less effective evidence through hierarchical reasoning for vague regions.

3.6.1 A hierarchical representation for vague regions

On vague regions also a hierarchy like an image pyramid can be built. For this task one has to decide how to treat mixed nodes that cover non-homogeneous areas in the original image. In the simplest approach, coarser elements (nodes) containing only exterior pixels at the finer level will be labelled as exterior, and nodes containing only interior pixels at the finer level will be labelled as interior. Finally, all other nodes will be labelled as maybe. That means that a parent node is labelled maybe if it covers children that are labelled maybe partly or in total, but also if it covers children belonging partly to interior and partly to exterior (which was called miXed in crisp context). Hence this rule generalizes the meaning of maybe. Definitions that diversify more specifically could improve the efficiency of the algorithm discussed later, by giving more specific information at coarser tree levels. An example of a hierarchy for a vague region is given in Figure 3.5.

F	F	F	F				
F	F	F	M	M	M	M	
F	F	M	M	**T**	**T**	M	
F	M	M	M	**T**	**T**	M	
F	M	M	M	**T**	M	M	
F	F	M	M	M	M		
F	F	M	M				
F	F	F	F				

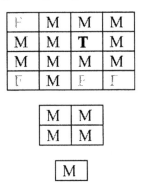

Figure 3.5 A pyramid for an 8 × 8 image containing a vague region.

3.6.2 Intersection of two hierarchical vague regions

With the vague hierarchy at hand, the hierarchical procedure for evaluating a topological relation is now applied. At the root level, two *maybe* nodes will be intersected, which results in an undetermined four-intersection with four wildcards. Traversing breadth-first through the hierarchy will replace wildcards step by step. With *maybe* labelled leaves, at the end of the hierarchical reasoning the four-intersection is not necessarily determined completely, leaving a set of topological relations as possible.

3.6.3 Comparison of the hierarchical approaches

So far the algorithm for hierarchical reasoning looks the same for crisp and for vague regions. One has only to replace the decision rules from Section 3.5 with the rules from Section 3.4.2, due to the different meaning of *miXed* and *maybe* pixels.

Consider first the set of decision rules for crisp regions. Overlaying two *miXed* nodes contributes no evidence for any intersection set. So if two *miXed* nodes are to be overlaid, than an incremental refinement at the next lower level at that position is necessary. However, overlaying a *miXed* node with an interior or an exterior node yields a completely determined four-intersection. At these positions it is no longer necessary to go deeper in the tree, and the process can stop at the actual level.

Consider now the set of decision rules for vague regions. Again, overlaying two *maybe* nodes contributes no evidence for an intersection set, and the four-intersection remains undetermined. At positions where two such nodes are to be overlaid, an incremental step to the next lower level in the hierarchy is necessary. But also overlaying *maybe* nodes with interior or exterior nodes leave some intersection sets in the four-intersection undetermined. That means, a hierarchical reasoning with vague regions needs incremental refinement if any of two corresponding nodes is labelled by *maybe*. In consequence the processing of a vague hierarchy takes more time, because indeterminacy needs to be fixed at lower hierarchy levels.

3.7 CONCLUSIONS AND DISCUSSION

We presented a hierarchical approach to determine the topological relation between two vague regions. For that reason, we adapted a hierarchical approach for crisp regions, which fits well because the semantic difference of a coarse (generalized) representation of a region and a vague region leads only to modified decision rules for local four-intersections, preserving the complete procedure.

Frequently, the result of the hierarchical reasoning will be a set of possible topological relations. This will be the minimal set given the vagueness of the location of the considered regions, with the extreme relations of the lower and the upper approximations of the regions, and the other relations of the set in between. One can interpret the set of possible relations as a two-dimensional connected interval, where the true relation (if a crisp realization exists) must be included. Generally, refinement of spatial resolution refines the set of relations, and decreases the uncertainty about the relation.

Vague regions are an interesting concept that fit better to the needs of representing spatial phenomena in digital systems than crisp representations. For that reason it is useful to develop reasoning methods dealing with vague regions, to include uncertainty of positional knowledge and propagate it to uncertainty of the reasoning results. It seems to be a natural approach to represent the uncertainty about the exact topological relation by a set of likewise possible relations. Additionally, the hierarchical reasoning is a more efficient procedure than the determination in image space.

REFERENCES

Baarda, W., 1967, *Statistical Concepts in Geodesy.* New Series, 2. (Delft: Netherlands Geodetic Commission).

Bittner, T., 1999, Rough Location. Ph.D. thesis, Department of Geoinformation, Technical University Vienna, Vienna.

Bittner, T. and Winter, S., 1999, On Ontology in Image Analysis. In *Integrated Spatial Databases*, edited by Agouris, P. and Stefanidis, A. Lecture Notes in Computer Science 1737. (Berlin: Springer), pp. 168–191.

Burrough, P.A. and Frank, A.U., 1996, *Geographic Objects with Indeterminate Boundaries.* GISDATA, 2,. (London: Taylor & Francis).

Casati, R. and Varzi, A., 1995, The Structure of Spatial Localization. *Philosophical Studies*, 82 (2), pp. 205–239.

Clementini, E. and Felice D. P., 1996, An Algebraic Model for Spatial Objects with Indeterminate Boundaries. In *Geographic Objects with Indeterminate Boundaries.* GISDATA, 2, edited by Burrough, P.A. and Frank, A.U., (London: Taylor & Francis), pp. 155–169.

Cohn, A.G., Bennett, B., Goodday, J. and Gotts, N., 1997, Qualitative Spatial Representation and Reasoning with the Region Connection Calculus. *Geoinformatica*, 1 (3), pp. 1–44.

Cohn, A.G. and Gotts, N.M., 1996, The 'Egg-Yolk' Representation of Regions with Indeterminite Boundaries. In *Geographic Objects with Indeterminate Boundaries.*

GISDATA, 2, edited by Burrough, P.A. and Frank, A.U., (London: Taylor & Francis), pp. 171–187.

Egenhofer, M.J., 1989, Spatial Query Languages. Ph.D. Thesis, University of Maine, Orono, Maine.

Egenhofer, M.J. and Al-Taha, K.K., 1992, Reasoning about Gradual Changes of Topological Relationships. In *Theories and Models of Spatio-Temporal Reasoning in Geographic Space*, edited by Frank, A.U., Campari, I. and Formentini, U. Lecture Notes in Computer Science 639 (Berlin: Springer), pp. 196–219.

Egenhofer, M.J. and Franzosa, R.D., 1991, Point-set topological spatial relations. *International Journal of Geographical Information Systems*, 5 (2), pp. 161–174.

Kong, T.Y. and Rosenfeld, A., 1989, Digital Topology: Introduction and Survey. *Computer Vision, Graphics, and Image Processing*, 48, pp. 357–393.

Pawlak, Z., 1982, Rough Sets. *International Journal of Computer and Information Sciences*, 11 (5), pp. 341–356.

Randell, D.A., Cui, Z. and Cohn, A., 1992, A Spatial Logic Based on Regions and Connection. In *Third International Conference on the Principles of Knowledge Representation and Reasoning*, edited by Brachmann, R., Levesque, H. and Reiter, R. (Los Altos: Morgan-Kaufmann), pp. 165–176.

Rosenfeld, A., 1979, Digital Topology. *American Mathematical Monthly*, 86, pp. 621–630.

Samet, H., 1990, *The Design and Analysis of Spatial Data Structures*, (Massachusetts: Addison-Wesley).

Sinowjew, A.A., 1968, *Über mehrwertige Logik. Deutscher Verlag der Wissenschaften*, (Berlin: Deutscher Verlag der Wissenschaften).

Smith, B., 1995, On Drawing Lines on a Map. In *Spatial Information Theory*, edited by Frank, A.U. and Kuhn, W. Lecture Notes in Computer Science 988, (Berlin: Springer), pp. 475–484.

Smith, B. and Mark, D.M., 1998, Ontology and Geographic Kinds. In *Proceedings of 8th International Symposium on Spatial Data Handling (SDH '98)*, edited by Poiker, T.K. and Chrisman, N., (Vancouver: International Geographic Union), pp. 308–320.

Tobler, W., 1984, Application of Image Processing Techniques to Map Processing, In *Proceedings of International Symposium on Spatial Data Handling*, (Zurich: International Geographical Union), pp. 140–145.

Winter, S., 1999a, Topological Relations in Hierarchical Partitions. In *Spatial Information Theory*, edited by Freksa, C. and Mark, D.M. Lecture Notes in Computer Science 1661 (Berlin: Springer), pp. 141–155.

Winter, S., 1999b, Distances for Uncertain Topological Relations. In *Geographic Information Research - Transatlantic Perspectives*, edited by Craglia, M. and Onsrud, H., (London: Taylor & Francis), pp. 449–459.

Winter, S., 2000, Location Similarity of Regions. *ISPRS Journal of Photogrammetry and Remote Sensing*, 55 (3), pp. 189–200.

Worboys, M., 1998, Imprecision in Finite Resolution Spatial Data. *GeoInformatica*, 2 (3), pp. 257–279.

CHAPTER FOUR

Topological Relationships between Spatial Objects with Uncertainty

Wenzhong Shi and Wei Guo

ABSTRACT

The research issue of topological relationships between spatial objects, particularly objects with uncertainties, has gained a lot of attention during the past two decades. However, the formal representation and calculation of topological relationships is still an open issue and needs to be further developed. This chapter presents a study on formal representing and calculating of the topological relationships between uncertain spatial objects in two- or higher-dimensional space. This is under the assumption that the core of a geographic object is certain geometrically, and its boundary area is uncertain. First, a unified structure for representing certain or uncertain spatial objects is proposed. Second, a formal framework to represent topological relationships between uncertain spatial objects is presented. Finally, the decision algorithms for determining topological relationships are provided.

4.1 INTRODUCTION

In GIS, the term spatial data model refers to the representation and organization of spatial data. Such a model should be able to represent various kinds of spatial objects in the real world, and to be used to answer various kinds of questions related to the objects. Topological relationships are defined as invariant under geometric distortion, and form a subset of spatial relationships. The representation of topological relationships between certain or uncertain spatial objects is one of the major issues in spatial data modelling for GIS. By applying topological relationships, it is potentially possible to speed up spatial access and to design consistency-checking tools. As a result, spatial analysis and spatial reasoning can be more efficient.

Topological relationships between spatial objects have gained much attention by researchers during the past two decades. Several significant formal models of topological relationships between spatial objects with precisely defined boundaries have been developed. Egenhofer (1993) provided a 4-intersection model and then a 9-intersection model for analysing binary topological relationships between point sets. Cohn and Gotts (1996) developed a theory of regions and their relationships based upon single primitive contact relationships, called connections, between regions. Smith (1996) introduced a mix of metrology and topology to provide a theory of parthood and boundary for spatial objects. However, these models cannot be directly applied to the topological relationships of spatial objects with uncertain boundaries that may appear in many cases, such as in geographical analysis or image understanding.

In recent years, research has been conducted for representing topological relationships between regions with indeterminate boundaries by extending the 9-intersection model. Clementini and Felice (1996) proposed an algebraic model by extending a region with a crisp boundary to a region with a broad boundary. Chen (1996) proposed a fuzzy 9-intersection model based on fuzzy set theory and mathematical morphology. Zhan (1998) defined an indeterminate boundary approximately by a set of α-boundaries, each of which are approximated by a set of α-cut level regions that can be treated as crisp regions. The binary topological relationships between fuzzy regions can thus be analysed by approximate computation. Cohn and Gotts (1996) constructed a set with 46 relationships that formed the basis of the "egg-yolk" representation model. They classified all possible cases according to the relationships between the spatial objects into the five basic relationships holding between the pairs egg-egg, yolk-yolk, and yolk-egg. Basically, this approach uses the same relationships between simple regions with sharp boundaries to describe topological configurations between the objects with uncertain boundaries by egg-yolk pairs.

The formal representation and calculation of topological relationships is still an open question to be further studied. In the past, most research has focused on formalizing two-dimensional topological relationships. Therefore, methods for formally representing uncertain topological relationships in three- or higher-dimensional space need to be further developed. Furthermore, for two given spatial objects it is necessary to design an algorithm to determine the appropriate topological relationship. In fact, a comprehensive theory and methods for spatial search are very important in the area of large spatial databases and also are essential for inferring new information in spatial reasoning.

This chapter presents a study on the formal representation and calculation of the topological relationships between uncertain spatial objects in two- or higher-dimensional space. There are three different types of uncertain geographic objects: (1) geometrically the core area is uncertain, and it is also surrounded by an uncertain boundary; (2) geometrically the core area is uncertain, but it is surrounded by a certain boundary; and (3) geometrically the core area is certain, and it is surrounded by an uncertain boundary. In this chapter, we find solutions for case (3), although geographical objects in both cases (1) and (2) exist in the real world. The rest of this chapter is structured as follows. Firstly, a unified definition and representation of uncertain spatial objects is proposed. Then, a framework to represent and calculate the topological relationships between uncertain spatial objects is presented. Lastly, the chapter includes conclusions and an outlook for further research in this area.

4.2 REPRESENTATION OF SPATIAL OBJECTS

Spatial objects can be defined based on the features that exist in the real world. Each spatial object is distinguishable from others, and phenomena in reality can be used to classify spatial objects into different types, where each type is a set of spatial objects with the same or similar properties. The definition of spatial object types is dependent on logical properties and has nothing to do with their representation and implementation within a computer environment. In general, spatial object types can be identified in n-dimensional space according to their degrees of freedom. For example, four types of spatial objects, such as point, line, surface, and body, can be identified in a 3-dimensional space. These spatial objects can be either certain or uncertain objects. A certain spatial object is an object whose spatial extents are clear and universally recognized, while an uncertain spatial object is an object that does not have well-defined boundaries or whose position and shape are unknown or cannot be measured exactly.

A spatial object can be defined based on topology. In this section, we will first explain the concepts of manifold, cell and cell complex. Then, we will give the definition of a spatial object in n-dimensional space.

4.2.1 *N*-Manifold

With respect to the conceptions of generic topology and algebraic topology, an n-dimensional manifold or n-manifold M^n is a topological space X in which any point $x \in X$ has a neighbourhood homeomorphic to an open unit n-dimensional disk D^n, where D^n is defined as in Equation 4.1:

$$D^n = \{(x_1, x_2, ..., x_n) \in R^n \mid (x_1^2 + x_2^2 + ... + x_n^2)^{1/2} < 1\} \qquad (4.1)$$

An n-manifold M^n with boundary is defined in exactly the same way as the excepted points on the boundary that have neighbourhood homeomorphic to an open unit n-dimensional hemi-disk $D^n_{1/2}$, where $D^n_{1/2}$ is defined as in Equation 4.2:

$$D^n_{1/2} = \{(x_1, x_2, ..., x_n) \in R^n \mid (x_1^2 + x_2^2 + ... + x_n^2)^{1/2} < 1 \; and \, x_1 \geq 0\} \qquad (4.2)$$

To simplify, it is assumed in this study that the boundary of an n-manifold with boundary has the properties of an $(n-1)$-manifold when it is considered in isolation (Pigot, 1992).

4.2.2 *N*-cell and *N*-cell Complex

An n-cell is a bordered subset of an n-manifold and thus homeomorphic to an n-manifold with $(n-1)$-manifold boundary. Depending on the dimension of the space, the cells of dimension 0, 1, 2 and 3 are defined respectively as follows:

    0-cell $\equiv$ point
    1-cell $\equiv$ line segment
    2-cell $\equiv$ area
    3-cell $\equiv$ sphere
    The geometric representation of these cells is shown in Figure 4.1.

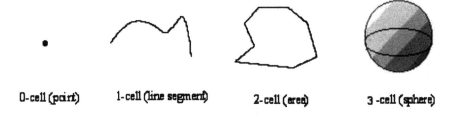

0-cell (point)   1-cell (line segment)   2-cell (area)   3-cell (sphere)

Figure 4.1 Cells of dimension 0, 1, 2 and 3.

An n-cell complex is the composition of all n-dimensional and lower-dimensional cells and satisfies the following conditions:
- Different elements of the n-cell complex have disjoint interiors;
- For each cell in the complex, the boundary is a union of elements of the complex; and
- If two cells intersect with each other, they also do so in a cell of the complex.

4.2.3 The definition of spatial object

Any certain n-dimensional spatial object with precise boundaries K can be represented as an n-cell complex. N-dimensional space can be subdivided into three components – the interior K°, the boundary ∂K and the exterior K^-. Let $C(p,r)$ denote an n-dimensional disk with radius r and centred at p, see Equation 4.3:

$$C(p,r) = \{q \in \Gamma | d(p,q) < r\} \qquad (4.3)$$

The interior, boundary and exterior of K can be defined as follows:

$$K^0 = \{p \in K : \exists \varepsilon > 0 : C(p,\varepsilon) \subset K\}$$
$$\partial K = \{p \in K : \forall r > 0 : C(p,r) \cap K \neq \varnothing \wedge C(p,r) \cap \Gamma - K \neq \varnothing\}$$
$$K^- = \{p \in \Gamma - K\}$$

where Γ represents the n-dimensional space 3^n.

There are two types of factors that may cause uncertainties of an object: randomness and imprecision. Randomness is an assessment of our belief (or doubt) about an outcome, based on the available data. Imprecision is a feature of the data itself. It refers to data expressed as a range of possible values (Altman, 1994).

Uncertainty due to randomness mainly caused by imprecise measurements of spatial data, such as the errors of scanning, digitizing, selecting, projection, overlaying *etc.* (Goodchild and Dubuc, 1987). Because of errors, spatial objects stored in a spatial database may not correspond with their actual locations in the real world. The true location of a spatial object is only known within a certain area around the represented location in the spatial database. A number of probability-based error modelling studies have been conducted, such as those by Shi (1998) and Leung (1997).

Uncertainty due to imprecision may occur, for example, in the recognition of spatial objects and their classification. The boundaries of imprecise objects may be fuzzy or indeterminate, such as the boundaries of a city or an area of forest. Fuzzy sets have been used for describing imprecise objects in GIS (Altman, 1994; Leung and Yan, 1998). For describing an imprecise object, we can take a precise region with a precise boundary as a reference. A fuzzy or indeterminate boundary region can be defined as a fuzzy region that surrounds the precise region.

By invoking methods to characterize certainty, imprecision and randomness of spatial objects, a unified framework for representing spatial objects can be developed as follows. Let Γ denote the n-dimensional space 3^n with the usual metric d, and suppose K is an n-dimensional spatial object, $\bar{x}$ is an arbitrary point in K and represented as $\bar{x} = (x_1, x_2, ..., x_n)$, where $(x_1, x_2, ..., x_n)$ is the spatial co-ordinate of $\bar{x}$. We can decompose K into three regions: (1) the core RK, (2) the crust BK; and (3) the exterior EK. The core region RK is the certain part of spatial object K which is a homogeneous n-dimensional manifold in Γ. It can be subdivided into two components: the interior RK° and the boundary ∂RK. The crust region BK is the uncertain part for spatial object K. For a certain object, the crust region is empty. It can be the indeterminate boundary of an imprecise object or the confidence region of a random object. The exterior region EK of spatial object K is the set Γ-K, that is, EK=Γ-K. The relations between the three regions can be concisely represented as follows:

$$K = RK \cup BK$$
$$RK \cup BK \cup EK = \Gamma$$
$$RK \cap BK \cap EK = \varnothing$$

According the method of fuzzy statistics, the result of statistical and probabilistic models can be translated into a membership function. Fuzzy set theory is prominent as an appropriate means of modelling both randomness and imprecision. Thus, an uncertain spatial object can be represented as a fuzzy object K. The fuzzy membership of K is denoted by Equation 4.4:

$$\mu_K(\bar{x}) = \begin{cases} 1, & \bar{x} \in RK \\ \mu_K, & \bar{x} \in BK \\ 0, & \bar{x} \in EK \end{cases} \tag{4.4}$$

where $0 < \mu_k < 1$, $\bar{x}$ is an arbitrary point in n-dimensional space.

For example, in 3-dimensional space, the core, the crust, and the exterior regions of a point object P*, a line object L*, a surface object S* and a body object V* can be represented as in Figure 4.2.

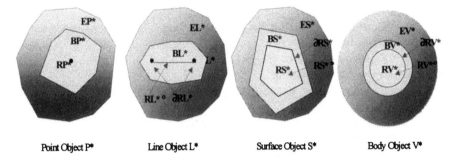

Figure 4.2 The core, crust, and exterior of uncertain spatial objects.

4.3 TOPOLOGICAL RELATIONSHIPS BETWEEN OBJECTS

Topological relationships become more complex under uncertain situations than under certainty, because any uncertain spatial objects K can consist of three components, that is the core RK, crust BK, and exterior EK. Meanwhile, the relationships between EK and other components can be determined by the relationships between RK and BK due to the existing relation $RK \cup BK \cup EK = \Gamma$, that is, $EK = \Gamma - RK \cup BK$. Thus, we propose our framework based on the relationships between core and crust components of two objects. The topological relationships between any two objects K_1 and K_2 can be concisely represented as a logical representation and thirteen basic relationship types as defined below:

1) The *disjoint* relationship
$$< K_1, disjoint, K_2 > \Leftrightarrow (K_1 \cap K_2 = \varnothing)$$

2) The *meet* relationship
$$<K_1, meet, K_2> \Leftrightarrow (RK_1 \cap RK_2 = \varnothing) \wedge (BK_1 \cap BK_2 \neq \varnothing) \wedge (RK_1 \cap BK_2 = \varnothing) \wedge (BK_1 \cap RK_2 = \varnothing)$$

3) The *nearly meet* relationship
$$<K_1, nearly\ meet, K_2> \Leftrightarrow (RK_1 \cap RK_2 = \varnothing) \wedge (BK_1 \cap BK_2 \neq \varnothing) \wedge ((RK_1 \cap BK_2) \cup (BK_1 \cap RK_2) \neq \varnothing)$$

4) The *nearly overlap* relationship
$$<K_1, nearly\ overlap, K_2> \Leftrightarrow (RK_1 \cap RK_2 = \varnothing) \wedge (BK_1 \cap BK_2 \neq \varnothing) \wedge ((RK_1 \cap BK_2) \neq \varnothing) \wedge (BK_1 \cap RK_2) \neq \varnothing)$$

5) The *overlap* relationship
$$<K_1, overlap, K_2> \Leftrightarrow (RK_1 \cap RK_2 \neq \varnothing) \wedge (RK_1 \not\subset K_2) \wedge (RK_2 \not\subset K_1)$$

6) The *nearly covered by* relationship
$$<K_1, nearly\ covered\ by, K_2> \Leftrightarrow (RK_1 \subseteq K_2)$$

7) The *nearly covered* relationship
$$<K_1, nearly\ covered, K_2> \Leftrightarrow (RK_2 \subseteq K_1)$$

8) The *covered by* relationship
$$<K_1, covered\ by, K_2> \Leftrightarrow (RK_1 \subseteq RK_2) \wedge (K_1 \subseteq K_2)$$

9) The *covers* relationship
$$< K_1, covers, K_2> \Leftrightarrow (RK_2 \subseteq RK_1) \wedge (K_2 \subseteq K_1)$$

10) The *inside* relationship

$$<K_1, inside, K_2> \Leftrightarrow (K_1 \subseteq RK_2)$$

11) The *contains* relationship

$$< K_1, contains, K_2> \Leftrightarrow (K_2 \subseteq RK_1)$$

12) The *nearly equal* relationship

$$< K_1, nearly\ equal, K_2> \Leftrightarrow (RK_1 \subseteq K_2) \wedge (RK_2 \subseteq K_1)$$

13) The *equal* relationship

$$< K_1, equal, K_2> \Leftrightarrow (K_1 \subseteq K_2) \wedge (K_2 \subseteq K_1)$$

For example, under uncertainty, the topological relationships between two spheres in 3-dimensional space can be described as shown in Figure 4.3.

The topological relationship types described above are mutually exclusive. That is to say, it is not possible that any of the above two different relationship types can hold, between K_1 and K_2, simultaneously. Furthermore, these relationship types make a complete set of all possible topological relationships. Thus, the relationship types can be taken as a partition of the topological relationships set. The topological relationships between any two given objects must be one of these relationship types. This conclusion can be proved by using Figure 4.4. Suppose the rectangle with maximum size represents the topological relationships set. It can be partitioned into thirteen connective and non-overlap components according to the conditions indicated by the arrowheads. Each component corresponds to just one of thirteen topological relationship types. To simplify, it is further assumed that the crust region of any object is much smaller than its core region. The close relationship proposed by Clementini and Felice (1996) and Cohn and Gotts (1996), thus, is not possible. Other topological relationships can be divided into one of the thirteen topological relationship types.

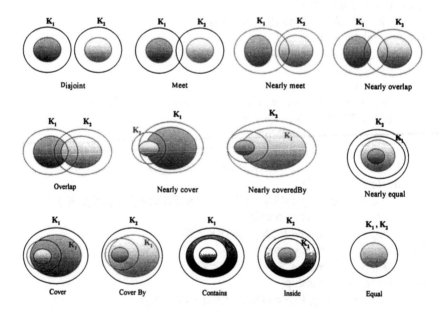

Figure 4.3. The topological relationships between two uncertain spatial objects.

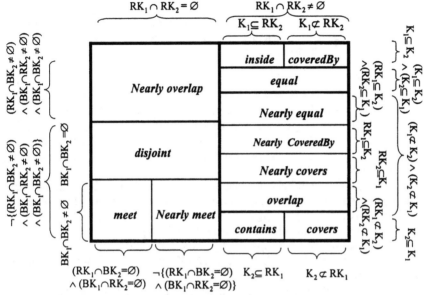

Figure 4.4. Mutual exclusiveness and full covering of topological relationships.

4.4 A DECISION ALGORITHM FOR TOPOLOGICAL RELATIONSHIPS

Suppose the fuzzy memberships of spatial objects K_1 and K_2 are μ_{k1} and μ_{k2}, which are denoted by Equations 4.5 and 4.6:

$$\mu_{K_1}(\bar{x}) = \begin{cases} 1, & \bar{x} \in RK_1 \\ \mu_{K_1}, & \bar{x} \in BK_1 \\ 0, & \bar{x} \in EK_1 \end{cases} \tag{4.5}$$

$$\mu_{K_2}(\bar{x}) = \begin{cases} 1, & \bar{x} \in RK_2 \\ \mu_{K_2}, & \bar{x} \in BK_2 \\ 0, & \bar{x} \in EK_2 \end{cases} \tag{4.6}$$

where $0 < \mu_{k1} < 1$, $0 < \mu_{k2} < 1$, and $\bar{x}$ is any point in n-dimensional space with co-ordinates $\bar{x} = (x_1, x_2, ..., x_n)$.

The topological relationships between two spatial objects K_1 and K_2 can be determined by the fuzzy memberships of Equation 4.7:

$$\mu(K_1, K_2) = \bigvee_{\bar{x}_i} [\mu_{K_1}(\bar{x}_i) \wedge \mu_{K_2}(\bar{x}_i)] \tag{4.7}$$

where the signs $\wedge$ and $\vee$ correspond to the minimum and maximum operations in fuzzy-set theory.

If $\mu(K_1, K_2) = 0$, K_1 and K_2 are *disjoint* with each other.

If $\mu(K_1, K_2) = 1$, it indicates that the interiors of K_1 and K_2 intersect. K_1 and K_2 may have one of the relationships: "nearly covered by", "covered by", "inside", "nearly covers", "covers", "contains", "overlap", "equal" or "nearly equal". They can be determined by the following methods.

For $\forall \overline{x}_i$, if $\mu_{k1}(\overline{x}_i) = \mu_{k2}(\overline{x}_i)$, K_1 is *equal* to K_2,

For $\forall \overline{x}_i \in K_1$, if $\mu_{k2}(\overline{x}_i) = 1$, K_1 is *inside* K_2, else if $\mu_{k2}(\overline{x}_i) \neq 0$, K_1 is *covered by* K_2. If the reverse is true, then K_1 *contains* K_2 or K_1 *covers* K_2.

For $\forall \overline{x}_i \in K_1$, if $\mu_{k1}(\overline{x}_i) = 1$, we have $\mu_{k2}(\overline{x}_i) \neq 0$, and for $\forall \overline{x}_j \in K_2$, if $\mu_{k2}(\overline{x}_j) = 1$, we have $\mu_{k1}(\overline{x}_i) \neq 0$, and we can say that K_1 *is nearly equal to* K_2. Otherwise, for $\forall \overline{x}_i \in K_1$, if $\mu_{k1}(\overline{x}_i) = 1$ and $\mu_{k2}(\overline{x}_i) \neq 0$, we can say that K_1 *is nearly covered by* K_2, and if the reverse, then K_1 *nearly covers* K_2. Otherwise, we say that K_1 *overlaps with* K_2.

If $\mu(K_1, K_2) = \mu$, it indicates that the interiors of K_1 and K_2 do not intersect. K_1 and K_2 may have the relationships "*meet*", "*nearly meet*" or "*nearly overlap*".

If $\exists \overline{x}_i \in K_1, \mu_{k1}(\overline{x}_i) = 1, \mu_{k2}(\overline{x}_i) = \mu_{k2}, 0 < \mu_{k2} < 1$, and $\exists \overline{x}_j \in K_2, \mu_{k2}(\overline{x}_j) = 1, \mu_{k1}(\overline{x}_j) = \mu_{k1}, 0 < \mu_{k1} < 1$, then K_1 *nearly overlaps with* K_2.

If $\exists \overline{x}_i \in K_1, \mu_{k1}(\overline{x}_i) = 1, \mu_{k2}(\overline{x}_i) = \mu_{k2}, 0 < \mu_{k2} < 1$, but $\forall \overline{x}_j \in K_2$, if $\mu_{k2}(\overline{x}_j) = 1, \mu_{k1}(\overline{x}_j) = 0$, we say K_1 *nearly overlaps with* K_2. If the reverse, we say K_2 *nearly overlaps with* K_1.

For $\forall \overline{x}_i \in K_1$, if $\mu_{k1}(\overline{x}_i) = 1$, we must have $\mu_{k2}(\overline{x}_i) = 0$, and for $\forall \overline{x}_j \in K_2$, if $\mu_{k2}(\overline{x}_j) = 1, \mu_{k1}(\overline{x}_j) = 0$, we say K_1 *meets with* K_2.

The value of $\mu(K_1, K_2)$ can be used to describe the magnitude of the common region of intersection. In general, the larger the value, the larger the common region of intersection. When $\mu(K_1, K_2)$ is equal to 1 or 0, we can define another metric function $\mu'(K_1, K_2)$ to describe the magnitude of the common region of intersection. For different application cases, the function $\mu'(K_1, K_2)$ may take different forms.

By applying the algorithms proposed in this chapter, we can determine which of the topological relationship types holds between any two given uncertain spatial objects. Spatial objects that have precise boundaries can be considered as a special case of those with indeterminate boundaries. The algorithms can be directly implemented in a raster-based or a vector-based GIS.

The concepts and methods can be further demonstrated with the following examples. Suppose we have two objects K_1 and K_2 with the fuzzy memberships listed in Figure 4.5. The spatial co-ordinates of the lower left corner for the two regions are $P(x_1,x_2,x_3)$ and $Q(x_1,x_2,x_3)$. If $P(x_1,x_2,x_3) = (1,1,0)$, $Q(x_1,x_2,x_3) = (18,1,0)$, we can calculate that $\mu(K_1, K_2) = 0$, and the relationship between K_1 and K_2 is *disjoint*. We now move the positions of the two regions. After this movement, we have $P(x_1,x_2,x_3) = (1,1,0)$, $Q(x_1,x_2,x_3) = (12,1,0)$, we can calculate that $\mu(K_1, K_2) = 0.6$. In this case, we can determine that the relationship between K_1 and K_2 is *meet* (see Figure 4.6). We further move the two regions, until $P(x_1,x_2,x_3) = (1,1,0)$, $Q(x_1,x_2,x_3) = (7,1,0)$ (see Figure 4.7). In this case, we have the result that $\mu(K_1, K_2) = 1$. Now, the relationship between K_1 and K_2 is *overlap*.

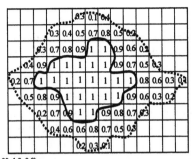
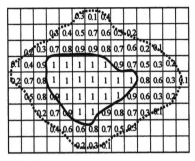

P (1.00, 1.0, 0.0) Q (18.00, 1.0, 0.0)

Figure 4.5 K₁ is *disjoint* with K₂.

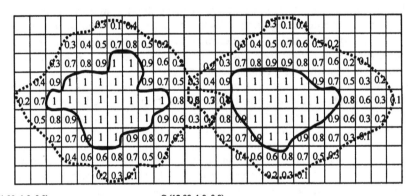

P (1.00, 1.0, 0.0) Q (12.00, 1.0, 0.0)

Figure 4.6 K₁ meets K₂.

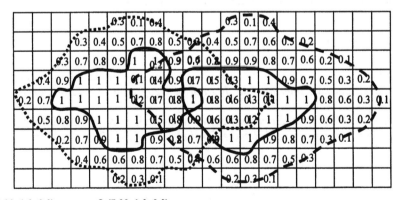

P (1.00, 1.0, 0.0) Q (7.00, 1.0, 0.0)

Figure 4.7 K₁ is *overlap* with K₂.

4.5 CONCLUSIONS

In this chapter, we presented a study of the formal representation of the topological relationships between spatial objects with uncertainty, in two-, three- or higher-dimensional space. The major difference between our approach and previous studies, such as the egg-yolk theory of Cohn and Gotts (1996) and the algebraic model proposed by Clementini and Felice (1996), is in the definition of spatial objects and the classification of topological relationships. In our approach, a spatial object is geometrically defined according to the specified topological relationships, and with the appropriate mathematical formula. As a result, these topological relationships can be quantified and calculated, which is very essential for practical implementation of the proposed method on a GIS platform. In addition, a unified structure for representing certain or uncertain spatial objects was proposed. Second, a formal framework to represent topological relationships between uncertain spatial objects was presented. This framework can be applied to solve topological and geometric problems using formal logic and algebraic methods. Third, decision algorithms for determining topological relationships were provided. An example was also given to demonstrate how the proposed framework was able to determine topological relationships on the uncertain objects.

This study provides a solution for determining topological relationships between objects with indeterminate boundaries, from a theoretical unified structure, and a framework of practical decision algorithms for implementing them in a GIS environment. This solution is not only for two- or three-dimensional objects, but is a generic solution for *n*-dimensional cases. The solution can potentially be applied to speed up spatial access. In addition, it is useful for the design of consistency checking tools. Other potential applications of the proposed solution are for higher efficiency in spatial analysis or for automated spatial reasoning.

This study can be further extended in the area of formal presentation of topological relationships between complex uncertain spatial objects. Furthermore, based on the presented theoretical models and algorithms, new tools for spatial query, analysis and reasoning can be developed for GIS.

REFERENCES

Altman, D., 1994, Fuzzy set theoretic approaches for handling imprecision in spatial analysis. *International Journal of Geographical Information Systems*, **8** (3), pp. 271–289.

Chen, X., 1996, Spatial relations between uncertain sets. In *International Archives of Photogrammetry and Remote Sensing, Vol. XXXI, Part B3*, Vienna, pp.105–110.

Clementini, E., and Felice, D.P., 1996, An algebraic model for spatial objects with indeterminate boundaries. In *Geographic Objects with Indeterminate Boundaries*, edited by Burrough, P. A. and Frank, A. U., (London: Taylor & Francis) pp. 155–169.

Cohn, A.G. and Gotts, N. M., 1996, The 'egg-yolk' representation of regions with indeterminate boundaries. In *Geographic Objects with Indeterminate Boundaries*, edited by Burrough, P. A. and Frank, A. U., (London: Taylor & Francis) pp. 171–187.

Egenhofer, M., 1993, A model for detailed binary topological relationships. *Geomatica*, **47** (3) & (4), pp. 261–273.

Goodchild, M. F., and Dubuc, O., 1987, A model of error for choropleth maps, with applications to geographic information systems, *Proc. Auto–Carto 8*, p. 165.

Kruse, R., Schwecke, E., and Heinsohn, J., 1991, *Uncertainty and Vagueness in Knowledge Based Systems: Numerical Methods*, (Berlin: Springer-Verlag).

Leung, Y., 1997, Point-in-polygon analysis under certainty and uncertainty. *Geoinformatica, No. 1*, pp. 93–114.

Leung, Y., and Yan, J., 1998, A locational error model for spatial features. *International Journal of Geographical Information Systems*, **12** (6), pp. 607–629.

Pigot, S., 1992, A topological model for a 3D spatial information system. In *Proceedings 10th International Symposium on Computer Assisted Cartography (AUTOCARTO 10), Technical papers, ACSM-ASPRS, Annual Convention*, **6**, Baltimore, Maryland, USA, pp. 369–391.

Shi, W., 1998, A generic statistical approach for modelling error of geometric features in GIS. *International Journalof Geographical Information Systems*, **12** (2), pp.131–143.

Smith, B., 1996, Mereotopology: A theory of parts and boundaries. *Data & Knowledge Engineering*, **20**, pp. 287–303.

Zhan, F.B., 1998, Approximate analysis of binary topological relations between geographic regions with indeterminate boundaries. *Soft Computing*, **2**(2), pp. 28–34.

CHAPTER FIVE

Error-aware GIS Development

Matt Duckham and John E. McCreadie

ABSTRACT

While techniques for handling error in geographical information are common in the research literature, such techniques are largely absent from today's commercial GIS software. This chapter highlights some of the barriers to the establishment of error handling in commercial GIS, and explores the development of practical error handling software that aims to overcome these barriers. The software developed has three distinct components, each of which is designed to make maximal reuse of existing technology already used in association with GIS, including object-orientation, artificial intelligence, distributed systems and the Internet. First, an *error-sensitive GIS* provides a flexible core of basic error handling functionality. Second, an *error-aware GIS* offers intelligent, domain specific extensions to the core error-sensitive functionality, so helping users to understand and make better use of spatial data quality. Third, a distributed systems bridge between error-sensitive and error-aware GIS enables developers to make a range of error handling tools available for users to access. To illustrate the effectiveness of the approach, the chapter outlines the application of the architecture to a legacy data capture project in the telecommunications industry.

5.1 INTRODUCTION

The inclusion of error handling capabilities within GIS is often viewed as crucial to the future commercial and legal viability of the technology. Error handling has remained high on the geographical information science research agenda for close to two decades. By contrast, commercial GIS have arguably not benefited significantly from this research. Systems capable of storing, managing and manipulating spatial data quality are a rarity outside the research laboratory. There are a variety of contributory factors to this undesirable situation, including:

- the lack of agreement on exactly what constitutes "error" and how it should be collected and reported;
- the cursory nature of most metadata, providing only limited raw materials upon which error-handling routines might operate;
- the highly technical, statistical or specialised nature of many error handling routines, thereby precluding their practical use for many general GIS applications and users;
- the negative connotations of the word "error", outside the research literature, which may deter software and data producers from emphasising error-handling capabilities.

It is certainly not possible to provide a technical solution to all these difficulties. Indeed, Goodchild (chapter 2, this volume) questions whether the problems posed by error handling may in fact require a complete rethink of GIS design. However, in this chapter we argue that it is possible to develop practical error-handling software using existing technology that at least eases some of the difficulties sketched above. An architecture capable of practical error handing is presented in three stages. Section 5.2 reviews the development of an *error-sensitive GIS*, which provides core error handling functionality in a form flexible enough to be widely applicable to error in spatial data. Section 5.3 introduces the concept of an *error-aware GIS*, which aims to provide intelligent, domain specific extensions to the core error-sensitive functionality, enabling users to better use and understand spatial data quality. Section 5.4 explores the deployment of distributed systems technology as a stable bridge between the flexibility of the error-sensitive GIS functionality and the specificity of the error-aware extensions. The resultant system, applied to a telecommunications legacy data capture example in Section 5.5, is used to indicate the suitability and robustness of the approach.

5.2 ERROR-SENSITIVE GIS

The term "error-sensitive GIS" (Unwin, 1995) is used to denote a GIS offering the basic error handling functionality needed to manage spatial data quality. In the past, attempts to develop error-sensitive GIS have tended to adopt one particular data quality standard as a starting point (e.g. Guptill, 1989; Ramlal and Drummond, 1992; van der Wel *et al.*, 1994). On first inspection this seems a reasonable decision. Most of the 22 national data standards included in Moellering (1997) are based largely on the U.S. National Committee for Digital Cartographic Data Standards proposed data quality standard (NCDCDS, 1988). The NCDCDS standard is now incorporated into the U.S. Spatial Data Transfer Standard (SDTS) and defines five elements of spatial data quality: lineage, logical consistency, completeness, attribute and positional accuracy. Reusing such standards is desirable, as it should ensure a degree of compliance and uniformity in error handling.

On closer inspection, however, the exact composition of data quality standards and the definition of individual elements within different standards exhibit wide variation. Standards are constantly evolving and new standards being developed. No one standard is necessarily better than another; it seems plausible that certain data standards will be well suited to particular spatial data types, and that certain spatial data may not even be adequately covered by current standards. The implication is that data quality standards can never be flexible enough for every user's needs. Before error-sensitive functionality can be included as a basic feature in GIS, a more flexible approach to data quality is needed that supports, but is not driven by standards.

5.2.1 Conceptual model of data quality

Happily, an approach to data quality that does not rely on any particular data quality standard does exist within the literature. The conventional conceptual model of information systems assumes that the complexities of the physical world can only usefully be handled following the abstraction of the physical world to a 'practically adequate' (Nyerges, 1991) idealised data set, often termed the *terrain nominal*. An actual data set or database constitutes a representation of this idealised data set. In contrast to geographical objects, however, there exists no meaningful concept of data quality in the physical world. Quality is not a feature of the objects being observed, rather it is a feature of the process of observation. Data quality arises only as a side effect of deficiencies in process of observation (David *et al.*, 1995).

A conceptual model of spatial data quality, then, can be founded on the idea that all error in geographical information results from the imperfect abstraction and representation of geographical phenomena. Since this model does not privilege any particular data quality standard, software development based on this model should result in an error-sensitive GIS able to support any quality standard or set of quality elements that might reasonably be proposed. By severing the relationship between software and standards, it is hoped that the emphasis for describing the quality of a data set will shift away from the standards organisation and towards the database designer, the data provider and the data user. In the same way as GIS currently allow users to design their own spatial data models, so error-sensitive GIS should allow users to design their own spatial data quality models. While data quality standards remain an important resource for data quality managers, it is data producers and users, not standards organisations, who are best placed to determine the most appropriate methods for representing the quality of their data.

5.2.2 Object-oriented software development

Following the identification of the conceptual model of data quality in the previous section as a suitable basis for error-sensitive GIS development, the next stage is to translate that conceptual model into a working software implementation. Object-oriented (OO) development techniques were selected for this task for two reasons. First, OO has been shown to be particularly well suited to the geographical information problem domain (Kösters *et al.*, 1997; Worboys *et al.*, 1990), and is now an established technology for GIS. Second, OO is a semantic modelling technique that can minimise problem complexity by allowing the development of models that are closer to our intuitive understanding of the problem domain. Details of the development process do not form part of this discussion (but see Duckham, 1999), although a few of the key results are highlighted below.

5.2.2.1 Object-oriented analysis results

Object-oriented analysis (OOA) is the first stage of OO software development and aims to produce simple architecture-neutral OO schemata that address a particular problem. Details of the OOA process can be found in a wealth of literature on the subject (e.g. Booch, 1994; Rumbaugh *et al.*, 1991). The analysis of the conceptual model of data quality above resulted in the specification of two core data quality concepts, related to the twin processes of abstraction and representation identified in Section 5.2.1 above.

The concept of *abstractive quality* is used to model data quality elements that are the result of deficiencies in the process of abstraction. For example, in order to be topologically consistent contours must not cross, and in order to be logically consistent must have height values that fall somewhere between their topological neighbours, usually changing in regular increments. Such behaviour, once specified, will be identical for all contour objects in a GIS. Logical and topological consistency are examples of abstractive quality elements, since they result from problems of specification and abstraction.

In contrast, *representative quality* is used to model quality elements resulting from deficiencies in the process of representation. Positional accuracy is one example of representative quality, since different survey methods, data collection and data provenance will mean that positional accuracy may vary for different objects of the same type. For the vector representation of a river network, say, we would expect positional accuracy to vary spatially across different regions, different rivers or even different vertices, perhaps in a complex autocorrelated way.

Existing data quality elements usually fall neatly into either the abstractive or representative quality categories. Since the conceptual model of data quality is focused on the underlying processes which give rise to data quality, it is argued that representative and abstractive quality form a more basic, flexible approach to data quality than standards. As a result, the model should be able to adequately model any reasonable data quality element (see Section 5.5.1).

5.2.2.2 Object-oriented design and implementation results

Using the OO mechanism of inheritance it is possible to transmit the properties encapsulated by abstractive and representative quality to all geographical objects throughout the error-sensitive database. This has two important implications. First, using this approach means that each individual geographical object in the database possesses the basic functionality necessary to be able to be interrogated about its own data quality. This provides the flexibility to associate data quality elements with other database objects at any level in the database, from individual coordinate objects, through complex aggregated geographical objects to entire map or coverage objects. Second, the approach should be compatible with any OO environment, including existing OO GIS. By taking advantage of the OO inheritance mechanism, the OOA results can be slotted into a pre-existing OO spatial database without requiring significant modifications.

Based on the OOA results, an OO error-sensitive data model was designed and implemented in Laser-Scan Gothic OO GIS. Actually using the error-sensitive GIS is similar to using a conventional OO GIS. Using a conventional OO GIS, such as Laser-Scan Gothic, begins by defining classes of geographical phenomena of interest (e.g. new "contour" and "river" classes), and making these classes inherit from appropriate generic spatial classes provided by the system, such as a generic "line" class. In a similar manner, using the error-sensitive GIS begins by defining a number of abstractive and representative quality elements, such as "logical consistency" and "positional accuracy", based on the generic quality classes, abstractive and representative quality, provided by the system. The resulting database can then be populated both with geographical and data quality objects, with the error-sensitive database managing the relationships between the two.

5.3 ERROR-AWARE GIS

The error-sensitive GIS development process discussed above aims to provide the core functionality needed to store and manage both geographical information and its associated data quality. However, in itself this functionality is of limited value. Information on error is only useful if it can be incorporated into the geographical decision-making process. Most research into spatial error handling does not go as far as to help users understand and apply data quality information in a practical way (Agumya and Hunter, 1997). An error-aware GIS aims to extend the core functionality of an error-sensitive GIS by assisting in the provision, use, and understanding of information about error in geographical information. The need to represent knowledge about error and apply that knowledge in an intelligent manner entails the use of artificial intelligence (AI) techniques. As an illustration of the concept, two example error-aware applications are discussed in this section.

5.3.1 Expert quality schema definition tool

As suggested in Section 5.2.1, an important aim of this research is to shift the emphasis in error handling away from standards organisations and towards data providers and data users. However, the task of producing an OO data quality schema matched to and integrated with a particular data set can itself be complex and rather subjective. In order to simplify the task, an expert system-based data quality schema definition tool was developed to assist non-specialist users with developing new quality schema for their spatial data. The tool models the data quality schema definition process as a mini-OOA of the factors likely to affect the quality of a particular data set. Tentative attempts to produce an expert system able to assist with the OOA process have already appeared in the literature (e.g. Belkhouche and Gamino, 1998). In fact, the problem domain for quality schema definition is comparatively narrow and consequently an expert system able to address that problem domain proved less problematic than attempts to build generic expert OOA tools.

The tool developed takes advantage of three sources of information. First, the OOA literature was used as a relatively rich source of rules and heuristics relating to the OOA process. Second, the spatial data quality literature was used both as a source of further rules and heuristics, and as the basis for a hierarchy of common data quality terms. Finally, the specific geographical object schema, which should be produced in the normal course of OO GIS database design, serves as an important fact-base for the expert system.

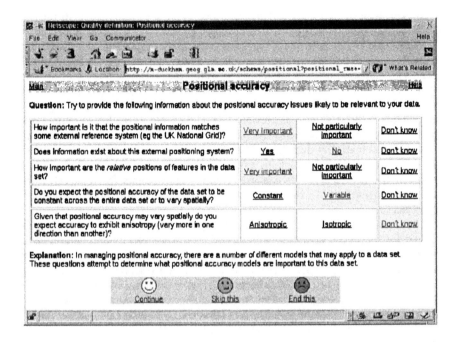

Figure 5.1 Expert quality schema definition tool interface.

The expert quality schema definition tool was implemented using the Java Expert System Shell (JESS), which is a Java-based version of the popular C Language Integrated Production System (CLIPS). The tool aims to encourage the database designer to answer questions about factors likely to affect the quality of data with respect to particular classes in the geographical object schema. The interface for the tool, shown in Figure 5.1, is hypertext-based to allow intuitive, non-linear navigation of questions. The information generated about the factors affecting the quality of the data set under investigation, combined with automatically generated facts about the geographical object classes, is used to build and integrate a new quality schema that hopefully meets the requirements of the data set in question.

5.3.2 Inductive quality capture tool

Despite growing acceptance amongst GIS companies and users of the importance of data quality, very often adequate data quality information for a given data set will simply not exist. Further, limited expertise and financial restrictions are likely to mean most data producers do not feel in a position to compile such quality information about their data. Consequently, an important component of the error-aware arsenal is a tool able to assist with the capture of data quality information in an efficient manner. To meet this need, a tool capable of assisting with the data quality capture process was developed based on an inductive learning algorithm.

Induction is not new to GIS and has been used, for example, to assist with geographical and habitat pattern analysis (Walker and Moore, 1988; Aspinall, 1992). The induction process implemented within the quality capture tool operates in two stages. First, based on a quality assessment for a small representative sub-sample of a full geographical data set (the *training* set), the induction algorithm uses patterns in the relationships between spatial data and its quality to build a rule-base that describes the quality of the training set. Subsequently, this rule-base can be used to automatically deduce data quality for the entire geographical data set, even when the training set contains examples that are similar but not identical to objects in the full data set. In this way, quality assessments for relatively small representative portions of a data set can be used to derive quality information for an entire data set.

In common with all inductive learning algorithms, the size of the training set required for successful operation will depend on the complexity and heterogeneity of the data set as a whole. In order to check the veracity of the induction process, the accuracy of the rule-base is cross-validated upon a reserved proportion of the training set, called the *test* set. This accuracy is reported at the end of any induction process as the percentage accuracy of induction. Low accuracy values are an indication to the user that the training data set needs to be extended and the process should be repeated. Conversely, high accuracy values are an indication that the induction should produce reasonable results when applied to the entire data set.

5.4 DISTRIBUTED SYSTEMS

While highly flexible, an error-sensitive GIS is an unsuitable tool for practical geographical decision-making. Error-aware GIS tools can assist in the use and understanding of data quality, but at the cost of domain specificity and concomitant loss of flexibility. Consequently, the isolation of the flexible error-sensitive data model from the domain specific error-aware applications is seen as crucial to the stability of this approach. Distributed systems allow the error-sensitive and error-aware GIS to seamlessly interoperate over a network, at the same time as safeguarding the flexibility of the error-sensitive GIS functionality.

There are a number of distributed systems software solutions currently available. The software used in this research was based on Java Remote Method Invocation (java.RMI). The advantage of java.RMI over similar distributed systems architectures, such as the Common Object Request Broker Architecture (CORBA) and the Distributed Component Object Model (DCOM), is that it is free, simpler and more powerful. The disadvantage is that java.RMI is currently a "Java only" solution, although java.RMI and CORBA are converging and it seems likely that a java.RMI-like architecture will be compatible with future CORBA standards.

The basic distributed systems architecture used in this study consists of three tiers, shown in Figure 5.2. A "middleware" java.RMI object server sits between the lower tier of the Gothic error-sensitive database and the upper tier of the error-aware applications.

5.4.1 Distributed architecture

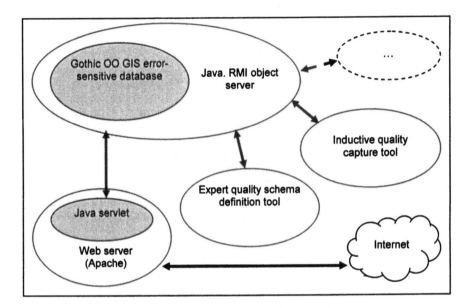

Figure 5.2 Three and four tier distributed systems architecture.

The java.RMI object server effectively implements a contract between error-aware and error-sensitive GIS, defining what services can be provided by the error-sensitive GIS and can be consumed by error-aware applications. The contract takes the form of a simple Java class definition file. Based on this file, an error-aware application can then make a request to the java.RMI server for an error-sensitive database object, either locally or over a network. Crucially, there are no other links between the error-sensitive and error-aware software, which as a result can be developed almost independently.

5.4.2 Java servlet architecture

While for security reasons it may not be appropriate to provide access to core database functionality over the Internet, some error-aware functionality, such as data quality exploration and visualisation tools, may need to be accessed over the Internet. Secure Internet access to the error-sensitive GIS is possible through the java.RMI object server using a four tier architecture. The fourth tier in this study was developed using the Java *servlet* application programming interface (javax.servlet API), which allows Java programs to be embedded within a web server such as Apache (illustrated in Figure 5.2). A Java servlet can access the java.RMI object server running on the error-sensitive host, and so provide a specific, restricted subset of distributed services more generally over the Internet. For example, a simple error-propagation and accuracy visualisation tool developed for use over the Internet is outlined in Section 5.5. In order to take advantage of familiar web-browser interfaces, such as Netscape, the expert quality schema definition tool, described in Section 5.3.1, is also based on the four tier architecture.

5.5 TELECOMMUNICATIONS APPLICATION

To be successful, error handling in GIS must be able to provide practical advantages over conventional GIS. Whilst the aim of this research has been the development rather than the application of error handling in GIS, the research would not be complete without an example application. The application chosen here is that of a telecommunications network in Kingston-upon-Hull, U.K. Utility applications generally are recognised as amongst the most important commercial uses of GIS, and telecommunications is one of the most active utility GIS application areas (Russomano, 1998). Advances in telecommunications use and technology coupled with the deregulation of the U.K. telecommunications industry have led to many telecommunications companies abandoning legacy paper-based mapping practices in favour of digital GIS. This study looked at how the proposed architecture would operate during this transition, with the assistance of one particular telecommunications company, Kingston Communications, U.K.

5.5.1 Error-sensitive application

Using documentation and data supplied with the consent of Kingston Communications, a version of an existing telecommunications database was implemented within the error-sensitive GIS, for a small region in Kingston-upon-Hull. The next step was to link a suitable quality schema to this database. To verify the flexibility of the error-sensitive database, three existing data quality standards were investigated in turn. The SDTS quality standard was chosen as it remains the *de facto* international standard whilst the International Standards Organisation (ISO) geographical information standard (ISO TC/211) continues development. The European draft standard (CEN TC/287), which contains a number of *meta-quality* elements (metadata about data quality), was also successfully tested. Finally, the Canadian Spatial Archive Interchange Format (SAIF) quality schema was successfully tested. Since SAIF is an explicitly OO standard, it was the most straightforward standard to support within the error-sensitive GIS. It is worth noting that the SAIF standard defines a number of quality elements as part of a broader approach to metadata. While the error-sensitive GIS is geared towards managing data quality information, the dividing line between quality and metadata is blurred, and there is no reason to suspect the error-sensitive GIS could not deal adequately with metadata more generally.

The study showed that all three quality standards investigated could be supported by the error-sensitive GIS database, including support for meta-quality. Just as importantly, however, the study also verified that hybrid and customised quality schemata could be used. A variety of such non-standard quality schemata were implemented, in particular those resulting from the use of the expert quality schema definition tool, discussed below. The ability of the error-sensitive GIS to deal with a range of data quality standards in addition to customised data quality schemata is taken to be a consequence of the in-built flexibility of the architecture, discussed in Section 5.2.2.1.

5.5.2 Error-aware application

Experiences during this study suggested that there exists a high level of informal awareness of spatial data quality issues amongst utility professionals. However, few utility companies see enough potential rewards in error handling to commit the high levels of additional resources needed to effect comprehensive spatial data quality management. The error-aware GIS architecture can help address this problem through the provision of simple, effective and cheap methods of data quality management. The error-aware expert schema definition tool described in Section 5.3.1 can be used to define customised quality schemata (discussed above in Section 5.5.1) automatically. In practice, the schemata produced by the tool were not vastly different to those found in data quality standards: the basic quality terms used in the fact-base were drawn from the same literature upon which data quality standards are usually based. However, the expert schema definition tool is intuitive and simple, automatically integrating the quality schema produced with the geographical object schema.

Following quality schema definition, the inductive quality capture tool can be used to efficiently capture data quality information, based on a pilot assessment of the quality of a small area of the legacy map data being captured. Perversely, a significant body of quality information associated with legacy paper maps will usually be lost during upgrading to digital mapping. Lineage information on the provenance of maps and map features is well known to engineers used to handling those maps. Levels of accuracy, precision and detail are often implied by the physical limitations of the map, limitations that do not apply once the map is digitised. Consequently, it proved entirely feasible to derive a picture of the history and accuracy of the pilot quality assessment without embarking on expensive resurvey. Once added to the telecommunications data within the inductive quality capture tool, this pilot quality assessment can be used as a training set for the induction process. The tool looks for patterns in the geographical data that imply patterns in the quality data, resulting in a rule-base tailored to the particular features of the telecommunications data being captured. Once created, this rule-base can then be applied to the remainder of the data, automatically populating the database with quality information deduced from the pilot quality assessment.

On balance, the patterns found in the data by the inductive algorithm should be indicative of more general processes. For example, complex or densely packed features on the paper maps were found to be more likely to be of lower accuracy once entered into the GIS, as they may be harder to understand and digitise or may be deliberately displaced on the paper map for cartographic reasons. The inductive algorithm is sensitive to data exhibiting these types of relationships and when shown a data set where low accuracy and high feature density are coincident is able to derive rules that reflect this relationship. However, the algorithm is far from infallible. In particular, if the training data set is unrepresentative or too small, the algorithm is much more likely to derive rules that relate to no particular processes or are entirely coincidental. When deciding whether a pilot assessment is sufficiently large and representative, the automatically produced cross-validation accuracy statistic provides an invaluable guide. The best results in this study were produced by pilot assessments covering between 5 and 10% of the total number of features. As a last line of defence, however, all automatically generated quality objects are associated with a meta-quality object that reports both the fact that the quality object was automatically generated and a record of the automated rules used to arrive at a particular quality object. This meta-quality information forms the basis of a "quality audit" that enables the original data set to be retrieved even after quality capture.

Finally, a simple error-propagation and accuracy visualisation tool was developed to provide access to the results of the data quality capture process over the Internet. The tool was aimed at site engineers in utility companies who need to know about both the location and accuracy of existing utility installations, for example when planning on-site work. Figure 5.3 shows an example visualisation, showing the likely intersection between existing telecommunication installations and a proposed new utility installation, based on the accuracies of the different features. The more blurred areas (red in the full colour visualisation) reflect the probability of an intersection between existing and proposed utility installations. The error propagation technique used to derive the visualisation is after Shi (1998). The error propagation itself is performed by a Java servlet using the four tier architecture described in Section 5.4.2, resulting in Graphics Interchange Format (GIF) files which can be accessed and displayed using any common web browser.

5.6 CONCLUSIONS AND FURTHER WORK

This research has attempted to sketch a blueprint for data quality handling in GIS. Importantly, the architecture proposed relies on reusing a wide range of existing technology, including OO, AI, distributed systems, servlets, hypertext and the Internet. The basic system architecture comprises three separate units. First, to address the need for flexible, generic error handling functionality, an OO error-sensitive spatial database was developed without relying on any particular data quality standard. Second, to address the need for domain specific, specialised error handling tools, the concept of an error-aware GIS is proposed, illustrated using two example error aware applications based on AI technology. To complete the architecture, distributed systems form a bridge between error-sensitive and error-aware GIS, enabling error-aware applications access to geographical information and its associated quality as well as safeguarding the flexibility of the error-sensitive database.

Arguably, adopting this architecture can ease some of the problems posed at the beginning of this chapter. Since there exists no widespread agreement on exactly what constitutes "error", the error-sensitive architecture is flexible enough to support a range of different data quality schemata suited to the needs of particular information communities. Despite the cursory nature of most metadata, the error-aware architecture is able to support AI-based tools, such as the inductive quality capture tool, which can make maximal use of minimal amounts of data quality information. Even though error-handling routines are often highly technical, statistical or specialised, the distributed architecture allows domain specific error-aware applications, such as the expert quality schema definition and accuracy visualisation tools, to be integrated with generic error-sensitive functionality at the same time as keeping the different components entirely separate. Finally, while the word "error" may always have negative connotations, integrating error-handling capabilities with GIS using existing technology should result in at least a more widespread acceptance of the endemic nature of error in geographical information. The hope is that by making spatial data quality information as easy, flexible and intuitive to use as other types of information within a GIS, use of data quality may become more commonplace, opening the door to far greater understanding, application and sharing of geographical information.

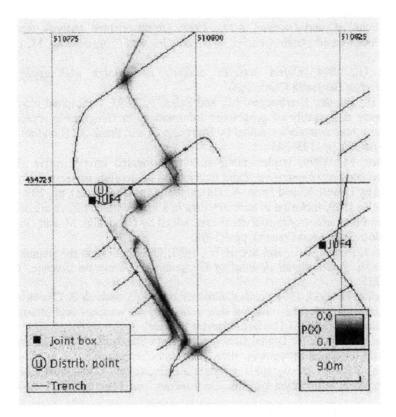

Figure 5.3 Example accuracy visualisation.

ACKNOWLEDGEMENTS

This research was funded by NERC and SDS U.K. under a CASE award at the University of Glasgow. Gothic software and support was generously supplied under a development license by Laser-Scan, U.K. Finally, the help, interest and openness of the employees of Kingston Communications and Informed Solutions is thankfully acknowledged. The author is currently funded by the EPSRC at the University of Keele under grant GR/M 55585 "Managing vagueness, uncertainty and granularity in spatial information systems".

REFERENCES

Agumya, A. and Hunter, G., 1997, Determining fitness for use of geographic information, *ITC Journal*, **2**, pp. 109–113.

Aspinall, R., 1992, An inductive modelling procedure based on Bayes theorem for analysis of pattern in spatial data, *International Journal of Geographical Information Systems*, **5**, pp. 105–121.

Belkhouche, B. and Gamino, A.M., 1998, Object-oriented analysis through a knowledge based system, *Journal of Object-Oriented Programming*, **11**, pp. 52–59.

Booch, G., 1994, *Object oriented analysis and design with applications*, (California: Benjamin-Cummings).

David, B., van den Herrewegen, M., and Salgé, F., 1995, Conceptual models for geometry and quality of geographic information. In *Geographic objects with indeterminate boundaries*, edited by Burrough, P. and Frank A., (London: Taylor and Francis), pp. 193–205.

Duckham, M., 1999, Implementing an object-oriented error-sensitive GIS. In *Spatial accuracy assessment: Land information uncertainty in natural resources*, edited by Lowell, K. and Jaton, A., (Michigan: Ann Arbor Press), pp. 209–215.

Guptill, S., 1989, Inclusion of accuracy data in a feature based object-oriented data model. In *Accuracy of spatial databases*, edited by Goodchild, M. and Gopal, S., (London: Taylor and Francis), pp. 91–98.

Kösters, K., Pagel, B-U. and Six, H-W., 1997, GIS-application development with GeoOOA, *International Journal of Geographic Information Science*, **11**, pp. 307–335.

Moellering, H. (Ed.), 1997, *Spatial database transfer standards 2: Characteristics for assessing standards and full descriptions of the national and international standards in the world*, (Oxford: Elsevier Science).

National Committee for Digital Cartographic Data Standards, 1988, The proposed standard for digital cartographic data, *American Cartographer*, **15**, pp. 11–142.

Nyerges, T., 1991, Geographic information abstraction: Conceptual clarity for geographical information systems, *Environment and Planning A*, **23**, pp. 1483–1499.

Ramlal, B. and Drummond, J., 1992, A GIS uncertainty subsystem. In *Archives ISPRS Congress XVII*, Volume 29.B3, pp. 355–352.

Rumbaugh, J., Blaha, M., Premelani, W., Eddy, F. and Lorensen W., 1991, *Object-oriented modeling and design*, (New Jersey: Prentice Hall).

Russomano, D., 1998, Utility network derivation from legacy source data for feature based AM/FM systems, *International Journal of Geographical Information Science*, **12**, pp. 131–143.

Shi, W., 1998, A generic statistical approach for modelling error of geometric features in GIS, *International Journal of Geographical Information Science*, **12**, pp. 487–495.

Unwin, D., 1995, Geographical information systems and the problem of error and uncertainty, *Progress in Human Geography*, **19**, pp. 548–549.

van der Wel F., Hootsmans, R. and Ormeling, F., 1994, Visualisation of data quality. In *Visualisation in modern cartography*, edited by MacEachren, A. and Taylor, D., (New York: Pergamon), pp. 313–331.

Walker, P. and Moore, D., 1988, SIMPLE: An inductive modelling and mapping tool for spatially-oriented data, *International Journal of Geographical Information Systems*, **2**, pp. 347–353.

Worboys, M.F., Hearnshaw, H.M. and Maguire, D.J., 1990, Object-oriented data modelling for spatial databases, *International Journal of Geographic Information Systems*, **4**, pp. 359–385.

CHAPTER SIX

A Theory for Communicating Uncertainty in Spatial Databases

Karin Reinke and Gary J. Hunter

ABSTRACT

With the increase in new users of geographical information systems (GIS) and the rapidly growing number of data sets available, the authors believe that the need for effective communication of spatial data quality is now vital. In particular, the novice user is often forgotten in much of our research, yet it is they who are in most need of new methods and tools to assist them in understanding the nature and importance of data quality. This chapter introduces the concept of communicating uncertainty in spatial data and describes the actions that take place during the communication process. Understanding the communication process assists in the development of specific uncertainty communication operations and the sequencing of these from data to visualisation. Previous work by Beard and Mackaness (1993) identified three key stages in the process of communicating data quality to user: that is notification, identification and quantification. A fourth stage, evaluation, has been identified and added to this list. This chapter focuses on each of these steps and provides a discussion on some conceptual and practical considerations.

6.1 INTRODUCTION

The communication or portrayal of spatial data quality has often been cited by researchers as being important in enabling users to better understand the inherent strengths and weaknesses of their data (Hunter and Goodchild, 1996; Leitner and Buttenfield, 1996). In addition data producers and users are becoming increasingly concerned about liability issues (Epstein *et al.*, 1998) and so the establishment of data quality communication guidelines and techniques have the potential to help facilitate the reduction of uncertainty by minimising doubt in the identification of spatial feature locations, forms and characteristics (Veregin, 1989).

While the methods employed to communicate data quality are dependent on the skill level of the user, the kind of data quality information, the type of data being used and the nature of the application, it is clear that communication techniques need to be flexible enough to meet individual user and project requirements whilst still providing guidance on the appropriate selection of communication methods. However, before this can occur we need to address the fundamental question of "How do we inform users that the data they are about to apply may contain error or uncertainty?". Accordingly, this chapter investigates a theory for communicating error in spatial databases and discusses how it might be applied in practice.

6.2 A MODEL FOR COMMUNICATING UNCERTAINTY

Communication in a cartographic context is considered to be the scientific process by which information is selected, represented (usually in map form), perceived and interpreted by the map user. Thus the ultimate goal has always been to improve maps or develop new representation methods to communicate more information, more accurately to the end user (Robinson, 1985). Taking this perspective further, it is possible to apply traditional cartographic communication models to the communication of uncertainty in geographical data. The communication structure therefore basically remains unchanged, but provides for potential sources of uncertainty in the representation of the data to be exposed.

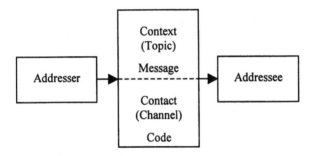

Figure 6.1 A model of communication recognised in linguistic and communication studies illustrating the different factors that must be present for communication (after Stubkjaer, 1990).

Stubkjaer (1990) applied a linguistic model of communication to provide a reference frame for communication and to serve as a paradigm for quality measures in spatial data (Figure 6.1). The model is interpreted in much the same way as other models of communication. The addresser sends a message that is received by an addressee but is extended to include the concept of context, that is something the message refers to other than itself. The message is transmitted via a contact that represents the physical and psychological connection between the addresser and the addressee. Finally, a code is incorporated with the contact as the shared meaning system that gives the message structure. Within cartography an example of this would be the legend provided with a map, where context is replaced by topic and contact is replaced by channel.

Stubkjaer uses this model as a structure to investigate concepts relating to the quality of the data, specifically those that relate to the topic, code and channel. He examines quality with the intention of providing a reference frame for communication in organisations. However, it can still be applied to the communication of uncertainty in spatial data. In particular, it is the measures that refer to the quality of the code used in the message and the relationship with the addressee or user that are of most interest. The measures that relate to code examine the quality of the meaning system by which the message is structured, for example the principles of the graphical variables. Using several of Dale and McLaughlin's (1990) characteristics of data quality, Stubkjaer identified particularity, clarity and compatibility as elements suitable for describing the quality of the code. When the code is considered in the context of geographical visualisation rather than the data itself, the elements remain suitable descriptors, albeit with some small adjustments.

The first element is particularity. This is the potential information content of a message, and the greater the number of objects or the more variables held by an object, the higher the particularity. The need for object definitions increases as particularity increases. Clarity is the second element and represents a measure of ambiguity. It concerns the probability of a message being unintentionally mistaken. The final element is compatibility, which measures the degree to which the code matches given standards or principles. Continuing with Stubkjaer's findings, further elements are identified as being suitable for measuring the relationship between the geographic visualisation message and the user. These are comprehensiveness and readability. Comprehensiveness is the ability of the user to understand the message, either intuitively or through explicit learning of the code. In some cases, the code is included as part of the message such as the legend on a thematic map. There is some expectation, however, of the user to be familiar with formalised codes. Finally, intrinsic to comprehensiveness is readability and this refers to the arrangement and organisation of the objects or signs in the message.

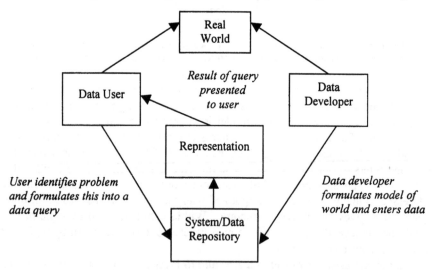

Figure 6.2 A general model of communicating uncertainty in spatial data
(after Gottsegen *et al.*, 1999, p. 178).

Gottsegen *et al.* (1999) put forward a general model of uncertainty in spatial data (Figure 6.2). The model illustrates the processes involved in the use and creation of data that may contribute to uncertainty. Uncertainty in the data is not considered within this model, which examines the quality of the message where the message is interpreted as a geographical visualisation. Gottsegen *et al.* conclude that there are two types of uncertainty related to the communication of data. The first is the degree to which the representation corresponds to the truth or what is taken to be the truth, while the second is jointly termed compatibility and consistency. This examines the match between the needs of the user and the representation.

The structure of the model is divided into two parts representing the data developer and the data user. It is both acceptable and possible for the data user to also be the data developer. The interaction with data and the development of the data progresses through a series of steps ranging from forming ideas and concepts to developing structured and formal specifications. Gottsegen *et al.* (1999) describe the processes of the data user and data developer components. Representation is the central component in this model and is considered to be the outcome of either final decisions or some problem-solving task. It may also arise from refinement of a problem in which the user proceeds with a new request of the data, new output and possibly new knowledge. This is a feedback loop between the data user, the data repository and the representation. Using the components of this model and incorporating the processes outlined in recent models of cartographic communication, a new model suited specifically to the communication of uncertainty in spatial databases is proposed (Figure 6.3).

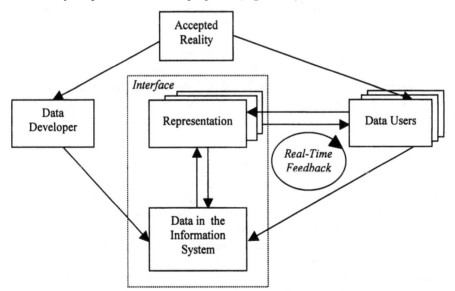

Figure 6.3 A proposed model for communicating uncertainty in spatial databases to users.

The arrows in the revised model represent a transfer of data or information, that is, some kind of message. This information can be in the form of a query or request between the user and the information system, or suggestive of the influence of real-world experiences of the user (for example, culture and knowledge). Observing the real world as accepted reality implies the importance and variation of context in spatial data tasks.

Unlike the general model offered by Gottsegen *et al.* (1999), this model allows for feedback between the data user and the representation. This is a key component of any communication paradigm used in geographical information science, and is illustrated by the direct two-way flow between the user and the representation. The interactive capabilities available as a result of new computing technologies has significantly altered the way in which data representations are used and understood. Pickett *et al.* (1995) observe there is enormous benefit to the user in being able to immediately observe changes in the data resulting from direct manipulation (for example, focusing). Direct interaction with data representations and the data itself are central to visualisation. This is indicated in the model by the double arrows and feedback loop shared between the user, the data and the representation.

Many researchers (such as MacEachren, 1992 and Monmonier, 1991) have argued against the search for a single optimum map. Advances in computing mean that users need no longer be confronted with only one type of representation, most commonly a map or textual table, but instead should be provided with a variety of representations in real time. Developments in computer visualisation, animation and multimedia demand that these new techniques be included. The model in Figure 6.3 extends the traditional view of cartographic communication to account for modern techniques of representation other than the traditional map. The possibility to select from a number of alternative representations is included as a feature of the model. This multiple representation approach is intended to minimize Type I and Type II visualisation errors although the actual effectiveness of multiple displays is yet to be confirmed (DiBiase *et al.*, 1992). Despite the lack of evidence to support the benefit of multiple representations it is assumed at this point, as aptly put by Monmonier (1991), that it is a case of "lying less by telling more".

A strong case for multiple representations is presented by MacEachren (1995) and used as a method for assessing the truthfulness of a display. He suggests that multiple views are the only practical way of gaining a measure of visualisation truth, and his case is founded upon the concept of geographic visualisation (map) stability. In turn, Muehrcke (1990) defines stability as the degree to which an observed characteristic of a representation remains consistent after changes are made in representation parameters or methods. Observations of a characteristic that appears in most of the representations equates to a dependable, and therefore truthful, property of the data. Whilst this feature does not directly assess the success of a visualisation it nonetheless facilitates the correct interpretation of the data. Another important adaptation of the proposed model is the suggestion of multiple users. This accommodates both ends of the private-public audience spectrum defined by MacEachren (1995) and suggests that different users can have quite different communication goals and ideas about the same data set.

At present the data user is required to interact with the data via an interface, usually a Graphical User Interface (GUI). The interface can affect the manner in which data users interact and think about the data or representation, and a well-designed interface should minimise the impact it has on the user. The ultimate goal is for the user to be unaware of the interface, and to think only of the data or problem at hand. The interface is an important factor in the communication process (Edsall and Peuquet, 1997; Davies and Medyckyj-Scott, 1994) and must be included as part of the model since it is a potential source of communication noise. A successful interface introduces minimal noise during the information transfer. There is now a trend towards both intuitive interfaces and 'invisible' interfaces, where users can become directly involved with the data and its representations.

This is evident in many desktop GIS where objects on a map can be graphically altered through point, click and drag operations using a cursor, and it is conceivable that this will soon be possible through touch-screens. Thus, the interface is the bridge between the user and the representation, and the user interface style is an issue that is central to the effective communication of uncertainty.

Table 6.1 The types of error that may occur during the visualisation of uncertainty in spatial data based on cartographic communication theories.

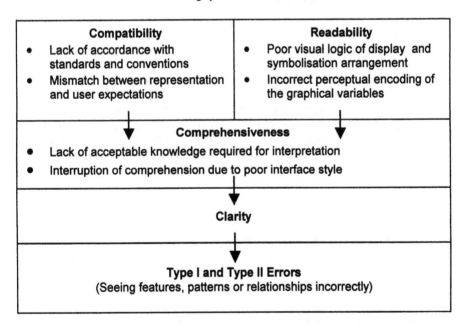

The research here is concerned with the transfer of information between the data, the representation and the user. The ultimate outcome is for the information to be accessed and received by the user as efficiently as possible, and for the output to achieve the optimal balance between a representation of the truth, and the user's expectations and perceptions. A breakdown between these components results in display error (Hunter, 1993) and may take the form of either Type I or Type II errors. Table 6.1 lists the main sources of error that can occur during the visualisation process and illustrates the flow-on effect that this has on map interpretation. The terminology used is consistent with that of Stubkjaer (1990).

The error sources are important because they affect the probability of a visualisation being correctly understood. A mismatch between the data and visualisation can result in perceptual error, while a mismatch between the visualisation and user expectations can result in cognitive error. However, features such as visualisation multiples, user customisation features and provision of marginalia can help improve message clarity if properly performed. The level and type of comprehension required to interpret a visualisation is governed by the degree to which a representation is iconic (or intuitive).

A highly intuitive display requires little knowledge by the user about how to read the display. In such cases, the perceptual encoding is a determining factor in the success of the display, although in special circumstances a perceptually correct display can still be misread by the viewer. As such, the importance of empirical studies and the provision of multiple views cannot be underestimated. It is important to note that this study is not concerned with the impact of conceptual error that may be conceived from poor problem definition or inappropriate abstractions of accepted reality. Instead, the proposed model identifies the agenda for future research, the underlying concepts to which successful communication should adhere, and the type of investigation necessary to evaluate the success of geographic visualisation.

6.3 FOUR STEPS TO COMMUNICATING UNCERTAINTY

Communicating uncertainty involves describing, presenting and explaining data quality information to the user, and Beard and Mackaness (1993) have proposed three core communication goals as being notification, identification and quantification. A fourth goal, evaluation, has been added by the authors. As a starting point for definition of each of these terms, a dictionary describes them as follows:

- Notification: to give notice of or report the occurrence of [quality];
- Identification: to establish the identity of or to determine the taxonomic position of [the quality parameter];
- Quantification: to determine, express, or measure the quantity of [a quality parameter]; and
- Evaluation: to determine or appraise the significance of [a data quality parameter].

Notification indicates the presence of data quality information in a data set, and Beard and Mackaness describe it as being the simplest of assessment techniques which offers the least detail to the user. It requires few system resources and does not need to interrupt current tasks. The analogy of a smoke detector is used here since it provides a warning that smoke exists, but yields no further information as to the characteristics or source of the smoke. So notification provides a basic nominal yes/no response to the question "Is data quality information available?"

Identification provides a description of the type of uncertainty or error that may exist and the spatial extent to which it applies. It has a nominal form because it distinguishes between data quality categories, for example, positional accuracy or attribute accuracy. At the level where location becomes identified it is similar to notification in that it provides a presence/absence response. With identification, the question that is asked is "What kind of data quality information exists here?".

Quantification describes the type of data quality information available, where it occurs, and its values. It is the most complex of all the levels of assessment and thus requires the most elaborate representations to correctly express the magnitude and variation of the quality measure (Beard and Mackaness, 1993). The type of measurement may be nominal (and even though nominal measurements have no magnitude it is still possible to represent different classes using appropriate visual methods), ordinal, interval or ratio depending on the quality parameter and its measurement. The question addressed in quantification is "What data quality values exist here?" It is suggested by the authors that quantification should be extended to include data simulations, given that spatial data may be inherently uncertain due to its representation, as often found in mapping the natural environment, for example in soil surveys.

The highest communication goal is evaluation. The communication activity is for the user to estimate or determine the significance that uncertainty has on the data set and given application. Evaluation acts as a visual aid for assisting in determining fitness-for-use of the data display. This may be done by viewing the changes that occur in the original data or by comparing the original data with the associated quality information. The primary representation issue then is how to present the data and its associated quality. Incorporating the actual data either as a separate or merged display is the only difference between evaluation and quantification where the communication tasks consists of comparison.

Each communication goal would normally require the user to have progressed through each of the proceeding communication tasks, even where some of these may have been bypassed. For example quantification, which requires the user to estimate magnitude, also requires the user to have detected, distinguished and located the relevant data quality parameter before proceeding with an estimation.

For each of these goals, the spatial extent to which the assessment applies may lie at the global, regional, feature class or individual feature levels. The spatial extent of the data quality information will affect the choice of representation only in the identification and quantification tasks. Communication of each of these levels of assessment relies on quality information being available for the given data. The potential for a data quality conflict is dependent upon the user's opinion of the significance of the error or uncertainty present in the data and the application concerned. This is likely to be determined at the quantification level but may also occur at the identification level. Each level presents progressively more information about the data and makes no assumptions of its significance to the user.

The methods employed to communicate data quality are dependent upon the skill level of the user, the type of data quality information, the type of data being used and the application. While it is difficult to prescribe a complete list of communication techniques suitable for every combination of application, data and user, it is clear that the higher levels of communication need to be flexible enough to meet individual user and project requirements whilst still providing some guidance on the selection of communication methods. Using MacEachren's (1992) map use cube model (Figure 6.4), the different communication goals can be placed according to the general trend exhibited along each axis. As would be expected, every subsequent communication goal increases in complexity and user flexibility. Because the intended communication system relies on prescriptive methods for assigning visualization techniques (although some user customisation and selection is possible) the higher communication goals do not extend well into the private domain of the user audience.

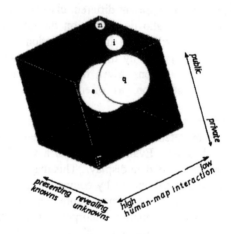

Figure 6.4 Siting the communication goals, that is, notification (n), identification (i), quantification (q) and evaluation (e) within MacEachren's (1992) map use cube.

Figure 6.5 establishes a framework that outlines the processes involved when communicating uncertainty. It describes the transformation of data quality information into data quality displays, and essentially represents a data-driven approach with acknowledgement of the considerations that influence the choice and interpretation of the final visualisation by the user. There is an upward move from individual preference factors to more global factors arising from the cultural setting of the user and the intended application. The level of communication (that is, notification, identification, quantification and evaluation) is shown in increasing complexity as the final stages of the process are reached. Overall, it describes the general process for communicating a single data quality parameter, and interaction is available to the user at each stage in the communication process.

The first step in the framework is to assume that data quality information is available for the given data set. At this point the user may be informed of the presence of data quality information and this comprises the notification phase.

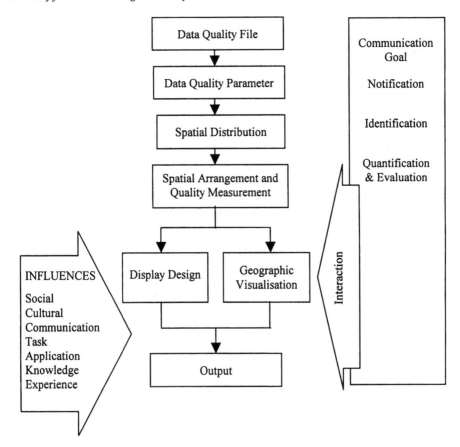

Figure 6.5 A framework summarising the systematic process that occurs during the communication of a single data quality parameter coupled with the potential user influences.

The next step describes the type of data quality parameter to be communicated, and the elements here include positional accuracy, attribute accuracy, logical consistency, lineage and completeness. This is where the identification phase begins. Continuing with this phase is the spatial existence of the data quality parameters. This describes the spatial occurrence of data quality throughout the data and results in a binary map of presence/absence. However, it is also at this stage that the quantification phase begins since the spatial distribution of the data quality information influences the type of visualization and interface design selected.

The type of information (that is, spatial variation as a binary field compared to one where the amount of variation is shown) and the communication purpose (that is, where is the data quality information available versus how it varies), form the distinction between spatial variation at the identification phase and spatial distribution at the quantification phase. The next step considers the spatial arrangement of the data model (that is, the degree of spatial continuity and spatial smoothness). The data quality measurement examines the type of measurement of the data quality parameter according to the levels of measurement as described by Chrisman (1999) and MacEachren (1995). The step prior to the presentation of the final output is that of interface design and visualisation.

The choice of how the data quality is to be represented is made based on the factors associated with the previous steps. Visualisation encompasses the familiar cartographic or graphic variables of saturation, hue, size, orientation and pattern, as well as new variables such as focus and techniques such as animation. The interface design considers the method employed to present and interact with the data. For example, in the form of side-by-side windows, toggling windows or bivariate displays. The display design and geographical visualisation together influence the effectiveness of the overall output and as such must be considered simultaneously. The quality output step ends the quantification phase by providing the user with an appropriate representation or multiple representations of the data.

6.4 NOTIFYING AND IDENTIFYING UNCERTAINTY

In the same way that computer users associate the question mark or light bulb with online help, tips or hints, the same should exist for communicating data quality to users of spatial data. However, preceding all communication is a fundamental step that is often overlooked and that is: "How do we let users know that the data they are about to use may contain uncertainty?". The answer lies initially in notification, and leads to the identification, quantification and evaluation levels. Notification also performs the important task of creating an awareness of the issue of uncertainty in spatial data and attempts to educate users on this issue.

As user knowledge and acceptance of the existence of data quality information increases, the need for notification decreases. However, until then it is a vital stepping stone in the communication of uncertainty in spatial data. Currently, there is little choice apart from visual or audible methods for gaining the attention of a user in a GIS environment. With respect to the notification task, the use of sound in desktop interfaces may be separated into two types: (1) realistic sounds referred to as mimetic sounds or 'earcons' (Gaver, 1986); and (2) abstract sounds. Earcons are sounds that resemble the true sound of the represented action, and Krygier (1994) gives the example of the 'thunk' sound emitted by a file that is successfully dragged (or 'dropped') into a digital trash can.

Alternatively, abstract sounds may be intuitive of a particular condition but do not or cannot imply the true sound, usually because one does not exist. Nevertheless, Krygier notes that abstract sounds can still be used to attract and direct the attention of users, for example, the 'ding' that sounds when an illegal operation is attempted. Sonic cues can exploit the fact that the human auditory system is constantly open. For example, humans can hear events or alarms even when not actively concentrating on the sound environment. Sonification can mark a specific condition and auditory alarms, such as military bugle calls, church bells and emergency vehicle sirens have long been used throughout history, testifying to their success in attracting the attention of people.

While the application of sound for notification purposes appears useful, there are problems associated with using it as the sole means of gaining a user's attention. For instance, when several software application packages or tasks are being used at the same time, a variety of sounds may be heard notifying the user of different activities potentially resulting in confusion of the true source. So the use of sound can be limited by the intrusion of environmental sounds and other practical limitations. For example, what happens once the sound has occurred, and what if it was not heard? As such, it is rare for software packages to use sound as the sole means of notification without the support of visual cues, and instead the sound serves to inform the user of a notification message (for example, incoming e-mail). In this regard, Beard and Mackaness (1993) suggest that visual techniques such as blinking or flashing icons be used to support notification, and it may be that several different but complementary techniques can be used to form a useful notification tool.

The purpose of notifying users of spatial data quality in GIS applications requires the method to be noticeable but not interruptive of user tasks; and comprehensible, small and quick for the user to respond accordingly. It is suggested that signs and symbols may be the most effective and efficient way of achieving this, and there is a noticeable preference for using symbology in current software, including GIS packages. Practically all commercial GIS software packages have now moved from command line systems to a graphic user interface environment. Icons grab the attention of the user and are an effective means of communication, particularly for the purpose of notification and identification.

Foley *et al.* (1991) have identified the advantages of icons as being language-independent, requiring minimal screen space and allowing for quick recognition by the user. Figure 6.6 presents warning screens found in modern desktop applications, while Table 6.2 lists the functionality behind each communication method and the level of user interaction. These examples provide a collection of methods that may be useful for future implementation of similar tools in GIS software.

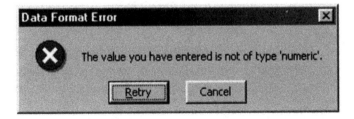

Figure 6.6 Common notification and identification techniques in modern computing environments.

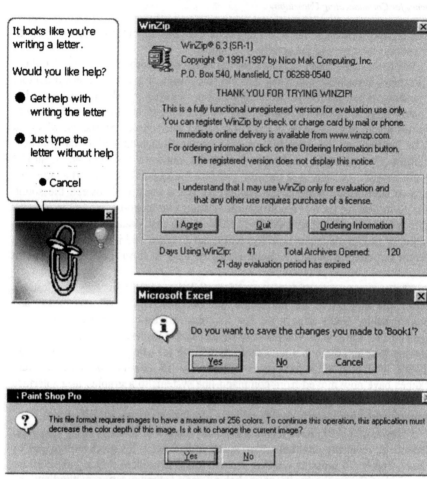

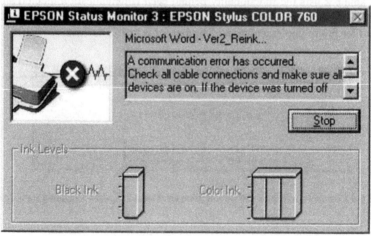

Figure 6.6 (cont.) Notification and identification techniques used in modern computing environments.

MESSAGE	EXAMPLE	METHOD	USER INTERACTION	COMMENTS
Warning to user of some condition, and cannot proceed with action NOTIFICATION OF CONDITION		Graphic appears	Passive / Active	Often used in conjunction with a message box. In the examples the user level becomes passive-active and the communication, level becomes identification.
Warning to user of some condition, and to proceed action with caution NOTIFICATION OF CONDITON		Graphic appears	Passive / Active	Often used in conjunction with a message box, in these examples the user level becomes passive-active and the communication level may become identification.
Alerts the user of new mail NOTIFICATION OF CONDITION		Graphic changes	Passive	Change in condition often accompanied by sound.
Alerts and directs the user to an existing hotlink IDENTIFICATION OF CONDITION		Graphic changes	Passive	Graphic changes as cursor moves over text that contains a hotlink
Task or action being processed NOTIFICATION OF PROCESS		Graphic changes and animation	Passive	Graphic changes into timer during processing, animates (processing) and returns to cursor graphic upon task completion
Indication that action (for example, help) is available NOTIFICATION OF FUNCTION		Graphic button	Passive	Button indicates that action is available. When grayed-out it indicates that the function is disabled It is at the user's discretion that the action is invoked
Task or action being processed QUANTIFICATION OF PROCESS		Graphic animation	Passive	Movement within box indicates status of processing task
Indication that help is available NOTIFICATION OF FUNCTION		Graphic appears	Passive / Active	Often used in conjunction with a message box, in these examples user level becomes passive-active and the communication level may become identification.

Table 6.2 Notification and identification examples in current computing environments.

Task/action and notification type	Symbol	Mechanism	Passive/Active	Description
Task or action being processed NOTIFICATION OF CONDITION	(● ●)	Graphic change	Passive	Indicates the completion of a process, ready for next task.
Notification of incorrect spelling. IDENTIFICATION OF CONDITION	Missnel word	Graphic appears	Passive	Alerts user to incorrect condition, and identifies where it occurs in the data.
Identification of a overshoot or undershoot of a linear feature IDENTIFICATION OF CONDITION	(graphic)	Graphic appears	Passive	A box appears around the feature (for example, an overshoot) that is potentially an error identifying the location and type of error.
Request for acceptance of conditions by user NOTIFICATION OF CONDITION	Window (a)	Button request	Passive / Active	User input required before continuation of program (graphic appears in the form of a window)
Notification of printing status NOTIFICATION OF PROCESS	Window (b)	Animation and button request	Passive	Controls are provided that enable the user to intervene with a process (graphic appears in the form of a window) but it still remains passive because it is optional
Notification of the condition that an update is available NOTIFICATION OF CONDITION	Window (c)	Button request and checkbox	Active	User input required before continuation of program but is given the option to 'turn' the message box off (graphic appears in the form of a window)
Notification of incorrect formula added, option to correct NOTIFICATION OF CONDITION	Window (d)	Button request	Active	User input required before continuation of program (graphic appears in the form of a window)
Query to initiate task (ie. file save) NOTIFICATION OF FUNCTION	Window (e)	Button request	Active	User input required before continuation of program (graphic appears in the form of a window)

Table 6.2 (contd) Notification and identification examples in current computing environments.

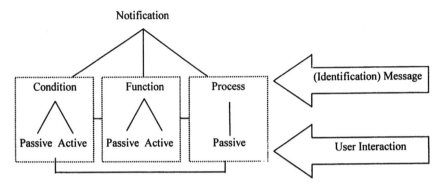

Figure 6.7 A classification of the characteristics describing common notification and identification techniques found in computing software.

From the illustrations in Table 6.2, it is possible to develop a general classification system of different communication techniques used for notification and identification purposes (Figure 6.7). The classification describes three types of messages as relating to a particular condition (some characteristic exists or has changed), a function (a capability to perform a particular task is available) or a process (computer resources are being utilised). Passive interaction is defined as being where the user has the choice of ignoring or reacting to a notification display, compared to active interaction where the user is forced to respond in some way to the display.

It is recognised here that these messages may not always be mutually exclusive. For example, most active notifications of a condition lead to notification or identification of functions or immediate processing. The type of message that is conveyed when notifying users of spatial data quality is either about the state of the data (condition) or about the availability of a utility that provides information about the quality of the data (function). Identification informs the user about the type of data quality parameter (condition). The main distinction being that notification is essentially hardcoded, whereas identification is more flexible in allowing users to select a form of representation.

Successful communication between the user and the notification or identification sign depends upon how well the concept of data quality has been portrayed, and how easily this is detected by the user. Figure 6.8 illustrates the interpretation phases that occur when information is presented by the computer to the user. Ito and Nakakoji (1996) refer to this as the listening mode. For identification, the communication task is firstly to distinguish the type of data quality parameter, and this involves recognising the data quality parameters as distinct entities. By distinguishing the different data quality parameters the user is able to draw conclusions about the general properties that define that parameter. The second communication task is to locate the spatial occurrence (that is, the absence or presence) of the data quality parameter. Whilst this is the most common

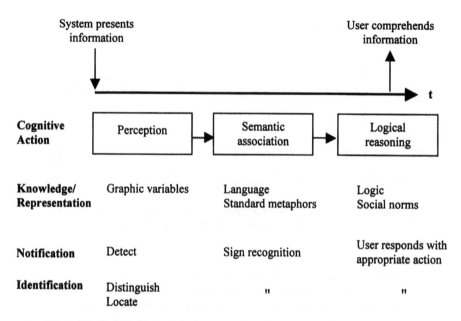

Figure 6.8 The listening mode in the communication of uncertainty for the purpose of notification and identification (after Ito and Nakakoji, 1996, p. 108).

type of locating, the ideal system should have the provision for location of data quality in the temporal and thematic domains.

The perception phase involves the user becoming aware of objects in the display that appear or have changed. This is notification. Usually graphic variables such as colour or shape are used, and the goal at this step is instant acknowledgment of the existence or alteration of the sign. The association step links the semantic meaning with what has been perceived, in other words the user is able to correctly identify the sign. The final step is logical reasoning, and the goal here is for the user to understand what has happened and to respond accordingly onto the higher levels of identification and quantification. It requires an understanding of the meaning of the object that is artificially portrayed by the sign.

6.5 QUANTIFYING UNCERTAINTY

Quantification is the third goal in the process of communicating uncertainty in spatial databases. It is the core of the entire process because it is the level where different visualisation techniques are assigned to the uncertainty based upon defined characteristics. It extends identification by providing a visualisation of the actual uncertainty values present in the data. The measurements may extend throughout the data as global, regional, feature class or feature level variations, and the user should be able to estimate the magnitude of a data quality measurement at any given location. This is dependent upon an appropriate presentation being created and a correct interpretation being made by the user. Identifying the requirements for mapping between the data and the visualisation is paramount to a successful quantification tool.

The communication model or paradigm used for communicating uncertainty in spatial databases is a useful starting point for recognising the requirements of quantification. Three main components for the quantification process can be identified as: (1) data specifications; (2) user specifications; and (3) visualisation categories. The data specification formally describes the data uncertainty characteristics that can affect the choice of visualisation. User specifications describe the interpretation aims of the user to be gleaned from the visualisation. This is an important component of quantification because it eliminates the issue of context-specificity for visualization by separating applications into visualisation goals and tasks. The visualization methods are broadly classified to formally identify the different (perceptual) characteristics of these methods and to use this classification as a means of mapping the different visualisation methods to different types of data.

The dominant component in assigning visualisations to uncertainty relies on describing the properties of the data and uncertainty values. These consist of the spatial variation of the uncertainty, the level of measurement of the uncertainty values, and the underlying data model (that is, the spatial arrangement). Many authors have identified the level of measurement with data quality (van der Wel *et al.*, 1994) or the underlying model with data quality (Buttenfield, 1991), as the framework for a communication system but there is a lack of models that account for all the data specifications. Figure 6.9 outlines the general approach for accommodating the different data specifications for communicating uncertainty.

The aim of the approach is to broadly identify the different characteristics of uncertainty necessary for choosing visualisation methods. It is not meant to be a complete mapping/assignment system but rather an example of how such a system could operate and the components required for such a system. It assumes appropriate visualisation methods are applied according to their perceptual characteristics. The final stage in this approach links the data specifications with user communication goals and tasks, and this last step attempts to find the corresponding visualisation method as defined by the visualisation task. If no common visualisation method occurs then the question arises "Which approach takes precedence, or should both recommended visualisations be simultaneously portrayed as a bivariate map?"

It has previously been defined that data quality is a function of application (in part). However, it is impossible to think of every application and it is limiting to restrict visualisations to particular types of applications. In this communication system, the application is replaced by user communication goals and tasks. In doing so, it is hoped that a more manageable set of operations can be defined based on communication tasks rather than application. Keller and Keller (1993) support this approach, arguing that most techniques that perform a particular communication task can be applied to any data regardless of application or discipline. The importance of user context (that is, communication goal and visualisation task) has also been emphasised by others (such as Goodchild *et al.,* 1994; Leitner and Buttenfield, 1997; MacEachren *et al.,* 1999; Beard and Buttenfield, 2000). The main visualisation tasks identified for the communication of uncertainty in spatial data following the four communication goals includes detection, identification, location, estimation of magnitude (or order) and comparison.

Data Quality Domain

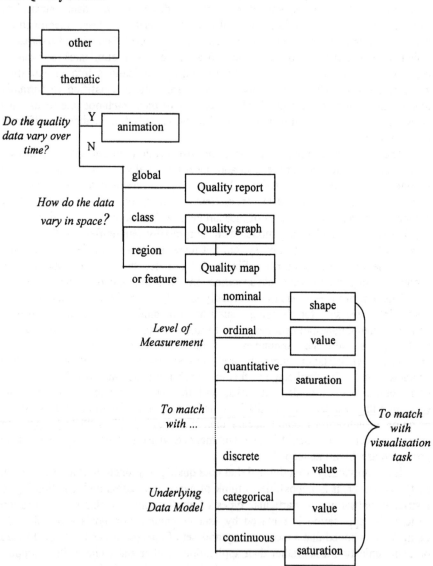

Figure 6.9 A data-driven approach for the communication of uncertainty in spatial databases using example visualisation and symbolisation techniques.

It is proposed to classify the different visualisation and symbolisation techniques according to the perceptual properties they carry and then associate these with the above tasks. Such a classification begins with categorising techniques according to their level of intuitiveness in conveying the concept of uncertainty. Three visualisation categories are proposed as being represented by icon, index and symbol forms based on categories of representation distinguished by Peirce (1985). Each visualisation form is a function of the level of interpretation required to reach an understanding of uncertainty. An iconic representation directly indicates or implies uncertainty; an index form of representation infers or suggests uncertainty through a familiar association between the user and the representation; and a symbolic representation is one where uncertainty is represented through the use of convention or agreement and requires explanatory supplementary information. Figure 6.10 ranks the different visualisation techniques according to their level of intuitiveness and need for supplementary information in forms such as marginalia, legends and textual descriptors.

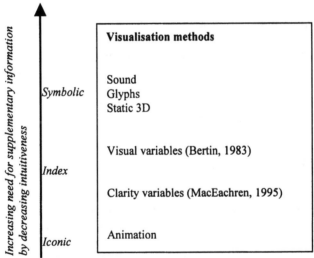

Figure 6.10 Proposed semiotic hierarchy of some example visualisation techniques that can be used to portray uncertainty.

The visual variables sit between index and symbolic form because different variables incite different perceptual responses with most requiring some type of explanatory key to enable the correct interpretation of uncertainty, but less so than symbolic methods. The next step is to determine the techniques that will be most effective in perceptually communicating uncertainty according to the different visualisation tasks identified within each of the communication goals. However, it may be that for some tasks such as quantification, users cannot accurately identify and quantify uncertainty types and values using intuitive visualisations, and other forms of visualisation may be more suitable. Establishing the most effective way for the user to identify, estimate and compare local, regional and global uncertainty measures is an area for further investigation.

The greatest problem faced with encoding visualisations with data specifications and visualisation tasks is the difficulty in developing and implementing a universal set of visualisation standards. It is doubtful that a complete standardisation of visual techniques is possible or even desirable (Gahegan, 1999; Morrison, 1977). Green and Horbach (1998) believe the number of available options alone is too difficult to code, and variation between the cognitive abilities and conventions used by viewers suggests that a rule-based system should not be imposed on the user. A perceptual-based approach used in conjunction with multiple representations (and corresponding marginalia or interpretation assistance) can partially overcome this. By providing the user with some freedom over what they are seeing they can match the visualisation with their expectation whilst reducing the likelihood of incorrect cartographic encoding. Muehrcke (1996) argues that because visualisations rely partly on intuition and partly on analytical reasoning, it is impossible to derive a prescriptive solution to a holistic problem.

The quantification system encodes the visualisation with the data type and synthesises the outcomes with the communication tasks of the user. User interaction during the quantification process is limited to interactions with the type of visualisation used for portraying the uncertainty. Bringing together each of the components in a working system is currently in the design and implementation phase, but it is founded upon matching common visualisation methods that are shared between user communication tasks and data specifications.

6.6 EVALUATING UNCERTAINTY

Evaluating uncertainty follows quantification and is a form of visual assessment rather than visual observation. The inclusion of the actual data in user interpretation is the essential difference between quantification and evaluation. It gives context to the uncertainty and is considered to offer the most utility to the user. The communication of uncertainty in spatial data for the purpose of visual evaluation consists of one of two types of visualisation tasks.

The first type of task estimates the significance the data quality has upon the given data set for a given application. The question being addressed here is "How do the data values vary according to the data quality?". In other words, what is the significance of change in the data caused by the uncertainty? This involves altering the original data according to the uncertainty measurements and may result in a series of realisations or as a single output created by computationally combining the data and uncertainty measures. The resulting representations would usually be portrayed as animations (showing possible realisations of the data) or by applying appropriate visualisation techniques identified in the quantification process.

The second type of visualisation task uses comparisons between the data and the uncertainty measurements to assist in visually evaluating the quality of the data for an application. Unlike the first task, this does not physically combine the original data and uncertainty measures but visually combines them. This task may take the form of separated (for example, side-by-side windows) or merged displays (for example, bivariate map). Display composition and selective association and differentiation visualisation properties are the main objectives at this level, depending upon whether separated or merged displays are used.

An example of the first type of evaluation task is being conducted in research carried out by the University of Melbourne in collaboration with the Country Fire Authority (CFA) in Victoria, Australia. The application aims to locate areas at risk of wildfires in outer metropolitan and semi-rural areas. Accordingly, the CFA has developed a wildfire threat model. This has many inputs such as digital elevation data combined with weather conditions, vegetation maps, fuels loads as well as demographic information. The digital elevation model has known errors in elevation, and these errors introduce uncertainty into CFA operational wildfire risk maps. Providing visual representations of the magnitude of error is limited for use by policy and decision-makers as a tool for assessing the confidence in their decisions (that is, locating and distributing resources to areas of higher risk). However, a visual representation that highlights areas where there is a change in risk class as a result of the uncertainty (for example, from a moderate to a high bushfire risk or *vice versa*) is expected to provide a more meaningful display to these users (pers. comm. Jones, S., December 2000).

Representations can show the location of where the class changes occur and the amounts of change. User-enabled modification of data quality amounts allow the user to investigate the sensitivity of the data to variation in uncertainty levels. This type of evaluation is an extension of the quantification task since it allows the user to investigate the direct impact uncertainty has upon the data rather than view the uncertainty in isolation. Both the visualisation tasks required for evaluation build upon previous visualisation tasks (that is, to distinguish, locate and estimate magnitude) in order to estimate the significance of the observed uncertainty.

Another method by which evaluation can be performed is through comparisons between the data and the quality information. This differs from the first approach by presenting the uncertainty values (that is, quantification) concurrently with the data. The aim here is for the viewer to observe similarities and differences between the data and uncertainty, and comparison is achieved by using the different display options (for example, merged displays or multiple displays) available to the viewer. This involves interaction between the viewer and the display composition.

The need for interaction between the viewer and the data is also apparent, and the ability to query and focus on user-defined data ranges (in theme, time or space) is considered useful for evaluation tasks. A working example can be found in Paradis and Beard (1994) who developed a data quality filter that allowed users to define and subsequently visualise their data quality requirements in conjunction with the actual data. Uncertainty thresholds should be able to be changed without changing the visualisation system, and interaction with the data quality measures is considered a common requirement for evaluation activities. During the evaluation phase the user is concerned with evaluating the significance the uncertainty has on the data, and therefore requires increased flexibility and interaction with the actual measurements.

Another type of interaction with the data that is an important utility during evaluation procedures is re-expression. Re-expression or re-classification of the data uncertainty classes can enable the data to be summarised or grouped rather than showing each individual value. A simple example can be found in Faiz *et al.* (1996), and these types of interactions with the data can reduce visual complexity and improve viewer comprehension (Gahegan, 1999). Interaction with the data quality visualisation and display can also occur during evaluation tasks. In particular, user access to how the data and quality should be displayed with each other is a primary function of evaluation.

6.7 FUTURE RESEARCH AND CONCLUSIONS

The future directions of this research are dictated by the following key questions relevant to each level in the uncertainty communication process, and the immediate action for the research program is the development and implementation of a working model of the communication system using a subset of proven visualisation techniques. Subsequent evaluation of the system will provide valuable confirmation about the actual requirements necessary for the success of tools for communicating uncertainty in spatial databases. The key questions identified are:

- Notification: What is the most effective way to notify users that data quality information exists without interfering with current mental tasks?
- Identification: What is the most efficient way to enable users to (1) identify a data quality parameter (either system or user activated), and (2) locate its occurrence within the temporal, thematic and spatial domain of the data?
- Quantification: What are the influencing factors to be considered in a quantification tool, and how should these be weighted? What visualisations correspond perceptually to which types of uncertainty data and visualisation tasks? Are intuitive visualisations more appropriate for quantification than abstract methods? How should this type of information be presented to the user, particularly where multiple outputs are available?
- Evaluation: How should the uncertainty and data be presented to the user for accurate visual comparisons? What level of interaction is required by the user?
- General: How does this uncertainty communication tool translate in terms of utility in decision-making processes? Are the identified communication tasks sufficient and complete?

Clearly, the development of new computer and spatially-related technologies has led to a shift from communication being a linear model where the map is a product of the cartographer's reality and abstraction, to one that incorporates user interaction as a key component in the communication process. While experts are still employed to design the data models, implement user interfaces and to create maps, the map creation process is no longer the exclusive domain of such people. Indeed, anyone can choose what they want to map and the way it should be done, and this needs to be reflected in current cartographic communication models. The transfer of scientific visualisation techniques into cartography and GIS has altered, or rather extended, the communication paradigm in digital mapping and most cartographic models now subscribe to a more user-dominated model and identify different types of visual tasks.

It is also apparent that there is very little in the way of exhaustive guidelines for communicating uncertainty in spatial data, and as a minimum the model proposed here should help coordinate and incorporate past, current and future evaluations. The communication process described here also permits the implementation of a systematic and practical approach for choosing visualisations for uncertainty. It is based on matching the visualisation capabilities to data uncertainty characteristics and user communication tasks. Finally, the communication model used here identifies the processes and components involved in the communication of uncertainty and assists in the design of ongoing experimentation.

ACKNOWLEDGEMENTS

The authors wish to acknowledge the support received for this research through Australian Research Council Large Grant No. A49601183 titled "Modelling Uncertainty in Spatial Databases".

REFERENCES

Beard, M.K. and Buttenfield, B.P., 2000, Detecting and Evaluating Errors by Graphical Methods. In *Geographical Information Systems: Principles, Techniques and Applications*, Vol. 2, edited by Longley, P.A., Maguire, D.J., Goodchild, M.F. and Rhind, D.W. (New York: Wiley), pp. 219–233.

Beard, M.K. and Mackaness, W., 1993, Visual Access to Data Quality in Geographic Information Systems. *Cartographica*, **30** (2 & 3), pp. 37–45.

Bertin, J. 1983, *Semiology of Graphics: Diagrams, Networks, Maps*. (Madison: The University of Wisconsin Press).

Buttenfield, B.P., 1991, Visualizing Cartographic Metadata. In *Visualization of the Quality of Spatial Data: Report on the Specialist Meeting for NCGIA Research Initiative 7*, edited by Beard, M.K., Buttenfield, B.P. and Clapham, S.B. (Santa Barbara: National Center for Geographic Information and Analysis), pp. c17–c26.

Chrisman, N., 1999, A Transformational Approach to GIS Operations. *International Journal of Geographical Information Science*, **13**, pp. 617–637.

Dale, P.F. and McLaughlin, J.D., 1990, *Land Information Management: An Introduction with Special Reference to Cadastral Problems in Third World*

Countries. (New York: Oxford University Press).

Davies, C. and Medyckyj-Scott, D., 1994, Introduction: The Importance of Human Factors. In *Visualization in Geographic Information Systems,* edited by Hearnshaw, H. M. and Unwin, D. J. (New York: Wiley & Sons), pp. 189–199.

DiBiase, D.A, MacEachren, A.M., Krygier, J B. and Reeves, C., 1992, Animation and the Role of Map Design in Scientific Visualisation. *Cartography and Geographic Information Systems,* **19** (4), pp. 201–214.

Edsall, R.M. and Peuquet, D.J., 1997, Graphical Query Techniques for Temporal GIS. In *Proceedings of AUTO-CARTO 13,* Seattle, (Seattle: ACSM/ASPRS), pp. 182–189.

Epstein, E., Hunter, G.J. and Agumya, A., 1998, Liability Insurance and the Use of Geographic Information. *International Journal of Geographical Information Science,* **12**, (3), pp. 203–214.

Faiz, S., Nzali, J.P. and Boursier, P., 1996, Representing the Quality of Geographic Information Depending on the User Context. In *Proceedings of the Second Joint European Conference and Exhibition on Geographic Information,* Vol. 1, Barcelona, (Amsterdam: IOS Press), pp. 73–77.

Foley, J.D., van Dam, A., Feiner, S.K. and Hughes, J.F., 1991, *Computer Graphics: Principles and Practice,* (Reading: Addison–Wesley).

Gahegan, M., 1999, Four Barriers to the Development of Effective Exploratory Visualisation Tools for the Geosciences. *International Journal of Geographical Information Science,* **13**, pp. 289–309.

Gaver, W., 1986, Auditory Icons: Using Sound in Computer Interfaces. *Human Computer Interaction,* **2** (2), pp. 167–177.

Goodchild, M.F., Buttenfield, B.P. and Wood, J., 1994, Introduction to Visualizing Data Validity. In *Visualization in Geographic Information Systems,* edited by Hearnshaw, H. M. and Unwin, D. J. (New York: Wiley & Sons), pp. 141–149.

Gottsegen, J., Montello, D. and Goodchild, M. F., 1999, A Comprehensive Model of Uncertainty in Spatial Data. In *Spatial Accuracy Assessment: Land Information Uncertainty in Natural Resources,* edited by Lowell, K. and Jaton, A. (Chelsea, Michigan: Ann Arbor Press), pp. 175–182.

Green, D.R. and Horbach, S., 1998, Colour - Difficult to Both Choose and Use in Practice. *The Cartographic Journal,* **35**, pp. 169–180.

Hunter, G.J., 1993, *Handling Uncertainty in Spatial Databases.* Ph.D. Thesis. Department of Surveying and Land Information, University of Melbourne, Melbourne, Australia.

Hunter, G.J. and Goodchild, M.F., 1996, Communicating Uncertainty in Spatial Databases. *Transactions in Geographic Information Systems,* **1** (1), pp. 13–24.

Ito, M. and Nakakoji, K., 1996, Impact of Culture on User Interface Design. In *International User Interfaces,* edited by del Galdo, E. and Nielsen, J. (New York: Wiley & Sons), pp. 105–126.

Keller, P.R. and Keller, M.M., 1993, *Visual Cues: Practical Data Visualization,* (Los Alamitos, California: IEEE Computer Society Press).

Krygier, J.B. 1994, Sound and Geographic Visualization. In *Visualization in*

Modern Cartography, edited by MacEachren, A.M. and Taylor, D.R.F. (Oxford: Pergamon), pp. 149–166.

Leitner, M. and Buttenfield, B.P., 1996, The Impact of Data Quality Displays on Spatial Decision Support. In *Proceedings GIS/LIS '96,* Denver, (Denver, Colorado: ASPRS), pp. 882–894.

Leitner, M. and Buttenfield, B.P., 1997, Cartographic Guidelines on the Visualization of Attribute Accuracy. In *Proceedings of AUTO-CARTO 13,* Seattle, (Seattle, Washington: ACSM/ASPRS), pp. 184–193.

MacEachren, A.M., 1992, Visualizing Uncertain Information. *Cartographic Perspectives,* **13,** pp. 10–19.

MacEachren, A.M., 1995, *How Maps Work: Representation, Visualization and Design,* (New York: Guildford Press).

MacEachren, A.M., Wachowicz, M., Haug, D., Edsall, R. and Masters, R., 1999, Constructing Knowledge from Multivariate Spatio-Temporal Data: Integrating Geographical Visualization with Knowledge Discovery in Database Methods. *International Journal of Geographical Information Science,* **13** (4), pp. 311–334.

Monmonier, M., 1991, Ethics and Map Design: Six Strategies for Confronting the Traditional One-Map Solution. *Cartographic Perspectives,* **10,** pp. 3–8.

Morrison, J.L., 1977, The Science of Cartography and its Essential Processes. *Cartographica,* **19,** pp. 58–71.

Muehrcke, P., 1990, Cartography and Geographic Information Systems. *Cartography and Geographic Information Systems,* **17** (1), pp. 7–15.

Muehrcke, P.C., 1996, *The Logic of Map Design,* (New York: Wiley and Sons).

Paradis, J. and Beard, M.K., 1994, Visualization of Spatial Data Quality for the Decision-maker: a Data Quality Filter. *Journal of the Urban and Regional Information Systems Association,* **6** (2), pp. 25–34.

Peirce, C.S., 1985, Logic as Semiotic: the Theory of Signs. In *Semiotics: An Introductory Anthology,* edited by Innis, R.E. (Bloomington: Indiana University Press), pp. 1–23.

Pickett, R.M., Grinstein, G., Levkowitz, H. and Smith, S., 1995, Harnessing Preattentive Perceptual Processes in Visualization. In *Perceptual Issues in Visualization,* edited by Grinstein, G. and Levkowitz, H. (Berlin: Springer-Verlag), pp. 33–46.

Robinson, V.B., 1985, About Different Kinds of Uncertainty in Collections of Spatial Data. In *Seventh International Symposium on Computer-Assisted Cartography,* Washington, (Washington D.C), pp. 440–449.

Stubkjaer, E. 1990, Communication Theory - A Basis for Data Exchange and for Measures of the Quality of Data. In *Third Scandinavian Research Conference on Geographical Information Systems,* edited by Hermansen, B., Balstrøm, T. and Frederiksen, P. (Helsingor, Denmark), pp. 213–229.

Van der Wel, F.J.M., Hootsman, R.M. and Ormeling, F., 1994, Visualisation of Data Quality. In *Visualization in Modern Cartography,* edited by MacEachren, A.M. and Taylor, D.R.F. (Oxford: Pergamon), pp, 313–331.

Veregin, H,1989, *A Taxonomy of Error in Spatial Databases.* Technical Report 89-12, National Center for GIS and Analysis, University of California, Santa Barbara.

Introduction to Part II: Methods for Handling Spatial Data Quality

Wenzhong Shi

This part includes seven chapters that describe methods for handling uncertainties in spatial data. These involve determining sample size in sampling techniques for spatial data quality assessment in GIS; new descriptions on resolution and precision –detect-ability and tuning digital precision to geographical resolution; visualization of uncertainties in spatial data and analysis; uncertainties in the overall spatial decision process based on a GIS; and finally registration of quality for spatial data.

The quality of spatial data in GIS will affect decisions that are made based on information generated from the spatial data set. Therefore, it is essential to test and control the quality of spatial data and make sure that the GIS data provided to users reach a certain level of standards. Practically, it is expensive if we evaluate the quality of spatial data according to an assessment of the overall data set in GIS; indeed, it is scarcely possible in many cases. As an alternative, a feasible solution is assessing and controlling the quality of a whole set of spatial data in GIS by using selected representative samples from the overall data set, where sampling technique is a critical issue. Traditionally, statistical quality control has been used as a method of assessing and controlling the quality of products. Many sampling techniques have been developed, such as random sampling, stratified sampling, systematic sampling and integrated sampling techniques. One of the targets of developing various sampling techniques is to try to find an appropriate method that can reflect the quality of the population data set as precisely as possible, by using relatively fewer number samples.

Wilhelm Caspary has been working in the area of surveying and geodetic data processing based on statistical and other approaches. In recent studies, he applies statistical methods from the area of surveying and geodetic data processing to spatial data quality measurement in GIS. In Chapter Seven, Wilhelm Caspary and Gerhard Joos introduce a method to determine sample size for testing and controlling spatial data quality in GIS while using sampling techniques. With the sample size, statistically significant statements can be made and the required acceptable quality level can be certain. First, the main parameters for a sampling plan are derived. Second, the central behavior of a sampling plan is formulated by the operating characteristic. Third, the operating characteristic – a curve – is used to determine the proper sample size. With this proposed solution, sample size can be reduced, especially for situations with very bad or good data sets. Finally, the quality level is determined by taking a cost model into consideration. With a smaller number of sample sizes, the cost of evaluating the quality of spatial data in GIS can be reduced. One assumption of the proposed technique is that the population of the data set meets the requirement of homogeneity in terms of error behaviour.

Representing the infinite and continuous real world as finite and discrete objects in GIS is an approximation process. Because of the differences between the entities in the real world and what is represented in GIS, many GIS models, using both field-based and object-based solutions, have been developed. However, the open issue of representing the real world perfectly in GIS remains a challenge. A number of concepts and corresponding methods have been developed to indicate the difference between what is represented in GIS and the real world, such as scale, completeness, resolution and others. Both Chapter Eight and Chapter Nine discuss the issues of resolution, scale and precision.

In Chapter Eight, Thomas K. Windholz, Kate M. Beard and Michael F. Goodchild propose a new concept, *detect-ability*: the ability to detect a specific object within a given field representation. Following this, a model to detect this capability is provided. The resolvability of a spatial object in GIS is determined by a function of its spatial extension and attribute value, and the three-dimensional relief of the inherent inaccuracy of the thematic representation. The proposed solution integrates spatial and thematic dimensions, and also resolves "objects" in "fields". With the proposed solution, a GIS user can determine whether the quality of a given field representation is sufficient for detecting a category of target objects.

In Chapter Nine, François Vauglin also presents a study related to precision and resolution, however from a practical point of view. In contrast with the proposed *detect-ability* for raster data sets in Chapter Eight, this chapter focuses on spatial data in vector format. Vauglin proposes the finer definitions for resolution and precision from the perspectives of geography and computer technology. Digital and geographical definitions are given for both resolution and precision. The new definitions lead to an opportunity for the optimization of geographical data sets: tuning digital precision to geographical resolution. A solution is also proposed for tuning digital precision to geographical resolution: routing. The solution is simpler and easier to industrialize, but may cause problems, such as changes of topology (inversions or merges), violations of specifications, or displacements.

The significance of the proposed solution is in its capability for dramatically reducing data volume in a GIS. This is essential for a data provider, such as a national centre with GIS data sets for a whole nation. GIS data users, particularly those who access data through the Internet, can benefit from being able to access the same amount of information in less time, due to the decrease in data volume.

Visualization of quality in spatial data is another key topic in the research field of spatial data quality. Chapters Ten and Eleven address this issue. There have been many developments in visualization techniques, such as three-dimensional, dynamic visualization and virtual reality techniques. In the past, the quality of spatial data was reported mostly by error indicators, such as standard deviation for the positional error of points on a digital map, and errors of commission and omission for a classified image. However, the spatial distribution of errors in GIS data is not represented very efficiently using only error indicators. A more efficient method of representing the spatial distribution of errors in spatial data is spatial distribution of the errors by visualization techniques. Several visualization techniques have been developed. Static solutions may include, for example, a three-dimensional uncertainty surface, integrated three-dimensional and colour presentation, color transformation, focus, blending and dazzling. On the other hand, the dynamic solutions may include blinking, quality slicing, moving and others.

In Chapter Ten, Igor Drecki presents a comprehensive review and analysis of the existing methods for visualizing uncertainties in spatial data. As a further step, Dr Drecki proposed two new methods for visualizing uncertainties in spatial data:

opacity and *squares*. The proposed methods are compared with the existing solutions for visualizing quality in spatial data. From the experimental study, it was concluded that the *squares* technique is the most effective means for visualizing uncertainty.

In Chapter Eleven, Tinghua Ai and Zhongliang Cai present a method for visualizing uncertainty in polygon filling, based on a Monte Carlo simulation. The proposed method is to animate uncertainty in polygon filling by means of a/the particle system model. An algorithm is proposed based on pixel random movement, including stages of birth, action and death. Each pixel in the raster map is described in terms of its position, movement speed and direction, color, brightness and life span. Uncertainty within a polygon can thus be reflected by randomization change of the variables.

An information system is a system to support decision making, while a GIS can be considered as a spatial-related decision support system. Uncertainties in a spatial decision-making process are thus a very essential issue, since they affect the quality of the final output of a GIS – decision making. The quality of using GIS for decision making may be affected by many factors, such as uncertainties in the spatial data and the information based on which the decision will be made, and in decision taking, decision analysis, and decision implementation. Therefore, uncertainties in spatial data, which has been the research focus, are only part of the overall research framework concerning uncertainties in GIS.

In Chapter Twelve, Bernard Cornélis and Sébastien Brunet give a comprehensive summary of the uncertainties in spatial decisions, from a policy-maker point of view. These uncertainties and their consequences are described based on a generic model of decision processes, the decisional fountain model. These include uncertainties in documentation/ information (with data, issue and context), decision analysis, decision taking, decision implementation and decision evaluation.

Instead of analyzing uncertainties from a data perspective, this study addresses uncertainties from the decision-making process point of view. This is actually a wider angle and more practical view of looking at uncertainties in GIS. A policy-maker needs to deal with many types of information with various types of associated uncertainty, while in scientific research we often limit the area of discussion on uncertainties and make certain assumptions. Therefore, the research issues concerning uncertainties in GIS need to be further extended to cover a wider range.

Registration of the quality of spatial data in a GIS is an important issue. Quality indicators are marks of the quality of GIS products. A spatial data user can thus select to use or not use a particular set of data based on the particular GIS application he/she has in mind and the quality of the available GIS data. Error indicators are defined as part of the metadata for GIS data. Two major issues need to be considered in presenting the quality of spatial data in a GIS: (1) what kinds of quality indicators to select as part of the metadata; and (2) how to present the indicators in a GIS.

In Chapter Thirteen, Henri J.G.L. Aalders presents a design of quality registration for spatial data in a GIS. First, identification of the quality indicators for spatial data in a GIS is conducted based on an analysis and comparison of the existing data standards proposed by ICA, CEN, and ISO. Second, there is a discussion on possible methods for adding metadata to a GIS based on a relational database solution.

Methods for handling the uncertainties of spatial data in GIS involve many research topics. Part II of this book addresses a number of these. Others, such as methods of assessing error accumulation and propagation in spatial analysis, and methods for determining uncertain relationships among GIS objects, which are described partially in Part I of this book, are not addressed in Part II.

CHAPTER SEVEN

Statistical Quality Control of Geodata

Wilhelm Caspary and Gerhard Joos

ABSTRACT

Statistical quality control as applied to industrial output is a method that can also be used for the control of geodata. This chapter discusses the requirements for applying statistical methods and it shows how the main parameters for a sampling plan can be derived. The central behaviour of a sampling plan can be formulated by the operating characteristic (OC). The OC-curve is used to determine the proper sample size required to achieve significant results and the corresponding acceptance number. The appropriate quality level can be determined by taking a cost model into consideration.

7.1 INTRODUCTION

Unless the quality of given geodata is well known, the data should not be used for any application. Certain types of errors in a database can be detected and sometimes eliminated automatically by programs. These are mainly violations of rules of integrity and consistency of the data concerning topology, attributes, and relationships between objects (Guptill and Morrison, 1995; Joos, 1999). To detect a second kind of error concerning accuracy, completeness, and correctness checking can only be done by comparing the data with either analogue or digital data of higher quality or with the original data sources (Caspary and Joos, 1998). Since this comparison requires usually visual searching for differences, it is a very laborious and expensive task (Claussen, 1996).

For economical reasons a producer or purchaser of a large database cannot perform this manual check for the complete amount of data. Nevertheless, he or she should carry out some type of quality control at the stage of taking over the data in order to get to know whether the data fulfils the stated quality specification. This inspection can be done by drawing random samples, checking the objects within each sample very carefully and draw conclusions from the result on the complete dataset using statistical methods. This chapter describes a method of statistical quality control as applicable prior to input into the database.

Very often sampling methods are used for randomly checking purchased geodata, but these checks are very seldom based on solid statistical theory. This chapter will show methods to determine the number of objects that have to be investigated in order to make statistically significant statements and to ascertain the required acceptable quality level (Joos, 2000).

7.2 REQUIREMENTS

By statistical quality control the error behaviour of the whole dataset is estimated based on a randomly sampled subset of that dataset. If the dataset does not meet the assumption of homogeneity, it has to be divided into different lots where homogeneity in terms of equally distributed erroneous objects can be assumed (Schilling, 1982; Uhlmann, 1982; Montgomery, 1991).

7.2.1 Assumptions about the data-set

A sample of geo-objects can be called a random sample, if the probability of choosing an erroneous object is independent of the location of this object. In other words there is no significant clumping of errors. Homogeneity in the data can only be assumed if all objects were captured:
- from equal data sources
- using the same capturing method
- by the same institution or even person
- and if the area has a similar characteristic (density of objects and distribution of errors).

If the dataset does not meet all criteria for its entire content, it must be split into different lots, where homogeneity with respect to these criteria can be assumed.

Statistical methods for the quality control of geodata are only sufficient if the data will never be used for decisions where the health or life of human beings could be endangered.

7.2.2 Quality criteria

Different object classes may have different error behaviour, therefore only one object class should be considered at a time when analysing the sample. An aggregation should only be applied if there is good reason to do so.

For the classification of errors in geodata three criteria should be considered:
- completeness,
- accuracy, and
- correctness

These are crucial for many applications and cannot be detected differently in contrast to consistency conditions. Data errors referring to these criteria can only be detected in direct comparison of the digital objects with their real world pendant or to their counterpart in the data sources or any other high level reference.

7.3 STATISTICAL CONTROL

7.3.1 Probability function

The theory on statistical control was originally developed for industrial production (Peach and Littauer, 1946; Mace, 1964).

Whether an object is faulty or not can be described using a random variable Y with two possible values 0 and 1. If there is a total number N of objects in a lot and if M of these objects shows at least one error, the probability of randomly choosing an erroneous object is given by the Equation 7.1.

$$\Phi := P(Y = 1) = \frac{M}{N} \tag{7.1}$$

The number N of total objects within a lot can easily be determined by the SQL-aggregate function count. The number M of faulty objects is unknown and subject to be estimated by evaluating the sampling results.

With a specific value y_i of the variable Y for the ith investigated object, the number of errors m in a sample of size n can easily be calculated by summation as in Equation 7.2.

$$m = \sum_{i=1}^{n} y_i \tag{7.2}$$

The random variable X holds for the number of errors in a given sample. The probability of drawing exactly m erroneous objects in a sample of size n is given by the hypergeometric distribution 7.3.

$$P(X = m) = \frac{\binom{M}{m} \cdot \binom{N-M}{n-m}}{\binom{N}{n}} \quad \text{with} \quad 0 \le m \le n \quad m \le M \quad \text{and} \quad n\text{-}m \le \tag{7.3}$$

The hypergeometric distribution is fundamental to much of acceptance sampling. It is applicable when sampling from a finite lot without replacement. It depends on the parameters N, n, and M. The random variable X has a hypergeometric distribution, or shortly

$$X \sim H(N,n,M) \quad \text{or} \quad X \sim H(N,n,\Phi) \tag{7.4}$$

For numerical evaluation the more stable product Equation 7.5 should be used (Joos, 2000)

$$P(X = m) = \prod_{i=0}^{m-1} \frac{n-i}{m-i} \cdot \prod_{i=0}^{m-1} \frac{M-i}{N-n+m-i} \cdot \prod_{i=0}^{n-m-1} \frac{N-M-i}{N-i} \tag{7.5}$$

7.3.2 Acceptance quality control

The decision from a thoroughly investigated sample whether the error-behaviour of a lot is satisfying stated requirements can be carried out by hypotheses tests. The Null hypothesis gives the quality a customer would expect, whereas the Alternative hypothesis refers to the lowest quality a customer can use for his particular application.

$$H_0: \quad M = M_0$$
$$H_A: \quad M = M_A \tag{7.6}$$

The Null hypothesis H_0 will be rejected in favour of the Alternative hypothesis H_A, if a certain number Ac of erroneous objects in a sample of size n is exceeded.

Since sampling is a random procedure it can easily happen that the number of detected errors exceeds the limit acceptance number (Ac) although the total number of erroneous objects in the lot falls below M_0. This leads to a wrong decision with probability of error of 1st kind or producer's risk. The probability at which this kind of error occurs is called error probability α. On the other hand, the number of erroneous objects in a sample can randomly be less or equal the limit Ac, although the true total number of errors M exceeds the stated limit M_A. The lot gets accepted despite its lack of quality. The risk of this wrong decision, also called risk of 2nd kind or consumer's risk, is given by the probability β. In all cases where the number of errors lay between M and M_0 producer and consumer share the risk of rejection or acceptance. Both cases of mistakenly making the wrong decision has influence on the producer as well as on the customer as summarized in Table 7.1.

Table 7.1 Risk of mistakenly making the wrong decision.

	Producer	Consumer
Good lots rejected	Good product lost (producer's risk)	Potential higher cost
Bad lots accepted	Potential customer dissatisfaction	Paid for bad product (consumer's risk)

7.3.3 Sampling plans

A simple sampling plan is characterised by the number n of inspected objects and the acceptance number Ac. There are several strategies to reduce the sample size in order to avoid unnecessary expense. The investigation can be truncated, if the number of detected errors exceeds the acceptance number Ac, even if not all objects in the sample have been investigated. The investigation can be carried out in multiple steps where three decisions are possible in every step: accepting the lot, rejecting the lot, or continue sampling. If a producer has delivered only excellent data in several consecutive lots, the customer can switch from normal sampling to reduced sampling, where the probability of accepting a bad lot increases, but the effort of investigating the samples reduces. If the customer has to reject a lot during reduced sampling, this is an indication of a significant drop of quality, so that he should switch to tightened sampling, even though it yields an increase of making an error of 1st kind.

7.3.4 Operating characteristic (OC)

The behaviour of a sampling plan can be expressed in terms of its operating characteristic. It gives the relation between the percentage of erroneous objects within a lot and the probability of accepting this lot. For a simple sampling plan defined by $Ac \parallel n$ the OC denoted by $L(\Phi)$ can be calculated using the hypergeometric distribution as given in Equation 7.7.

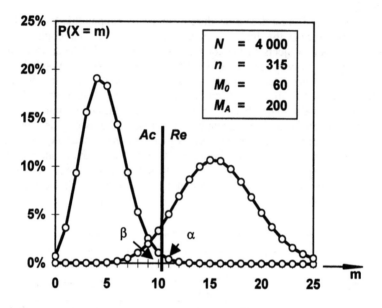

Figure 7.1 Example for the probability of drawing m erroneous objects under the Null and the Alternative hypothesis for a simple sampling plan.

$$L_{N,n,Ac}(M) = L_{N,n,Ac}(\Phi) = \sum_{i=0}^{Ac} P(X = i) \tag{7.7}$$

The operating characteristic curve shown in Figure 7.2 applies for the given parameters, N=10,000, n=100 and Ac=2 defining the sampling plan, i.e. a sample of 100 objects is drawn from a lot of 10,000 objects, and this lot is rejected if more than 2 objects of the sample are erroneous.

The OC curve has three particular points with probabilities which are of special interest for the application of quality control. These points and their associated names are listed in Table 7.2. The probability levels are chosen by convention and standardised in ISO 2859.

If the error rate of the lot is 0,8 % ($= \Phi_{95\%}$), than we have the probability of 95% of accepting H_0 under the sampling plan with OC of Figure. 7.2, i.e. the risk of 1st kind is 5%. For an error rate of 2,7% ($= \Phi_{50\%}$), the probability of acceptance and of rejecting H_0 are equal to 50%. The acceptance probability drops to 5% if the error rate reaches 6,1% ($= \Phi_{5\%}$).

7.3.5 Required sample size

Assuming the acceptable and limiting quality level to be known the parameters defining a sampling plan, namely the sample size n and the acceptance number Ac, can be derived. For this purpose the hypergeometric distribution is transformed into a normal distribution.

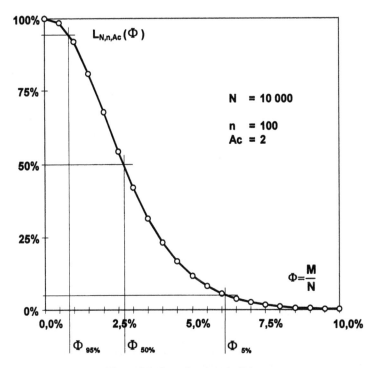

Figure 7.2 Operating characteristic curve.

Table 7.2 Special probability levels for accepting a lot.

Probability	Term	Abbr.	Probability point
95 %	Acceptable quality level	AQL	$\Phi_{95\%}$
50 %	Indifference quality	IQ	$\Phi_{50\%}$
10 % or 5 %	Limiting quality Lot tolerance percent defective	LQ LTPD	$\Phi_{10\%}$ or $\Phi_{5\%}$

The normal distribution has the advantages that it has a continuous density function and that the quantiles of this distribution are very well known.

Equation 7.8 transforms the discrete variable X into a random variable U with standardised normal distribution.

$$U = \left(2 \arcsin \sqrt{\frac{X}{n}} - 2 \arcsin \sqrt{\frac{M}{N}} \right) \sqrt{n} \; \sim \; N(0;1) \tag{7.8}$$

The operating characteristic curve for the wanted sampling plan has to meet two conditions simultaneously:

$$P \big|_{M=M_0} (X > Ac) = \alpha$$
$$P \big|_{M=M_A} (X \le Ac) = \beta \tag{7.9}$$

After applying the transformation this yields the relations 7.10:

$$
\begin{aligned}
1 - P(U < u_{1-\alpha}) &= \alpha \\
P(U < u_\beta) &= \beta
\end{aligned}
\;\;\Leftrightarrow\;\;
\begin{aligned}
\left(2 \arcsin \sqrt{\frac{k}{n}} - 2 \arcsin \sqrt{\frac{M_0}{N}} \right) \sqrt{n} < u_{1-\alpha} \\
\left(2 \arcsin \sqrt{\frac{k}{n}} - 2 \arcsin \sqrt{\frac{M_A}{N}} \right) \sqrt{n} < u_\beta
\end{aligned}
\tag{7.10}
$$

$u_{1-\alpha}$ and u_β refer to the quantiles of the normal distribution at probability level $1-\alpha$ and β respectively. Eliminating the unknown k and substituting the less than sign with equality, since only the minimal number is of interest, the equation can be solved for the desired necessary sample size n.

$$n = \left(\frac{u_\beta - u_{1-\alpha}}{2 \arcsin \sqrt{\frac{M_0}{N}} - 2 \arcsin \sqrt{\frac{M_A}{N}}} \right)^2 \tag{7.11}$$

Inserting sample size n into the equation for the hypergeometric distribution 7.7 yields the maximum allowed number of erroneous objects Ac in a sample of size n (Equation 7.11).

7.3.6 Average outgoing quality

If a lot gets rejected, all objects in that lot have to be thoroughly investigated and all detected errors have to be corrected. Consistency control programs should protect the data from new errors introduced during the correction phase. Even if the controlling person works reliably, you cannot expect that he or she will find all errors. Whether the lot gets rejected or not it will contain erroneous objects. Therefore it is important for a user of the data to know the average amount of these errors. It can be calculated with the operating characteristic (Equation 7.7) and the unavoidable error rate $\Phi_u = M_u / N$.

$$E(\hat{\Phi}) = \frac{M_u}{N} + \frac{M - M_u}{N} \cdot L_{N,n,Ac}(M/N) \tag{7.12}$$

Figure 7.3 gives the graph of the average outgoing quality (AOQ) for a remaining error rate proportional to the previous error rate and with a constant remaining number of errors after inspection of the rejected lot respectively.

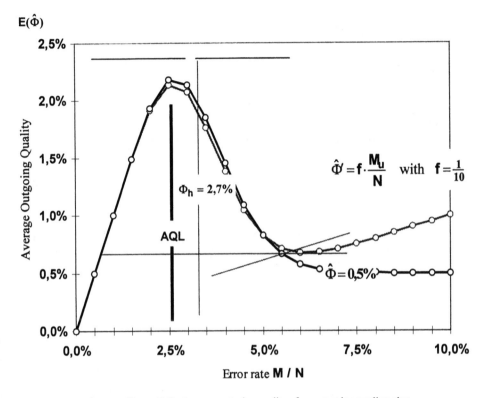

Figure 7. 3. Average outgoing quality of an example sampling plan.

7.3.7 Determination of limiting quality considering costs induced by erroneous data

If a user of a GIS works with erroneous data, he or she will make wrong decisions caused by wrong information from the GIS. These decisions can lead to tremendous cost, which could have been avoided by a rigorous quality control system.

The higher the level for the limiting quality (LQ) the more effort is needed to reach this level. For that reason the costs for quality control have to be on an equal level with the benefit of that quality control. The considered cost items are listed in Table 7.3.

The costs induced by erroneous objects increase with the frequency of access to these objects and with the error rate. The balance is given by Equation 7.13a or 7.13b, dependent on whether the user has to cover the costs for the correction of the data himself, or if he rejects the data so that the producer has to do the corrections with no extra charge to the customer.

Table 7.3 Different cost items.

C1*	Costs caused by erroneous objects without quality control
C1	Costs caused by erroneous objects after quality control with a remaining error rate independent of the error rate before the investigation
C1′	Costs caused by erroneous objects after quality control with a remaining error rate proportional to the error rate before the investigation
C2	Costs caused by the quality control
C3	Costs to correct detected errors
C3′	Costs to correct detected errors dependent on the error rate

$$C1^* = C1 + C2 \quad \text{or} \quad C1^* = C1' + C2 \tag{7.13a}$$
$$C1^* = C1 + C2 + C3 \qquad C1^* = C1' + C2 + C3' \tag{7.13b}$$

Figure 7.4 gives the cost development for the different models. The values with and without statistical quality control are compared and the vertical lines give the error rates for LQ, where effort and benefit are equal.

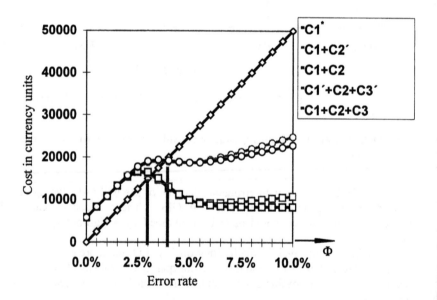

Figure 7.4 Comparison of different cost models for evaluation of LQ.

7.4 CONCLUSIONS

Statistical quality control is an inexpensive method to check whether a given data set fulfils the quality requirements for a particular application. In order to avoid wrong conclusions you have to make sure that the investigated population of objects meets the requirement of homogeneity in terms of error behaviour. Once the AQL and LQ and their reliability levels are fixed, the parameters of the sampling plan can be mathematically derived using the hypergeometric distribution. Whether for very bad or very good data sets, there are strategies to reduce the required sample size in order to keep the expenses for quality control as low as possible.

REFERENCES

Caspary, W. and Joos, G. 1998, Quality Criteria and Control for GIS Databases. In *Proceedings of Symposium on Geodesy for Geotechnical and Structural Engineering*, Eisenstadt, Austria, edited by Kahmen, H., Brückl E. and Wunderlich, Th.(Wien: Institut für Landesvermessung und Ingenieurgeodäsie at TU Wien), pp. 436-441.

Claussen, H., 1996, Qualitätsbeurteilung Digitaler Karten für Fahrzeugnavigationssysteme. *Geo-Informations-Systeme*, 9, pp. 23-29.

Guptill, S. C. and Morrison, J. L. (Eds.), 1995, *Elements of Spatial Data Quality.* (Kidlington, Tarrytown, Tokyo: Elsevier Science).

Joos, G., 1999, Assessing the quality of geodata by testing consistency with respect to the conceptual data schema. In *Geographic Information Research: Trans-Atlantic Perspectives*, edited by Craglia, M.and Onsrud, H. (London: Taylor & Francis), pp. 509-519.

Joos, G., 2000, *Zur Qualität von objektstrukturierten Geodaten.* Schriftenreihe des Studienganges Geodäsie und Geoinformation, Universität der Bundeswehr München, Heft 66. (Neubiberg: Studiengang Geodäsie und Geoinformation at UniBwM).

Mace, A. E., 1964, *Sample-Size Determination* (New York: Reinhold).

Montgomery, D. C., 1991, *Introduction to Statistical Quality Control.* (New York: John Wiley & Sons).

Peach, P. and Littauer, S. B. 1946, A Note on Sampling Inspection. *The Annals of Mathematical Statistics,* 17, pp. 81-84.

Schilling, E. G., 1982, *Acceptance Sampling in Quality Control.* (New York: Marcel Dekker).

Uhlmann, W., 1982, *Statistische Qualitätskontrolle.* (Stuttgart: Teubner).

CHAPTER EIGHT

Data Quality: A Model for Resolvable Objects

Thomas K. Windholz, Kate M. Beard and Michael F. Goodchild

ABSTRACT

The model developed in this chapter considers the combined effect of spatial and thematic dimensions. The objective is a metric to resolve "objects" in "fields". From a three-dimensional representation of the residuals (stored representation vs. higher accuracy) we obtain a relief map showing the minimal determinable variations—which can be used to detect the minimal size of a resolvable object. Thus, the resolvability of a spatial object can be determined by a function of the spatial extension of an object, its attribute value, and the three-dimensional relief of the inherent inaccuracy of the thematic representation. "Objects" in the context of this chapter are considered to be patches of higher concentration, density, etc. A large patch may not be resolvable if its attribute value is weak compared to the accuracy of the "field" representation. This chapter includes a case study of a sea surface temperature data set collected off the coast of Maine.

8.1 INTRODUCTION

The computer is a finite system. We cannot duplicate the infinite real world. Any representation stored in a geographical information system (GIS) is imperfect. The quality of the data within a GIS depends on several different components (e.g., inaccuracy, inconsistency and resolution). Consequently, any query result—generated by this GIS—introduces a level of uncertainty about the state of the quality of the outcome. It is important to provide the GIS user with the necessary awareness that these problems exist. Although there is a growing interest in improving data quality standards (CEN 1995, ISO 1998, SDTS 1994), commercial GIS packages put little or no effort in calculating and communicating the inherent imperfections to the user. In the literature (Chrisman 1983, Gopal 1989, Hunter and Beard 1992, Goodchild and Goodchild 1993, Heuvelink 1993, 1998, Guptill and Morrison 1995, Beard 1996, Parsons 1996), we can find several different approaches for handling either a single imperfection (e.g., inaccuracy) or a combination of imperfections (e.g., imprecision and inconsistency).

8.1.1 Problem statement

There are several methods of generating a GIS map. One of them is to generate a raster representation of a continuous variable (e.g., sea surface temperature). For example, we could sample the variable and then generate a kriged map. Then we could ask the question: "is the resulting map 'good' enough for the purpose of finding an object of a certain spatial extension and attribute value within a field representation (e.g., an area of warmer water with an extension of one square mile)". Thus, this section investigates a model that provides the GIS user with the necessary tools to judge the quality of a stored map with respect to its ability to identify a certain object in a continuous field representation.

8.1.2 Terminology—from scale and resolution to detect-ability

In general one can say that storing a certain representation within a GIS requires a model of the real world at a meaningful (for a specific purpose) scale and at a meaningful—in respect to a chosen scale—resolution. This representation cannot be identical with the real world and thus introduces imperfections (e.g., inaccuracies). However, in order to avoid any confusions with the terminology we want to clarify the terms resolution, inaccuracy and detect-ability.

The stated problem of deciding whether one is able to detect a specific object within a given field representation is dependent on the combined imperfections within the represented area—one of these components could be interpreted as the resolution of the stored field representation. Here the term resolution is a combination of some properties of the photographic heritage—the definition: the degree of discernable detail (Gonzales and Woods, 1993)—and some of the decisive properties inferred by the scale of a paper map—the users expectation to identify specific features at a certain scale. On the other hand, we could also use the term "level of geographic detail", which is discussed by Goodchild and Proctor (1997) as a possible augmentation of the term "scale" in the digital geographical world.

Another component of imperfections, namely inaccuracy, is also closely related to detect-ability. Inaccuracy is commonly associated with a certain feature, for example a point within a polygon or an edge of a thematic class. On the other hand, inaccuracy can be given as a single value for a complete layer (e.g., a DEM can have an attached inaccuracy information indicating that the representation has an accuracy of ±5m). However, these inaccuracy values neglect to include any information about their spatial distribution. The model in this section includes the spatial dependencies of the calculated inaccuracies and thus differs from simple inaccuracy values.

Since some of the terminology is used differently in different disciplines or might be misleading we do not want to reuse terms like resolution, inaccuracy, or scale for the model introduced in this section. Thus, we introduce another term: detect-ability, which combines properties of the field (that can be seen as aspects of resolution or inaccuracy) and properties of the object. Their distinct dependencies (e.g., sample size or object size) are explained in more detail in the following section.

8.2 DEPENDENCIES OF DETECT-ABILITY

This section discusses the parameters that influence the outcome of the question where (within the field representation) one can identify objects—we refer to this as the dependencies of detect-ability. An intuitive approach to this question suggests that there are two main components influencing the results. On the one hand, there is the field representation and on the other hand, there is the object. However, here we are interested in a more detailed list (Figure 8.1).

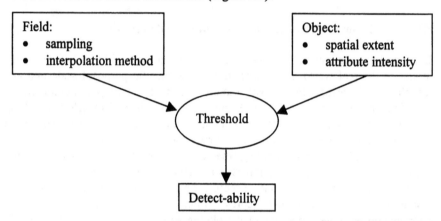

Figure 8.1 Dependencies of detect-ability.

First, let us take a look at the decisive parameters used to generate a field representation (Peuquet *et al.*, 1999), where we assume that we start with sampling the underlying variable. Sampling introduces two constraints: (1) the number of sample points; and (2) their distribution. The next step when aiming at generating a continuous coverage is deciding which interpolation process (e.g., universal kriging) to choose. The final field representation will differ if any of these three components vary. Some of the results are more accurate than others (e.g., more sample points) and some of them will be smoother compared to others—depending on the interpolation method. The accuracy or smoothness of the representation influences whether we can detect an object or not.

Second, we would like to focus on some properties of the object itself. There are two components that are of interest when formulating its detect-ability within a field representation: (1) the spatial extent of the object; and (2) the attribute height (or strength) of the object. Assuming that an object within a field has a distinct outline, its spatial extensions can be given by a single value, namely by its area in square units. The attribute height of the object is in the same units as the field representation and is a relative comparison to its neighbourhood. For an object showing a small spatial extent we can say that it will be detected more easily with increasing attribute height. An object having a small attribute height will be more detectable with a larger spatial extent.

The third dependency is given by the threshold. The threshold determines the percentage of the object that has to be visible for its detection and it could vary—

up to a certain degree of freedom—as a parameter specified by a GIS user. The determination of the visibility and thus, detect-ability is discussed in the following section.

8.3 THE MODEL—HOW TO DETERMINE DETECT-ABILITY

The applied method for generating a representation (e.g., sampling followed by kriging) introduces some constraints on the level of detail that one is able to provide within the GIS. In this section we discuss a model that results in a binary map that identifies areas where a certain object can be determined and where it can not.

8.3.1 Approach

The model is based on the residuals calculated by subtracting the generated field representation from the ground truth. For an implementation we can substitute ground truth with any layer that we accept as being true. This could either be a comparable representation of higher accuracy (if available) or multiple (e.g., n = 100) generations of simulated realizations using conditional simulation (e.g., Gaussian Simulation). Within this section both approaches are addressed. The residuals can then be seen as a result of (1) the sample method; and (2) the model effects inherent in the interpolation method used to generate the field representation (e.g., kriging). The residuals represent an indicator of how well the representation matches reality—where one could say that this is the accuracy of the map. This is one way of interpreting these residuals, however, here we are looking beyond the numeric information and consider the spatial distribution of the values of the residuals. These residuals can be used to determine the detect-ability (or resolution) of a given representation.

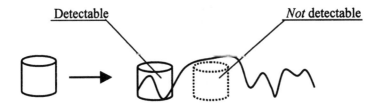

Detectable   *Not* detectable

Figure 8.2 Object representation and relief map of the residuals.

Within the field representation—at any given location—one cannot determine a feature occurring in the real world if the variations are smaller than the residuals. Let us take this idea a step further: if we generate a three-dimensional representation of the residuals we obtain a relief map (similar to a DEM) of the minimal determinable variations. Looking at this relief map we can now specify a representative object and compare it to the terrain outlines of the relief map. Figure

8.2 illustrates a profile through such a terrain and an example object we wish to detect in this terrain. If the object is hidden by the relief map then we say it cannot be detected from the kriged map representation. On the other hand, if the object is fully visible on the outside of the relief map than we would be able to detect the object within the kriged map.

Next we would like to discuss the generation of a representative object for the comparison mentioned above. We suggest a representative object in the form of a cylinder. This is a result of the fact that a circle is the most compact form and that the height is a parallel movement of the objects outline. The radius of the circle is determined by the spatial extension of the object (e.g., we want to identify an object that has an area of π square units than the radius of the cylinder would equal 1 unit). The height of the cylinder represents the attribute value.

Finally we combine the relief map with the cylinder. In order to determine the areas of possible detect-ability the cylinder is moved over the relief map. At each location of the cylinder (i.e., representation of the object) we now have to determine whether the top of the cylinder extends above the relief map (i.e., inherent inaccuracies/noise) or not. If the top of the cylinder is visible we can infer that an object located at this position would not be covered by the inaccuracies and thus be detect-able. However, we can say that if the spatial extent of the cylinder is represented by, for example, 100 pixels it is sufficient to see 99 pixels in order to detect the cylinder. Thus, the introduction of a threshold for the detect-ability allows a percentage (e.g., 5%) of the cylinder to be obscured by the relief map.

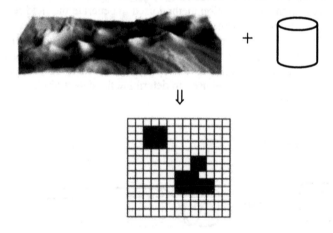

Figure 8.3 Schematic representation of the moving object and the resulting binary map.

Figure 8.3 shows a schematic representation of calculating the detect-ability from the relief map and a representative object. The result is a binary map, where areas of positive detect-ability (i.e., the object can be detected) are marked white and areas of negative detect-ability (i.e., the object cannot be detected) are marked black. The areas refer to the centre of the object. Thus, if parts of a given object are within a black area, but it is centred within a white area, we would be able to detect the object. Regarding the visualization of the resulting binary map, it might be better to represent areas of positive detect-ability green and areas of negative detect-ability red. These colour settings might improve the communication of the inherent imperfections to the GIS user.

	Certain about statement	Uncertain about statement
Object is in observed area	white pixel ☐	black pixel ■
Object is *not* in observed area	white pixel ☐	black pixel ■

Figure 8.4 Implied inferences.

The resulting binary map needs some more discussion in order to clarify the inferences we can make about the areas of positive (white) and negative (black) detect-ability. For the white areas we can say that whether an object is present or not, the field representation is "good" enough to state that we are certain about the represented facts (i.e., there is an object or not). Whereas, for the black areas we have to state that the field representation does not allow us to make any inferences about the existence or non-existence of the defined object. Thus, all inferences made about objects within a black area introduce uncertainty in any derivations made from these field representations. This relationship is shown in Figure 8.4.

8.3.2 Applications

In this section we take a closer look at some interesting applications of the discussed approach. In general one can divide the applications into two major categories. On the one hand, there are those applications where the whole area of interest is already sampled or where—in addition to sampling—the kriged map is already generated. Here the model would be able to tell the user if the quality of the representation is high enough to derive conclusions with a desired certainty. The model could also be used to determine the appropriate sample size for a specific purpose (i.e., detecting objects of a certain size).

First, we would like to discuss issues of examples where the whole study area has been sampled. Applications could invoke the identification of, for example, warm core rings (i.e., warmer water pools), which would lead to a different ecological system within a cold-water area. This phenomenon occurs in the Gulf of Maine when warm core rings get separated from the Gulf Stream. The sizes of these separations have to fulfil minimum requirements regarding their spatial extent in order to have an impact on the ecological system. The issue is to prove that the change in an ecological system was initialized by one of these pools. Thus, it is of interest to have the ability to say—with certainty—that there was no such object (i.e., pool) within a given field representation (i.e., map of sea surface temperature generated from sample points). Another application could be the detection of patches of high concentration of soil pollution in a rural area. This case introduces another interesting aspect, where operators of a chemical plant might have an interest to prove—with certainty—that there are no high concentrations of soil pollution in a specific sub-area. Thus, here we deal with a legal issue to prove that a map is fit for the specific purpose.

Second, a slight modification of the discussed model could be used to determine whether a proposed sample size and distribution is sufficient for detecting a certain object prior to sampling the whole area of interest. Here the problem is more focused on the determination of whether the combination of the applied methods (i.e., sampling and interpolation method) will yield a sufficiently accurate field representation. The first step would require collecting sample points within a predefined sub-area, where objects do not necessarily have to be located. Then, at arbitrary locations within the sub-area, perturbations of the size of the given object would be introduced. Finally, an application of the suggested model to determine the detect-ability would clarify if the applied methods (i.e., sampling and interpolation method) are sufficiently accurate. If there are any black areas in the resulting binary map, changes are necessary (e.g., increasing the sample size). This method would require the implementation of conditional simulations—as indicated earlier.

8.4 CASE STUDY

In this case study we want to determine whether we can detect pools (with a radius of about 10km) of different water temperature (e.g., ±2°C and ±5°C). Here we discuss two different approaches. First, we generate the accuracy information by subtracting a kriged map (generated with an isotropic variogram model and punctual kriging) from the satellite image (i.e., ground truth). The second approach uses 50 conditional simulations (Gaussian) to generate the required accuracy information. In the latter approach we use the difference between the lowest/highest simulated attribute value and the kriged map for each location within the study area.

8.4.1 The used data

We use a satellite image showing the sea surface temperature (Figure 8.5) in the Gulf of Maine and a set of 231 sample points within the area shown in Figure 8.5. The sample points follow a regular distribution with a spacing of about 20km between them.

For the analysis and subsequent processing of the data several different software programs are used—with the key products: ARC/INFO, ArcView, GS+, gstat and S+. Some intermediate steps and the comparison between the accuracy information and the objects are generated by custom applications written in C++.

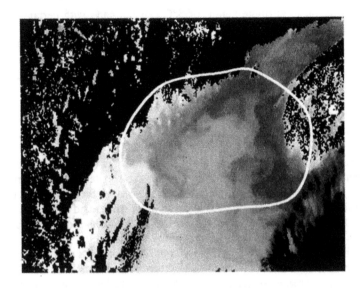

Figure 8.5 Satellite image, showing sea surface temperature.

8.4.2 Results using the satellite image

When applying the discussed model we investigate the detect-ability for two different objects. One of them with an attribute height of 2°C and the other one with an attribute height of 5°C, where the remaining dependencies (e.g., radius = 10km, threshold = 85%) of detect-ability are kept constant. The results can be seen in Figure 8.6a—for the 2°C object—and in Figure 8.6b—for the 5°C object.

a   b

Figure 8.6 Resulting binary maps a) for the 2°C object and b) for the 5°C object using the satellite image.

A comparison of the two results (shown in Figure 8.6) confirms the assumption that the areas where inferences about an object of an attribute height of 2°C can be made with certainty are clearly smaller than the areas where inferences about an object of an attribute height of 5°C can be made with certainty. These results lead to the following conclusions:

- If objects of 2°C attribute height need to be detected the applied method (e.g., sample spacing) is not sufficient.
- If objects of 5°C attribute height need to be detected the used method is sufficient.

8.4.3 Results using conditional simulations

For the second approach we used the same data set of 231 sample points extracted from the satellite image to generate a continuous representation. Furthermore, we also used the same objects (of 2°C and 5°C, respectively). The remaining dependencies were also kept constant at a radius of 10km and a set threshold of 85%. The difference for this approach, however, is the generation of the accuracy information and the subsequent calculation of the binary results.

To obtain the accuracy information required for this approach we have chosen to generate 50 Gaussian simulations (conditional upon data) of the given area. In order to increase the quality of the results we added 100 additional sample points. The resulting maps where then compared to the kriged map (using 231 sample points) consequentially generating 50 binary images showing the given detect-ability. In a final step all binary images were added. The pixels were assigned a "black" value if one or more of the 50 generations indicated a "black" value. On the other hand, they were assigned a "white" value only if all 50 generations resulted in a "white" value. If we would generate 100 simulations we would allow 2 generations to show a black pixel and still assign a white pixel to our final result. This approach is taken to gain independence of the number of generated simulation.

From the results shown in Figure 8.7a (for the 2°C object) and in Figure 8.7b (for the 5°C object) one can see that similar conclusions can be derived. However, as expected one can see an obvious smoothing effect when compared to the previous results.

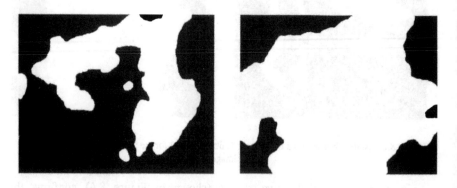

a   b

Figure 8.7 Resulting binary maps a) for the 2°C object and b) for the 5°C object based on conditional simulations.

8.5 CONCLUSIONS AND FUTURE WORK

The discussed model enables the GIS user to determine whether the quality of a given field representation is sufficient to detect a representative object. The result is presented via a binary raster representation where we can identify areas of positive and negative detect-ability. The user has to provide the spatial extent and the attribute height of the object. Furthermore, if required, the user should have the ability to vary—up to a certain degree (e.g., 0% to 20%)—the threshold for the determination of detect-ability.

In our case study we used two different approaches. At first, a satellite image was used as a reference (i.e., ground truth) to calculate the necessary residuals for the relief map. Subsequently we focused on including the model of conditional simulations to gain independence of a ground truth reference. In our approach we included additional sample points for the simulations to increase the quality of our results.

Another promising research area using the discussed model is the investigation of the influence on the binary result map of varying the dependencies of detect-ability. For example, we could reduce or increase the number of sample points and then analyse the relation between the number of sample points and the area of positive detect-ability. It would also be of interest to investigate variations in representative object shapes. Here we would like to look into the outcomes of replacing the cylinder by a line.

This section investigates a simple approach to communicate aspects of inaccuracy and resolution of a field representation to the GIS user. Future work will show aspects of an exploration of the effects of the dependencies of detect-ability on the results.

REFERENCES

Beard, M. K., 1996, A Structure for Organizing Metadata Collection. In *Proceedings 3rd International Conference on Integrating GIS and Environmental Modeling*, Santa Barbara, (CA: NCGIA), URL: http://www.ncgia.ucsb.edu/conf/sante-fe_cd_rom/main.html.

CEN / TC 287, WG 2, 1995, *Geographic Information - Data Description-Quality. draft for discussion* URL: http://forum.afnor.fr/afnor/WORK/AFNOR/GPN2/Z13C/PUBLIC/DOC/.

Chrisman, N., 1983, The Role of Quality Information in the Long Term Functioning of a Geographic Information System. In *Proceedings Auto Carto 6*, Ottawa, Vol. 1, pp. 303–312.

Gonzales, R. and Woods, R., 1993. *Digital Image Processing,* (Addison-Wesley Publishing Company).

Goodchild, M. F., 1993, Data Models and Data Quality: Problems and Prospects. In *Environmental Modelling and GIS*, edited by Goodchild, M. F., Parks, B. and Steyart, L. (New York: Oxford University Press), pp. 363–371.

Goodchild, M. F. and Gopal, S., 1989, *Accuracy of Spatial Databases,* (London: Taylor & Francis).

Goodchild, M. F. and Proctor, J., 1997, Scale in a Digital Geographic World. *Geographical & Environmental Modelling,* 1 (1), pp 5–23.

Guptill, S. and Morrison, J. L., 1995, *Elements of Spatial Data Quality,* (Tarrytown, NY: Elsevier Science).

Heuvelink, G.B.M., 1993, *Error Propagation in Quantitative Spatial Modeling,* (Utrecht: Drukkerij Elinkwijk).

Heuvelink, G.B.M., 1998, *Error Propagation in Environmental Modelling with GIS,* (London: Taylor & Francis).

Hunter, G. and Beard, M. K., 1992, Understanding Error in Spatial Databases. *Australian Surveyor,* **37** (2), pp. 108–119.

ISO / TC211 1998, URL: http://www.statkart.no/isotc211/

SDTS National Institute of Standards and Technology, 1994, *Federal Information Processing Standard Publication 173.* (Spatial Data Transfer Standard Part 1.Version 1.1), U.S Department of Commerce.

Parsons, S., 1996, Current Approaches to Handling Imperfect Information in Data and Knowledge Bases. *IEEE Transactions on Knowledge and Data Engineering,* 8 (3), pp. 353–372.

Peuquet, D., Smith, B. and Brogaard, B., 1999, The Ontology of Fields, *Report of a Specialist Meeting held under the auspices of the Varenius Project.* Bar Harbor, Maine, http://www.ncgia.ucsb.edu/pubs/pubslist.html#varenius.

CHAPTER NINE

A Practical Study on Precision and Resolution in Vector Geographical Databases

François Vauglin

ABSTRACT

Vector geographical databases are designed to represent the world by means of points, lines, polygons, and attributes. This fundamental information is highly structured in most geographical information systems, in which complex concepts like topology can be implemented. This chapter describes research work on precision and resolution of vector data. Precision and resolution can be defined and measured for raster data or even for classical paper maps. Their definition for digital vector data is presented as a consequence of the corresponding definitions for classical paper maps and as a result of production processes and human perception. Trying to make precision and resolution compatible is then presented as an optimisation problem for geographical databases. A tool providing a simple solution is proposed as well as a classification of the issues that are raised.

9.1 INTRODUCTION

Vector geographical databases are designed to represent geographical information. They include positional and attribute information and metadata. Positional information provides a description of the shape and position of geographical features, which are modelled by points, lines, and polygons.

From a user point of view, there seems to be no link between geographical databases and the concept of scale. One can always zoom in or out: points, lines and polygons will simply be displayed bigger or smaller. With geographical information systems (GIS), the representation of features may change by zooming. This can lead the user of geographical data to believe that data are relevant whatever the scale. As this is obviously not correct, it is necessary to study further the role of scale in digital geographical information. Multiscale functionalities in GIS have now become a specific issue requiring a strong theoretical background (Devogele, 1997; Goodchild and Proctor, 1997).

On classical maps, scale is the ratio that must be applied to convert a distance measured on the map into a ground distance. Points, lines and contours are drawn with a thickness varying between 0.1 mm to 1 mm or more. Smaller details than these sizes come from the printing device or paper irregularities and have no meaning on the ground: it is useless to make measurements on smaller details even with a very precise device. The fineness of drawing is called "graphical resolution" by Arkin *et al.* (1991).

Analogous concepts may be constructed with digital geographical databases. Sections 9.2 and 9.3 will present what can be called "precision" and "resolution" in a geographical database. The advantages and problems that arise when trying to tune precision to resolution are described in Section 9.4. An assisting tool for tuning precision to resolution is briefly presented in Section 9.5. This tool has been designed and implemented at the COGIT laboratory of IGN. Some results are shown and commented upon and Section 9.6 concludes this chapter.

9.2 PRECISION OF GEOGRAPHICAL DATABASES

On classical maps, sizes of features or distances between locations can be measured with a ruler and converted to ground distance or size. The result may vary if one uses a very precise linen tester with a magnifying glass or if one roughly and visually estimates the measure. This emphasises the importance of the precision of the tool used for measuring.

In a digital geographical database, there is no ruler. Features are stored with a description of their geometry as points, lines, and polygons. Each feature materialises as a list of coordinates. Geographical information systems simply use geometry to compute sizes and distances from coordinates. And just as happens on classical maps, results will vary according to whether one uses a very precise data set or not. The main difference comes from the tool that is used for measuring: distances are assessed in GIS by computations of differences and root squares in the coordinates. These operations are very accurate in a computer. The results will have at least as many digits as there are in coordinates. The precision of a database can thus be defined as follows.

Digital precision. Number of digits used to store coordinates. This definition is linked to the way geographical information is represented and stored in computers. For that reason, it will be called "digital precision".

Goodchild has given another and different definition for precision in spatial databases:

Geographical precision. Number of significant digits used to report a measurement (Goodchild, 1992). "Significant digits" is related to the geographical meaning of the information represented in the database. A global positioning system (GPS), for example may provide coordinates in millimetres, but there may be no reliability in the digits below 1 metre. In statistical words, the digits corresponding to decimetres, centimetres and millimetres of the coordinates are random. Therefore they cannot be considered as "significant digits".

A consequence of these definitions is that digital precision is always higher than geographical precision.

Other definitions for statistical precision may be found in the literature (see SDTS - Spatial Data Transfer Standard; FGDC - Federal Geographic Data Committee; or Bertin, 1983; Laurini and Thomson, 1992; ISO, 1996; David and Fasquel, 1997; Vauglin, 1997). They all present precision as a statistical parameter that has no link to the definitions above. This chapter will stick to the definition of "digital precision" and "geographical precision" as stated above.

Other works by the Güting group on the Rose Algebra could probably be used to strengthen theoretical definitions (Güting and Schneider, 1995). Theoretical frameworks on resolution have also been defined by Worboys (Stell and Worboys, 1998), (see also Skogan and Fimland, 2000). These are not included in this chapter, which focuses on a presentation of experimental problems.

9.3 RESOLUTION OF GEOGRAPHICAL DATABASES

Similar digital/geographical distinctions may be brought to the definition of resolution (Omrane, 1998; Vauglin, 1997). Precision is a number of digits for coordinates whereas resolution is a corresponding length on the ground.

Digital resolution. Smallest distance that is expected to exist between two different nodes stored in the database. Both digital resolution and digital precision are linked to the number of bytes used to store the coordinates.

Any two nodes that are closer than the digital resolution will be considered as equal. Any two nodes separated by a distance larger than the digital resolution will have different coordinates in the database. Digital resolution is the step of the magnetic grid that would be in a drawing application. It is the smallest detail that can be represented in the database.

Geographical resolution. Size of the smallest detail with a geographical relevancy. Some very small geometrical details that are stored in geographical databases are artefacts with no counterpart on the ground. They come from flawed production processes and should not be considered as geographically "relevant".

Taking a similar example as for geographical precision in Section 9.2, a GPS provides coordinates that are perturbed by a small random noise. Therefore, small variations appear when digitising a straight road with a GPS (see Figure 9.1). These variations have no geographical relevancy. The geographical resolution of the data set produced is larger than the size of these variations.

Figure 9.1 A straight road (straight line) and its digitisation by a GPS (polyline).
Each vertex of the polyline is a point as measured given by the GPS device.
R is the resolution of the data set

To ensure a high quality of their products despite the limits of human perception, mapping agencies often digitise geographical features at a larger scale than the scale of use. This enables the stereoplotter to reach beyond his/her own limits of perception and "see more" from aerial photographs. The result is of higher resolution, but too many details and details too small are digitised. Generalisation is then performed to keep the features compliant with the specifications of the database. This process leads to geographical features with too many digits to describe their geometry and to details smaller than the actual resolution.

Digital precision is too high and geographical resolution is not expected. Hence, the data set could be smaller with the same amount of geographical information. Section 9.4 develops on that.

9.4 OPTIMISING RAISES PROBLEMS

9.4.1 Digital precision versus geographical resolution

Section 9.3 has presented why geographical databases often have too high a digital precision and why they contain details smaller than the geographical resolution. This situation is a source of several problems such as:

- Larger data sets than necessary. When digital precision is higher than geographical precision, irrelevant digits are stored. If several digits can be dropped for each coordinate, the global volume of data is reduced. This helps to spare storage and makes applications go faster with fewer useless data.
- Illusion of high precision. The user can zoom in further than geographical precision, resulting in taking into account details that have no geographical meaning. This fake precision can be very embarrassing in some applications.

Example. The French cartographic database (BDCarto®) is digitised from the 1:50,000 scale French maps. According to its specifications, it is supposed to be used at scales between 1.100,000 and 1:500,000. Geographical resolution is around 10 metres (Spec, 1994). Digital precision is 12 digits and digital resolution is 1 micrometer. This means that 6 digits are dropped when digital precision is brought to 6 digits and digital resolution to 1 metre. 7 digits are dropped if digital resolution is tuned to geographical resolution (10 metres). Table 9.1 presents the size of such data sets. This example comes from the hydrography layer of the French BDCarto® on the Calvados département (4319 lines).

Table 9.1 Size of a geographical data set according to its digital precision.

Digital precision	Digital resolution	Size of data set (bytes)	Example of coordinate	Comment
12 digits	1 µm	2,605,451	356,292,656,250 m	Original data set from BDCarto®
6 digits	1 m	1,245,774	356,293 m	
5 digits	10 m	1,080,937	35629 x 10 m	data set where digital resolution = geographical resolution

Table 9.1 shows that non-significant digits can be very space consuming in computers. This is due to geometry that usually takes about 90% of data volumes. The remaining 10% is used by attributes and topology. Over-sized data sets are also time consuming in applications.

The second problem mentioned at the beginning of this section is the illusion of high precision. Naive users seeing that all coordinates are given in millimetres may think that such a data set is very precise. They may use it without caution at too high scales, even as high as 1:100 or higher. This is all the more likely as GIS zooming is often not restrained in scale. Such misuse of BDCarto® would be the source of lots of problems, especially when users need to overlay very precise data and muddle them up by trying to make them coincide with the cartographic background.

9.4.2 Problems occurring when tuning digital precision to geographical resolution

Tuning digital precision to geographical resolution is a matter of optimisation. From the example given in Table 9.1, one may just think that coordinates must be rounded to the geographical resolution to optimise the data sets. This comes down to putting the points and nodes on a grid where its step is equal to the geographical resolution. Rounding is a simple solution that raises several problems, especially with topology. The impact of rounding has been studied on a set of simple geometrical figures as proposed by Arkin *et al.*, (1991) and Omrane (1998). The results presented in Section 9.5 are on geographical examples.

Topological problems involving one feature. Topology can be changed when coordinates are rounded. When occurring within one feature, this change modifies the number of faces of a polygon or the number of intersections of a line with itself. Basic examples are presented in Figure 9.2.

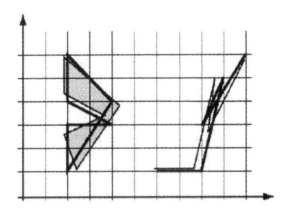

Figure 9.2 A polygon (thin line - grey polygon on the left) and its rounded version (thick line on the left). A line (thin line on the right) and its rounded version (thick line on the right). Each vertex has been moved to the grid coordinates.

Topological problems involving several features. Changes of topology can also occur between several features from one layer or from several layers. This kind of modification mainly causes inversions of spatial relations (e.g. a feature that is on one side of another feature moves over to the other side) or merging of features that are different. Figure 9.3 shows a road on the left of a railway. The road is moved over to the right of the railway when its coordinates are rounded.

Changes of topology lead to features that are inverted but also to features that are merged. This situation is presented in several examples in Figure 9.4.More complicated situations occur when problems are intricate, as shown in Figure 9.5.

Topological problems involving several features become conflicts between layers when at least two features from two different layers are involved. This implies cross-verification between layers when rounding is performed.

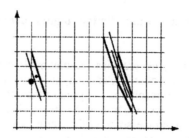

Figure 9.3 A line (thin line on the left), its rounded version (thick line on the left), and a vertex (small dot on the left) and its rounded version (big dot on the left).
A similar change of topology can occur between two lines (as shown on the right).

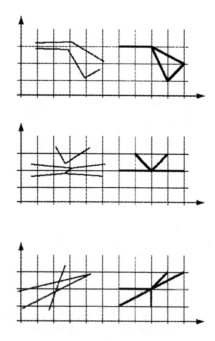

Figure 9.4 Lines (thin lines on the left) and their rounded versions (thick lines on the right).
Changes of topology can occur between two or more lines.

Violations of the database's specifications. Geographical databases provide a digital representation of the world. Their production requires specifications that describe the way the world is modelled and represented (Frank, 1997). Specifications include a description of the quality conditions that data sets must meet.

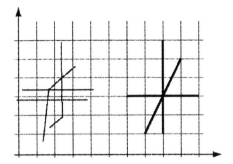

Figure 9.5 Merges of lines combined with inversion of topological relations, all due to rounding of coordinates.

The following examples are taken from the BDCarto® specifications on data quality for linear features (Spec, 1994). All linear features in BDCarto® must verify three criteria within a layer (see Figure 9.6):

A- Two end nodes of a line or two vertices from different lines cannot be closer than 20 metres;

B- Two successive vertices of an edge cannot be closer than 10 metres;

C- Any vertex must achieve a deviation greater than 5 metres within a line.

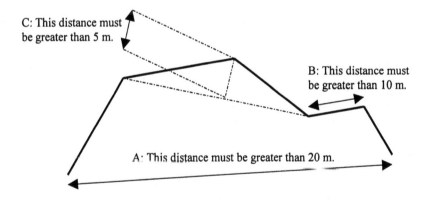

Figure 9.6 Specifications of BDCarto® include three criteria on vertices, nodes and lines (A, B, C).

Rounding of coordinates moves nodes and can cause problems with topology. It can also be the source of violations of the specifications when nodes and vertices are moved in a situation that does not comply any more with criterion A or B or C. An example is presented in Figure 9.7.

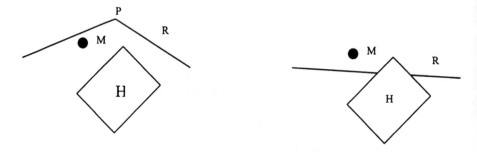

Figure 9.7 Example of violation of specifications. After rounding of coordinates of a vertex (P), its position is moved in such way that it comes to violate criterion C. This situation is complex: Node (P) cannot be dropped to comply with criterion C because this would create a topological problem as the road (R) would cross the house (H) and another node (M) would be on the other side of the road (R).

9.4.3 Toward a solution

The rounding of coordinates leads features to be displaced. These slight movements are limited by the digital resolution r used for rounding. The maximal distance D of displacement of nodes is given by Equation 9.1 (Omrane, 1998).

$$D = \frac{\sqrt{2}}{2}r \qquad\qquad (9.1)$$

The displacement due to rounding is always lower than or equal to D, and its spatial distribution is random. Hence, rounding does not change much the positional accuracy of the data set. As digital precision (often over-estimated - see Section 9.3) is not linked to tools used for data capture, positional accuracy is actually related to geographical resolution, not to digital precision.

Therefore, the important problems due to rounding are only those presented above: topological problems and violations of specifications. For that reason, a tool using rounding has been developed to optimise geographical data sets. This tool performs a rounding that must not be greater than the geographical resolution. If it detects a situation where topological problems or violations of specifications could arise, it asks the user what to do. In that situation, the user can allow a displacement higher than given by Equation 9.1 so that a solution with no change of topology nor violation of specifications can be reached.

Results are presented in the next section.

9.5 RESULTS

A simple tool has been developed in C++ and results can be displayed using Arc/Info. The data used for tests are three layers from BDCarto® covering the

French Département of Calvados. Layers used are roads and railways (20,564 lines), hydrography (4319 lines), and administrative boundaries (3425 lines). Other tests involve the region of Bretagne (roads and railways – 79,000 lines) and the whole of France (hydrography and roads and railways).

Topological problems. All changes of topology involving one or several features have been counted. As geographical resolution of BDCarto® is considered to be around 10 metres, three values for rounding have been tested: rounding of coordinates at 1 metre, 5 metres and 10 metres.

Table 9.2 presents the number of topological problems due to the truncation of coordinates. In this table, only one layer is examined.

Table 9.2 Number of topological problems involving one layer after rounding.

Digital resolution after rounding	Calvados: Administrative boundaries	Calvados: Roads and railways	Calvados: Hydrography
1 metre	0 *(0 %)*	0 *(0 %)*	0 *(0 %)*
5 metres	0 *(0 %)*	2 *(0.0097 %)*	2 *(0.046 %)*
10 metres	3 *(0.088 %)*	25 *(0.12 %)*	3 *(0.069 %)*

Digital resolution after rounding	France: Roads and railways	France: Hydrography
5 metres	96	40
10 metres	899	448

The first column of Table 9.2 contains the final digital resolution after rounding of coordinates. The other columns present the number of topological problems that arise during rounding. Values in brackets and italic are the percentages of problems among all the lines.

Figure 9.8 shows two real examples from BDCarto® where the topology of roads changes under rounding of coordinates.

Table 9.3 presents the number of topological problems due to coordinate rounding. In this table, only the number of problems involving two layers are counted: these problems are topological problems between at least two features from two different layers (roads - railways and hydrography in Bretagne). Values in brackets and italic are the percentages of problems among all the road and railway lines.

Violations of specifications. As shown in Section 9.4.2, rounding can also cause specifications to be violated. As for topology above, problems have been inventoried. Results are presented for each criterion A, B, and C (see Section 9.4.2). Results on a 1 metre rounding are presented in Table 9.4. The data set used to build this table is the roads and railways networks of the French Calvados département (20,564 lines).

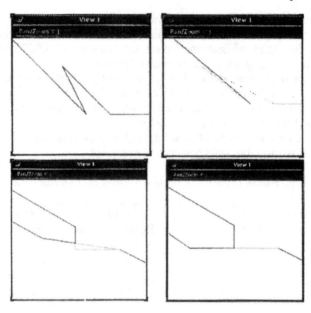

Figure 9.8. Two examples before (on the left) and after rounding (on the right).
Digital resolution is changed to 10 metres.

Table 9.3 Number of topological problems involving several layers after rounding.

Digital resolution after rounding	Bretagne: Roads and railways / Hydrography
5 metres	5 (0.0063 %)
10 metres	67 (0.085 %)

Table 9.4 Number of violations of specifications after rounding.

Criterion violated	Number of violations
Criterion A (minimal distance, inter)	2584 (12.57 %)
Criterion B (minimal distance, intra)	19000 (92.40 %)
Criterion C (minimal deviation)	723 (3.52 %)

In Table 9.4, the "number of violations" is referring to the number of couples of vertices that do not respect one of the three criteria. In brackets and italic this number is given as a percentage of the number of road and railway lines.

The high percentage for criterion B in Table 9.4 means that nearly all the resulting lines have two successive nodes that are closer than 10 metres. This is not that surprising since data sets are produced at a scale larger than the scale of use (see Section 9.3), inducing the stereoplotter to digitise too many nodes for capturing lines.

Then, filters may be applied to make the dataset comply with criteria A, B and C. This leads to lots of lines having nodes that barely comply with criterions A, B and C. Slight movements due to rounding make a lot of these nodes violate one of the criterions.

These violations, however, are not as severe as topology changes. Most of them will have no consequence at all on applications, whereas any small modification of topology can entail very strong problems in applications.

The tool that has been developed to help during rounding is able to warn the user when configurations may lead to violations of specifications or changes of topology. These configurations are highlighted in grey on Figure 9.9.

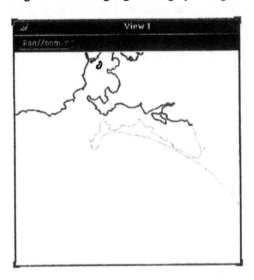

Figure 9.9 This example is taken from the see coast in Bretagne.
The line that may lead to changes of topology or violations of specifications after rounding is highlighted (in grey) in order to warn the user.

9.6 CONCLUSIONS

Resolution and precision are widely used in geographical information. This chapter proposes finer definitions according to the specificity of the two main components of GIS: geography and computers. A digital definition and a geographical definition can be given for both resolution and precision. These new definitions reveal an opportunity for the optimisation of geographical data sets: tuning digital precision to geographical resolution can decrease the size of data without loss of significant information.

A simple solution to that problem is the rounding of coordinates, but this can raise other problems, such as changes of topology (inversions or merges), violations of specifications, or displacements. Counting them shows that rounding is easy to industrialise because there are few big problems. Despite of problems, rounding is the easiest solution. Furthermore, the problems can be automatically selected for an interactive solution.

Currently, the tool counts merges together with inversions of topology. A future version should give separate results to provide a finer description of violations of topology. Further studies (involving the integration of simple generalisation algorithms to locally move around some nodes together) could lead to yet another future step for the tool: proposing an automatic solution for all situations.

From that study on BDCarto®, very few serious problems arise when rounding the coordinates to 1 metre. This means that setting the digital precision to 10% of the digital resolution may be a relevant choice in regard to topological quality. Higher values for digital precision lead to more problems, which however may be considered as minor for some quality standards.

The examples show that such optimisation leads to a data volume 210% lighter. This improvement makes the packaging of large databases easier: The size of a BDCarto® layer is currently about 10 Mbytes for one Département, for the whole of France, it is about 800 Mbytes. Reducing its size is a great opportunity for users. It is more easily tractable and also makes applications faster.

As further research, the notion of accuracy could be examined with a similar state of mind. This would bring it closer to users' interests. A more formal link between resolution, precision, and positional accuracy would be a significant contribution to the development of a coherent theory of geographical data quality. Using tools and concepts of the ROSE algebra could be a good contribution to further research (Güting and Schneider, 1995) as well as using concepts developed by Worboys on resolution (Stell and Worboys, 1998).

REFERENCES

Arkin, E. M., Chew, L. P., Huttenlocher, D. P., Kedem, K. and Mitchell, J. S. B., 1991, An Efficiently Computable Metric for Comparing Polygonal Shapes. *IEEE Transactions on Pattern Analysis and Machine Intelligence*, **13** (3), pp. 209–216.

Bertin, J., *Semiology of Graphics, 1983: Diagrams, Networks, Maps,* (Madison, Wisconsin USA: University of Wisconsin Press).

CEN/TC287, 1996, Geographic Information - Data Description Quality. European Standard - Draft for Public Enquiry, Comité Européen de Normalisation.

David, B. and Fasquel, P., 1997, Qualité d'une Base de Données Géographiques : Concepts et Terminologie. *Bulletin d'Information de l'IGN*, Vol. 67. (Paris, France: Institut Géographique National).

Devogele, T., 1997, Processus d'Intégration et d'Appariement de Bases de Données Géographiques : Application à une Base de Données Routières Multi-échelles. PhD, University of Versailles, France.

Frank, A. U., 1997, Metamodels for Data Quality Description. In *Data Quality in Geographic Information: From Error to Uncertainty*, edited by Goodchild, M. and Jeansoulin, R., (Paris: Hermes), pp.15-29.

Goodchild, M. F., 1992, Dealing with Uncertainty in Spatial Data. *Workshop on Data Quality, Fifth International Symposium on Spatial Data Handling,* Charleston, August 1992.

Goodchild, M. F. and Proctor, J., 1997, Scale in a Digital Geographic World. *Geographical & Environmental Modelling,* vol. 1, no. 1, pp. 5–23.

Güting, R. H. and Schneider M., 1995, Realm-based Spatial Data Types: The ROSE Algebra. *Very Large Data Bases (VLDB) Journal* 4, pp. 100–143.

ISO, 1996, Geographic Information - Quality Evaluation Procedures. Draft, International Organisation for Standardisation.

Laurini, R. and Thompson, D., 1992, Fundamentals of Spatial Information Systems. *The Apic Series,* vol. 37. (San Diego, California: Academic Press).

Omrane, N., 1998: Étude de la Résolution de la BDCarto®. Master thesis, University of Marne-la-Vallée, Paris, 63pp.

Skogan, D. and Fimland, M., 2000, A Multi-resolution Approach for Simplification of an Integrated Network and Terrain Model. *Proceedings of Spatial Data Handling 2000,* pp. 4b26–39.

Spec-94, 1994, Spécifications de Contenu de la BDCarto®. (Paris: Institut Géographique National).

Stell, J. G. and Worboys, M. F., 1998, Stratified Map Spaces: A Formal Basis for Multi-resolution Spatial Databases. In *Proceedings of the 8th International Symposium on Spatial Data Handling (International Geographical Union),* edited by T. K. Poiker and N. Chrisman, pp. 180–189.

Vauglin, F., 1997, Modèles Statistiques des Imprécisions Géométriques des Objets Géographiques Linéaires. PhD, University of Marne-la-Vallée, Paris.

CHAPTER TEN

Visualisation of Uncertainty in Geographical Data

Igor Drecki

ABSTRACT

This chapter reviews developments in visualising uncertainty in geographical data. It explores, develops and evaluates specific visual techniques to represent uncertainty in such data. The review points out the significance of visualising uncertainty and the specific challenge of uncertainty in raster data is identified. Existing visualisation methods are explored and attention is focused on examining current techniques and suggesting new ones. A satellite scene, classified in terms of land cover classes, is used to derive a measure of uncertainty suitable for the examination of uncertainty patterns. A number of already documented visualisations, as well as the new techniques, referred to as *opacity* and *squares*, are defined and constructed based on the land cover classification. To evaluate these methods World Wide Web (WWW) technology is used to deliver an assessment exercise, and the administration of this is described.

The analysis of the assessment exercise reveals a variation of responses to the methods offered to the respondents, with the *squares* technique being shown to be the most effective means of visualising uncertainty among the methods considered. The preferred methods, as determined by the respondents, are shown not necessarily to be the most effective ones for communicating patterns of uncertainty to the user.

10.1 INTRODUCTION

Traditionally, hard copy maps were the main source of spatial information for geographical interpretation and analysis. In the past two decades this situation has changed rapidly due to a significant growth in the processing and analysis of digital data. Hard copy maps are now often regarded as no longer of the required standard of accuracy and are being replaced by digital spatial databases, which become the ultimate source of geographical data. However, these databases often contain spatial information of inadequate quality. Error is frequently introduced by data compilation methods and later propagated through various spatial operations. This has an impact on the reliability of spatial databases and creates an urgent need for data quality visualisation.

In the last decade there has been an increasing demand for appropriate recognition of data quality information to be associated with spatial databases. This need is of particular importance when spatial data are used in decision-making processes, where there may be significant implications from using inadequate data. Data quality information is also an important consideration when assessing suitability of data for particular applications. Veregin (1995) points out the importance of the data quality component in data transfer, where some standards, such as the Spatial Data Transfer Standard (SDTS)(NIST, 1992), need to be applied in order to maintain data integrity. Effective representation and interpretation of data is also dependent on information about their quality, as this affects the credibility of any geographical information systems (GIS) operations.

Geovisualisation provides a valuable approach for representing data quality information in a spatial environment. Rapid changes in spatial information technology now allow fast processing and display of large volumes of data. This has prompted the exploration of visual methods for representation of data reliability and spatial variability. The application of appropriate visual techniques has also been found very useful in providing immediate and easy to understand information about uncertainty and directly relating data reliability with its location (Kraak *et al.*, 1995).

This chapter explores, develops and evaluates specific visual techniques to represent uncertainty in geographical data. The exploration stage begins with a review of existing and documented techniques to establish a number of methods that are considered successful in representing uncertainty. An area for the potential development of new techniques is also identified. The development stage consists of two parts. In the first part, an existing SPOT satellite image, classified in terms of land cover classes, is used to derive a measure of uncertainty suitable for the examination of uncertainty patterns. In the second part, new techniques, referred to as opacity and squares, are defined and actual implementations are designed and constructed based on the land cover classification. A number of existing methods are selected and developed for evaluation purposes. The evaluation stage also consists of two parts. It begins with the development of the electronic assessment exercise utilising World Wide Web (WWW) technology. The final part focuses on processing and evaluation of the results and the performance of new and existing techniques is assessed.

10.2 VISUALISING UNCERTAINTY

Information on data quality is of paramount importance. It affects the reliability of data analysis and presentation, as well as impacting on the credibility attached to decision-making with GIS (Goodchild *et al.*, 1994a).

Geovisualisation is a powerful approach in addressing this problem, where the application of appropriate visualisation techniques could provide immediate and more understandable information (spatial and temporal) about data uncertainty (van der Wel *et al.*, 1994). It plays an important role in communicating, interpreting and understanding spatial phenomena in any discipline of earth sciences. Geovisualisation is seen as a recognised method of scientific research, where maps and visual displays are part of a research sequence (DiBiase, 1990; DiBiase *et al.*, 1992). It uses visual means to present spatial relationships that can not be detected by other forms of expression (textual or verbal) and engages the principal human information-processing abilities, in particular, those associated with vision (Buttenfield and Mackaness, 1991; MacEachren *et al.*, 1992).

10.2.1 Requirements for visualising data quality

Apart from visual clarity, based on principles of graphic design and acceptable aesthetic appeal, design requirements for visualising data quality also contain a number of considerations specific to spatial data. They include issues of scale, spatial and temporal characteristics, local variations and computational efficiency (after Beard and Mackaness, 1993).

Data quality information is often generated and visualised at a particular scale, at which the level of detail is optimised and spatial patterns can be identified. Manipulation of scale, however, can affect that balance and misinterpretation of quality information may occur when data is viewed at inappropriate scale. Therefore, visual techniques should be applied to relate the level of detail to an adequate scale of display, i.e. the relevant information will be displayed at one scale and not shown at the other scales.

Visualisation of data quality is also concerned with its spatial and temporal location. Information about quality at specific locations can be very useful for the evaluation, analysis and interpretation of spatial data. Therefore, visual techniques should be employed to construct appropriate graphic displays capable of relating the quality information to its spatial and temporal location.

The general quality of an entire data set is often of satisfactory standard. However, it does not mean that the quality is evenly maintained throughout the data set. In some areas/locations significant variations in quality may exist. These areas or "hot spots" are usually of special interest to the user. Therefore, the design of visual displays should be capable of showing and attracting the attention of the user to the local variations in data quality.

Recently, the issue of computational efficiency in the visualisation of data quality has become increasingly important. Visual techniques require computing resources for generating and display of quality information. High consumption of those resources may cause some computational limitations that can lead to poor graphics and miscommunication. Therefore, it is desirable to apply visual techniques that are computationally efficient, retain good design principles and still provide the desired data quality information.

Taken together, the above requirements provide a general framework for visualising uncertainty effectively. Application of appropriate visualisation techniques within this framework would provide immediate and easily understood information (spatial and temporal) about data quality.

10.2.2 Computer technology in visualising uncertainty

The development of computer technology, high performance hardware and sophisticated software solutions creates an environment where the design and generation of spectacular and useful visualisation tools and effects can be used to represent uncertainty effectively. The visualisation of data and quality can be specified as either separated or combined (after MacEachren, 1992). In a separated relation, data and quality can be represented by either a static map pair (data and quality shown as separated maps but at the same time), or a succeeding map pair (first data, then quality), or an alternating map pair or map sequence (data, quality, data, quality...)(MacEachren, 1992). Examples of a separated relation between data and quality information are *double* and *multiple displays* for static representations, and *toggling/flicking* and *fading* for dynamic representations. A bivariate map represents a combined relation, where data and quality are shown by one visualisation. This chapter examines the use of the latter approach in more detail below.

10.2.2.1 Bivariate displays

This approach, although it can be complex and difficult to interpret, provides an immediate and compact view of uncertainty. Bivariate displays play an important role in the current exploration and development of data quality visualisations, where a significant number of techniques have been proposed and applied. They can be static or dynamic. Examples of static techniques are *colour transformation, blending, focus, three-dimensional uncertainty surface* and *dazzling*. Dynamic techniques cover *quality slider* or *slicing, blinking, moving* and *zooming*.

   Colour transformation allows the conversion of colours from the RGB (red, green, blue) system to IHS (intensity, hue, saturation) components. Colour hue is then used to convey nominal information, while intensity and/or saturation to represent quality information. By converting IHS values back to RGB, the original thematic data and their uncertainties are readily available in the RGB image as a bivariate display. An example of the *colour transformation* technique is given by Jiang *et al.* (1995) in developing a modified IHS colour system to apply in fuzzy overlay operations for the visualisation of uncertainty. This method is often called *colour saturation* technique, when saturation is used to represent reliability of data.

   Blending is particularly useful in showing transitional zones between polygons representing positional and attribute ambiguity of boundary location and attributes. Carefully mixed colours/greyscale in blending zones can correspond to fuzziness of vegetation or soil type boundaries as proposed by McGranaghan (1993). Application of this technique can create an impression of continuous variability between natural features, rather then showing sharp boundaries.

   In principle, *focus* is a similar technique to blending, but instead of colour mixing it uses sharpness or contrast to determine the quality of the data. Sharper objects represent more reliable data, while blurry or fuzzy objects show ambiguity. McGranaghan (1993) proposed this method to visualise the uncertainty associated with a polygon boundary location by creating a fading out vignette (buffer) around possible border line. Focus can also be expressed as the variation in the display resolution to convey variability of data quality (MacEachren, 1994).

   If uncertainty values are plotted along the *z* axis, a *three-dimensional uncertainty surface* can be generated to represent the positional variability of data quality. This method was employed by Rejeski (1993) to visualise ecological risk by modelling an uncertainty surface. It can be further refined by draping thematic information over the quality landscape to create a bivariate representation and communicating attribute uncertainty (Beard and Mackaness, 1993).

Combinations of different patterns, textures or colour hues can produce a *"dazzle painting" effect* to camouflage or prevent one from seeing the areas of low reliability. This technique was widely used on military maps during World War I to hide strategically important objects (van der Wel *et al.*, 1994). Although it is an old and time-consuming manual method, it can be revitalised using computer graphics technology for generating quality displays in a fraction of the time.

In a dynamic approach, uncertainty can be represented by *quality slider* or *slicing*. In this method different threshold values can be set to control the display of data quality. The data fades out successively as the quality slider is moved up to the higher values of data reliability. An example of this technique is given by MacEachren *et al.* (1993) to visualise the uncertainty in Chesapeake Bay monitoring data.

The principle of the *zooming* technique in visualising uncertainty is based on the zoom tool available in most of the software applications. The quality of the data can be viewed from different perspectives. A distant view can be obtained for uncertain data, while for highly reliable data a detailed view or close-up can be used.

Variations in the display time of map features, usually proportional to their reliability, can create a *blinking* or *twinkling* effect. Frequently blinking objects on the screen indicate high uncertainty values, while more stable or constantly displayed information convey reliable data (Fisher, 1994a). Refinement of this technique is proposed by Fisher (1993, 1994b), where a displayed soil grid cell changes its colour according to the proportion of it being assigned to one of the existing soil classes. In this case blinking (displayed/not displayed) is replaced by colour change, however the principle of this technique remains the same.

Animation can also be applied to create *multiple positions*, or *moving* effects. While a *multiple positions* method usually refers to positional uncertainty, a *moving* effect can also communicate attribute uncertainty. In the former case, Fisher (1994b) gives an example of an animated dot density population map where continuously changing positions of dots communicate the spatial distribution of population, rather than the particular location of population. Another example of a *multiple positions* method can be represented by a sequence of displays showing ambiguity in location of a coastline, where the coastline is shown by a number of thin lines each obtained from various surveys representing a different degree of accuracy (McGranaghan, 1993). In the other technique moving objects, such as "fog" or "mist", can drift across the display and obscure uncertain areas, while more certain parts of the display would remain unchanged (McGranaghan, 1993). Manipulating colour saturation or opacity of the original display can create the obscuring effect.

Bivariate displays can be complex and difficult to understand especially for inexperienced users (van der Wel *et al.*, 1994). For example, the *multiple positions* method as discussed using the example of a coastline location, could suggest the movement of the coastline rather than communicate its positional uncertainty. However, bivariate displays provide an immediate and compact view of uncertainty, which is a significant advantage over other visual techniques.

10.3 DESCRIPTION AND MEASUREMENT OF GEOGRAPHICAL DATA

Derivation of data quality measures may be difficult, especially when available sources of information have been heavily processed. Generalisation, approximation and interpolation are among standard procedures frequently applied to the original data. Uncertainty measures obtained from already processed data may not be adequate or of low quality. However, remote sensing imagery provides suitable data for derivation of reliability measures that are spatially uniform and free of processing error. For this project a 1994 SPOT HRV XS satellite image was obtained for the Albany Basin located in the western part of North Shore City (Metropolitan Auckland, New Zealand). The location is centred on latitude 36°44'S, longitude 174°42'E and covers an area of 36km² (6km × 6km). It represents an area where there has been substantial land cover changes in the last decade associated with urban expansion.

10.3.1 Classification procedure and data reliability measures

The classification procedure involved classifying the SPOT scene, using the maximum likelihood decision rule, into eight spectral classes: residential, industrial and commercial, transportation, transitional, pasture and cropland, mixed scrub, mangrove and water. The resultant image formed a land cover map that was later used as the original spatial data component in the design and implementation of bivariate reliability displays (Figure 10.1 in the colour plate section). Another product of the classification process was a distance image file that contained information on the reliability of the performed classification.

The maximum likelihood algorithm calculates probability values for every pixel in the image and stores them as a distance image file. This file contains spectral distance information, based on the Mahalanobis distance between the measurement vector and the class mean vector (ERDAS Inc., 1994). Pixels that are at a greater distance from their class mean are more likely to be misclassified, while those at a smaller distance are more likely to be classified correctly. The resulting distribution of pixels is often characterised by a *chi-square* distribution.

The process of identifying the pixels in a classified image that are likely to be misclassified is called thresholding (ERDAS Inc., 1994). The uncertain pixels are grouped in the "tail" of a distance image histogram. Different levels of reliability can be calculated by thresholding the "tail" of the histogram at various points. The threshold point can be determined statistically by numerical input of the *chi-square* parameter or distance measurement. When maximum likelihood classification is performed, the *chi-square* statistics are used to determine the threshold by calculating distance values (ERDAS Inc., 1994). In this case the *chi-square* value is a function of the number of degrees of freedom and the confidence level (Siegel, 1956). The degrees of freedom refer to the number of the input bands (three for SPOT data) and the confidence level to the percentage (%) of pixels that are believed to be incorrectly classified (ERDAS Inc., 1994) (Table 10.1).

In order to visualise uncertainty associated with the classification of the SPOT satellite data, 100 single band images with confidence interval of 1% were

Table 10.1 The critical values of *chi-square* for the three input bands of the SPOT satellite data at various confidence levels.

Degree of freedom = 3	Confidence level (%)												
	99	95	90	80	70	60	50	40	30	20	10	5	1
Chi-square	0.1	0.4	0.6	1.0	1.4	1.9	2.4	3.0	3.7	4.6	6.2	7.8	11

calculated. The *chi-square* statistics were calculated automatically for each of the images once a particular confidence level was specified. These images contained information about the number of pixels distributed among eight land cover classes within the specified confidence range (e.g. >0 to <100%, 1 to <100%, 2 to <100%, up to 99 to <100%) and were later used in the data processing stage for construction of reliability displays.

10.4 DESIGN AND IMPLEMENTATION OF RELIABILITY DISPLAYS

The requirements for visualising uncertainty discussed earlier need to be carefully considered when selecting and designing visual techniques. It also needs to be realised that a single technique approach is not the only solution when visualising data quality. The heuristic and hybrid approaches have proved to offer some interesting and valuable alternatives (MacEachren *et al.*, 1993; Hunter and Goodchild, 1996; Drecki, 1997). Finally, specifics of the raster data model used in this study influenced, to a certain degree, the selection and design process.

10.4.1 Selection of visualisation techniques

Selection of visualisation methods that provide immediate, easily understood, and accurate information about uncertainty could be a complex task. Early studies in the area of data quality visualisation have generally focused on the development of a single technique, often with several realisations, that conveyed uncertainty information in the optimal way (Fisher, 1993; Rejeski, 1993; Goodchild *et al.*, 1994b; Jiang *et al.*, 1995). However, Buttenfield and Beard (1994) found in their experiment related to the effect of symbolising annual and cumulative tree ring growth on an animated map, that users could easily associate gradations in symbol size with cumulative growth, but had some difficulties in associating variations in colour saturation with quality of the tree ring growth. The above experiment revealed that the single technique approach in visualising uncertainty might not always provide the best solution to data quality queries.

The alternative is to adopt a heuristic approach, in which users are provided with a selection of visualisation tools to create their model of uncertainty. The Reliability Visualisation System (RVIS) developed by MacEachren *et al.* (1993) is an example of this approach. RVIS uses an interactive environment offering a number of possible manipulations to convey uncertainty estimates associated with dissolved inorganic nitrogen in Chesapeake Bay (MacEachren *et al.*, 1993). However, manipulation of data by inexperienced users may lead to the construction of misleading displays of data reliability.

A hybrid approach offers another alternative. In this approach users are exposed to a variety of different visual techniques that represent the same uncertainty information, while the choice of the most informative methods depends on their individual preferences. Visualising uncertainty using this approach could be time consuming and ineffective in terms of computing resources. However, the hybrid approach, although not well documented in the literature, seems to have a number of distinct advantages. It gives a range of visual techniques that suit a wider group of users — experts as well as novices. It leaves the design and development of uncertainty displays to the professionals with extensive graphical skills, rather than to the users with limited or no experience in graphics design. Finally, it allows perceptual testing of individual techniques as well as their comparison (Drecki, 1997).

In order to provide an immediate and compact view of uncertainty, bivariate displays seem to be an obvious consideration despite their limitations. They allow an instant comparison of data and quality in relation to their spatial location (Kraak *et al.*, 1995). They are also capable of showing and alerting users to the local variations in data quality.

One of the objectives of this study was to develop new methods of representing uncertainty in geographical data. Two new techniques were designed and implemented including an *opacity* method and a *squares* method (Drecki, 1997). Adopting a hybrid approach allowed selection of both visual techniques.

10.4.2 Implementation of reliability displays

Visual clarity, conformity with the principles of graphic design, appropriate selection of visual variables, and a good balance between data and quality information are obvious graphical considerations when producing reliability displays. The specifics of spatial data also influence the implementation process.

The *opacity* technique was developed at a fixed scale to relate the level of detail to the scale of display. Although the interactive *squares* method had a zoom tool available to manipulate the scale of display, only the extreme use of this tool could affect the communication of uncertainty in this technique. The spatial distribution of error and its variations were retained with both bivariate displays. By definition they relate data quality to its location and provide immediate information about local variation and size of error. Finally, the implementation of both techniques was relatively simple and computationally efficient.

For evaluation purposes a further three existing and documented techniques were designed and implemented. They included *colour saturation* (MacEachren, 1992, 1995), *three-dimensional reliability surface* (Beard and Mackaness, 1993; Rejeski, 1993) and *blinking* methods (Fisher, 1993, 1994a, 1994b).

10.4.2.1 Opacity technique

The implementation of the bivariate *opacity* technique (Figure 10.2 in the colour plate section) required two components of data and quality. The quality component was established by processing 100 confidence images obtained from land cover classification of SPOT data and assigning the opacity values.

To maintain a relatively high accuracy of reliability estimates and computational efficiency these confidence images were merged into 20 single band images utilising the matrix overlay function. The loss of information for visual interpretation was minimal due to the limited human visual capacity, but the computation time in turn, decreased five times. The resulting images consisted of pixels in each of the eight land cover classes classified within the specified 5% confidence range (i.e. 5 to <10%, 10 to <15%, 15 to <20%, etc). They were later merged to produce a single band image with 20-step confidence scale.

The next phase involved assigning the opacity values for the confidence steps so as to convey data quality information. Because opacity is specified in the 0 - 100% range, each step was allocated the corresponding opacity value in 5% intervals. The original spatial data component was added by allocating the relevant colour hue to the pixels in the corresponding land cover classes, and the final bivariate appearance of the *opacity* technique was achieved.

10.4.2.2 Squares technique

The bivariate *squares* technique (Figure 10.3 in the colour plate section) was implemented primarily as a dynamic visualisation of uncertainty. The processing of reliability data included merging 100 confidence images to a single band image that contained information for every pixel based on 100-step confidence scale. By utilising the image to file conversion function, a text file was produced that stored x and y coordinates and integer confidence values in the 1 to 100 range for every pixel in the image. The original classified image was also converted into a text file. It contained x and y coordinates and the land cover class code for each pixel in the image. Both text files were later used in developing a customised implementation program to display the interactive *squares* technique.

The *squares* method uses visual variables of size and interaction for communicating uncertainty. In this technique each pixel of the original classified image was represented by a single square symbol, located in the same relative position as on the original image, colour coded (RGB) to the corresponding land cover class, and varied in size according to its confidence value. Smaller squares represent lower confidence in land cover classification, while larger squares refer to higher confidence.

The interactive mode of the *squares* method was achieved by writing a customised implementation program in Perl (Romaniuk, 1997a) to produce a VRML file. The legend was embedded in the display to allow the comparison of confidence values during interactive manipulation on the screen, when the size and position of symbols is changing. Interaction (especially zooming) provides a valuable tool for accurate quantitative estimates of reliability and detection of local anomalies in data quality.

Later, a static version of the *squares* technique was produced by modifying the original Perl program (Romaniuk, 1997b). This allowed the development of an appropriate map for hard copy output (Figure 10.3 in the colour plate section).

10.5 EVALUATION OF RELIABILITY DISPLAYS

The evaluation of reliability displays in visualising uncertainty concerns both users and producers of spatial data. While the users are looking for the right tool to assess the reliability and usefulness of data for various applications, the producers search for methods that convey the reliability of their products best. Furthermore, as Buttenfield and Beard (1994) note, software development and database solutions cannot be optimised due to the lack of appropriate evaluation of visual methods.

Many authors indicate a need for systematic and comprehensive empirical testing to evaluate the effectiveness of reliability displays in visualising uncertainty (e.g. Buttenfield, 1993; McGranaghan, 1993; Fisher, 1994a; van der Wel *et al.*, 1994; Hunter and Goodchild, 1996). However, to date there is very little experimental evidence on perceptual studies of data quality visualisations. Some tests were carried out (Goodchild, 1991; Buttenfield and Beard, 1994; Pang *et al.*, 1996), but their documentation is often limited to general statements, rather than detailed reports on the results and methods used to derive them.

A detailed evaluation of reliability displays, including a discussion of the results, was carried out by Evans (1997). The study provides valuable insight on the design and implementation of the experiments, but concludes that more research is needed to develop a better understanding of how well these visualisations are utilised and comprehended. It also points out that the search for new visual techniques in representing uncertainty should be continued (Evans, 1997).

In this study five different reliability displays were subject of the comprehensive assessment exercise to test their performance in visualising uncertainty. They included *opacity*, *squares*, *colour saturation*, *three-dimensional reliability surface* and *blinking* methods.

10.5.1 WWW technology in development of assessment exercise

The World Wide Web (WWW) technology provides an interesting environment for the design and implementation of questionnaires, where a broad accessibility, a standardised semi-platform independent format, and simplicity of use are among its main advantages (Drecki, 1997). The electronic answer sheets provide all responses in a digital format, so they are readily available for digital processing and analysis. In the case of this assessment exercise a wide spectrum of respondents in terms of number and personal characteristics was required. The questionnaire was designed primarily for individuals that have some experience or at least an interest in spatial data handling, and the WWW technology provided a means to control this. The ability to display static as well as dynamic visualisations was critical, as some of the techniques required the use of animation. Time efficiency was also an important consideration due to the large volumes of data for processing, which would otherwise need to be done manually. Therefore, the WWW technology provided an obvious solution to meet these requirements.

In order to publish the assessment exercise, as well as to enforce the correct answering order, a customised program was written in Perl (Glen, 1997). This program created the HTTP (Hyper Text Transfer Protocol) responsible for searching and retrieving the constructed WWW pages. It also utilised a cookie function to protect the respondents from answering the same section twice, responding in reverse order, or omitting some sections altogether.

Execution of the assessment exercise required the Netscape Navigator version 3.0 or higher WWW browser and QuickTime movie plug-in (for the animated *blinking* method). HTML itself provided a universal, largely platform-independent format accessible to a wide spectrum of respondents through the WWW.

10.5.2 Implementation of assessment exercise

To allow a comparison of responses, the same set of standard questions was asked for each method (Questions 1–5). However, for some visualisations, such as *opacity* or *squares*, additional questions were included to obtain more information about their performance (Questions 6–7). For each technique a different portion of the study area was used. This protected the respondents from being influenced by the answers from previous visualisations, as they applied to a different set of data. Also, to provide an adequate image for Question 1, the reliability displays were divided into a square matrix, nine for each display. However, due to its coverage the *squares* method was divided into four areas. The animated *blinking* technique was not altered, because it was believed that additional lines on the image could interfere with the blinking effect.

For ease of use, all questions were developed in a point and click environment with a few fields provided for longer answers or for entering specific values. The list of standard questions asked for each visual method, with available answers, is given below:

1. The image is divided into nine squares (in the case of the squares method there were only four squares). Click the corresponding box below and enter an estimate of the general reliability of data in that square using the following scale: 1 — most reliable; 2 — reliable; 3 — fairly reliable; 4 — unreliable; 5 — most unreliable.
2. Identify, which land cover classes are the most reliably classified ones (please tick no more than two): residential; industrial and commercial; transportation; transitional; pasture and cropland; mixed scrub; mangrove/water?
3. Identify, which land cover classes are the least reliably classified ones (please tick no more than two): residential; industrial and commercial; transportation; transitional; pasture and cropland; mixed scrub; mangrove/water?
4. Identify, which land cover class is more reliable: e.g. pasture and cropland; e.g. mixed scrub; they are equally reliable?
5. Can you identify any spatial patterns of error distribution on the image: yes, please fill the attached form (no more than 60 words); no?

The additional questions included:

6. Is the image showing a majority of: reliable pixels in confidence range 67 — <100; fairly reliable pixels in confidence range 34 – <67; unreliable pixels in confidence range >0 – <34?
7. How many land cover classes can you detect on the image: 3 – 5; 6 – 8; 9 – 11; 12 – 14; 15 – 17; >17?

The above Questions 1, 2, 3, 4 and 6 focused on the respondents' perception of reliability attached to the examined uncertainty displays, while Question 5 tested their ability to identify spatial patterns of error distribution in the data. Question 7 was specifically designed for the *colour saturation, opacity* and *blinking* methods, where variations in colour or twinkling effect could be mistakenly interpreted as the presence of additional land cover classes.

In the summary section of the assessment exercise, the respondents were asked to compare the effectiveness of all presented methods based on their opinion and expectations. They were also given the opportunity to comment on these techniques.

The last set of questions related to the respondents' profile and their experience with spatial data handling. It was believed that responses to all the above questions could provide valuable material to evaluate the effectiveness of the presented visual techniques, as well as give an opportunity to identify some answering patterns related to respondents' personal profiles.

10.5.3 Analysis and results

The implementation of the questionnaire on the WWW allowed collection of the relevant data in digital format. Analysis followed a relatively straightforward path. The first sequence of tabulation focused on the effectiveness of the visualisation techniques developed for representing uncertainty. The second stage compared these results with the respondents' opinion as to which method performs best or worst in visualising uncertainty. This includes their identification of which techniques are the most favourable for developing a good understanding of the reliability associated with the land cover classification performed in this study.

The nature of the data collected, including the limited sample size for fine subdivision of the sample has meant not all avenues could be pursued. Consequently the analysis presented reflects the priority of specific areas in terms of gaining a robust, general understanding of the findings.

In order for the evaluation process to be valid, the assessment exercise required knowledge of the range of respondents in terms of their background and personal profile. A total of 50 responses were received (Table 10.2). However, due to the inconsistency in answering the assessment exercise by the respondents, as well as to the design of the questionnaire that allowed respondents to complete the first few sections without answering the remaining ones, there were a number of invalid (not answered) records.

The analysis revealed that the majority of respondents were undergraduate students with a particular interest in geography or GIS, and little experience with spatial data handling and analysis. The above characteristics of respondents, usually referred to as novices (Hunter and Goodchild, 1996), suggested that they possess a uniform level of skill and similar ability to identify patterns in spatial data. Therefore, the expectation was that valid results for novice users could be derived.

10.5.3.1 Effectiveness of reliability displays

In order to assess the effectiveness of the reliability displays in visualising uncertainty, the results of the assessment exercise were compared with the true reliability measures. These true measures were obtained by extracting the relevant confidence values from the original land cover classification.

Table 10.2. The respondents' profile information and distribution of responses.

Respondents profile information	Number of responses
Sex	
• female	16
• male	28
Status	
• undergraduate student	32
• graduate student	5
• doctoral student	1
• a member of the academic staff	0
• researcher/scientist	2
• other	3
Main interest	
• cartography	0
• computer science	1
• geography (human)	11
• geography (physical)	15
• GIS	8
• psychology	0
• remote sensing	1
• spatial analysis	3
• other	3
Experience with spatial data handling	
• <1 year	16
• 1 – 2 years	24
• 3 – 5 years	2
• >5 years	2

Only the first four standard questions were considered, because they applied to all visual techniques and provided quantitative data. In Question 1, 2 and 3 the respondents were asked for general ranking of reliability levels, rather than precise reliability measures, and therefore the ranking system also had to be applied to the true values of uncertainty. The comparison of these rankings using the Spearman rank correlation coefficient (Siegel, 1956) allowed proper evaluation of responses for these questions. The responses in Questions 4 were analysed using standardised data values expressed in percentages (%).

The final ranking (Table 10.3) was determined by calculating the sum of ranks scored by each individual technique and dividing it by the number of questions (four).

The evaluation of reliability displays in terms of their effectiveness in conveying uncertainty was influenced by several factors, mainly associated with the design of the assessment exercise and the characteristics of the individual visualisations. For instance, each visual technique used different portions of the study area. Because these portions were not uniform, some variations in uncertainty patterns inevitably occurred. In order to systematically deal with the factors influencing the results, the following discussion is divided into the individual techniques.

Table 10.3. Rankings for all reliability displays.

Visual method	Question 1		Question 2		Question 3		Question 4		Final rank
	r_s	rank	r_s	rank	r_s	rank	%	rank	
Squares	1.00	1	0.77	1	0.60	3	46	2	1
Opacity	0.91	2	0.46	2	0.66	1	15	5	2
Blinking	0.75	3	0.18	5	0.61	2	82	1	3
3-D reliability surface	-0.32	5	0.46	2	0.46	4	39	4	4
Colour saturation	0.46	4	0.34	4	0.09	5	44	3	5

r_s - rank correlation coefficient (Siegel, 1956).

The *squares* method was found to be the most effective one (see Table 10.3). It was developed especially for this study as a new approach in visualising data quality information. The success of this method could lie in the simple design and clarity. The *squares* method was very consistent in terms of rankings in all four questions. This suggests that it was effective in all aspects of uncertainty visualisation raised in the assessment exercise. However, there are a number of issues that could have had an impact on such a result. Because of the computational demands of the *squares* technique, only a small portion of the study area was visualised. In Question 1 the respondents had only four squares to estimate the general reliability of data instead of nine as in the remaining techniques. Therefore, the chance of assigning the correct reliability values was much greater. In Questions 2 and 3 the respondents had a choice of six possible classes to indicate the most or least reliable land cover classes, instead of seven. In Question 4 it was easier to determine the more reliable land cover class from three possible choices, as there was less data on the display to analyse.

The *opacity* method was the second most effective technique (see Table 10.3). It was also developed particularly for this research as a new approach to visualise uncertainty. The results show that it was a very consistent method in terms of rankings across all four questions. This indicates that it conveys all aspects of uncertainty examined in the questionnaire very effectively. However, in Question 4 some confusion occurred in identifying the reliability of land cover classes. As for all the other methods there was a significant difference in the reliability between two classes, but in the case of the opacity method this difference was minimal. This resulted in an incorrect judgment by the respondents and eventually in the lowest overall ranking for this question. Another possible problem was associated with the appearance of this technique on a computer screen. Instead of the real *opacity* effect, the image was made up of little dots that formed a pattern that varied in density depending on data reliability values. This could mislead the judgement of respondents, but it is unclear in what way. Interestingly, the 20-step opacity scale seemed to have no adverse effect on the overall performance of the *opacity* technique (as opposed to the *colour saturation* method).

The animated *blinking* technique was the third most effective method in data quality visualisation. It was based on a similar technique developed by Fisher (1994b). Although Fisher did not document the effectiveness of this technique, he reported that it received very positive comments from a group of observers (Fisher, 1993). The results of this questionnaire revealed that the blinking method is very inconsistent in terms of its effectiveness in providing answers to all four questions. For example, it proved to be a valuable tool in distinguishing between reliability of two different land cover classes, where the difference between frequently blinking and stable classes was easily detectable. In Question 1 the dividing lines of nine squares to estimate the general reliability of data were not shown, which resulted in many missing answers (only 36 answers were valid, out of the possible 50). This could actually improve the performance of the *blinking* method, because only those respondents that were confident about the reliability values responded to this question. However, when it comes to distinguishing the reliability between more stable classes (Question 2), the blinking method did not perform well. Another issue associated with this method is the effect of blinking itself and its influence on the human visual system. Blinking could be very irritating for many users and have an adverse effect on the performance of this method.

The *three-dimensional reliability surface* method was ineffective in visualising uncertainty (Table 10.3). Although this method has been documented in literature by several authors (e.g. Beard and Mackaness, 1993; Rejeski, 1993; Pang *et al.*, 1996), there is no evidence of empirical testing of its effectiveness. There are two major issues associated with its poor performance. The view angle of the reliability surface made the data hard to interpret particularly in the background. It created pockets of viewshed where lower reliabilities were hidden behind the higher values. It also reduced the gaps between applied fishnet mesh lines causing problems in identifying the land cover classes. Another issue was that a fishnet mesh did not allow comparison of reliability between different areas of display. Contour lines would be more effective, as they allow the comparison of reliability values at any desired point of the reliability surface. However, the *three-dimensional reliability surface* technique was very effective in representing the most reliable data, as the interpretation of peaks in the reliability surface could be easily determined.

The *colour saturation* technique was the least effective one in representing uncertainty according to the results of the assessment exercise (Table 10.3). This method was based on the suggestion made by MacEachren (1992, 1995) that saturation is one of the most logical choices to represent data quality information. The *colour saturation* method was implemented by Buttenfield and Beard (1994) in their animated map to illustrate annual and cumulative tree ring growth. They reported that the viewers of the animation were confused by variations in colour saturation and its association with data quality information. It seems that the results of this questionnaire correspond with the findings of Buttenfield and Beard (1994). The *colour saturation* method was consistently ranked as ineffective in conveying uncertain information. The major disadvantage is the fact that in lower levels of saturation the colours representing land cover classes become greyish, which make them hard to distinguish between each other. Another reason for its poor performance could be the use of 20-step colour saturation scale, as opposed to the 15-step scale used by Schweizer and Goodchild (1992), or 3-step scale recommended by Brown and van Elzakker (1993). This could have a significant impact on the ability to interpret uncertainty values.

10.5.3.2 Effectiveness versus perceived effectiveness

The second stage of the evaluation analysis was to compare the above results with the respondents' opinions. The respondents were asked to share their opinion about which methods they believed were the best (Question 1) or worst (Question 2) in visualising uncertainty, and which techniques were the most favoured in developing a good understanding of the land cover classification (Question 3).

Table 10.4 presents the derived rankings for all three questions. The *colour saturation* and *squares* methods were the most preferable, while the *opacity* method was consistently determined as the respondents' last choice. These results were compared with the rankings scored by visual techniques (see Table 10.3). The comparison revealed that the respondents' opinions were in opposition to these rankings. Later, the Spearman rank correlation coefficient (Siegel, 1956) was calculated to determine whether there is any correlation between the most effective techniques and the most preferable ones in the opinion of respondents. The findings concluded that there was no significant correlation between any of the corresponding sets of rankings. The calculated values ranged between -0.35 and 0.10 for all possible combinations.

Table 10.4 Rankings from the summary section questions.

Visual method	Question 1	Question 2	Question 3
Colour saturation	2	1	1
Opacity	5	5 (worst)	4
3-D reliability surface	3	4	3
Squares	1 (best)	2	2
Blinking	4	2	5

This proves that the method preferred by the user is not necessarily the most effective one in conveying uncertainty information. It also means that the empirical tests to evaluate the effectiveness of various visualisations are absolutely necessary, as the users' guess of which method or methods are the best in visualising uncertainty can be very misleading.

10.6 CONCLUSIONS

In recent years there has been a significant growth in the development and utilisation of spatial databases. Since uncertainty is almost invariably a part of spatial data, the development of methods to visualise the data quality issues became a priority. This research acknowledges this demand through the exploration, development and evaluation of visual techniques to represent uncertainty in geographical data.

In the exploration stage, the examination of new digital techniques was carried out to provide valuable information on the current developments and trends in visualising uncertainty. This allowed identifying the effective and accepted methods, and determining the potential areas for the development of new techniques. Among the already developed and documented methods, the bivariate representations offered the most compact view. Methods that are generally accepted and regarded as effective include *colour saturation, three-dimensional reliability surface, quality slider*, animated *blinking* and *zooming* techniques. It was found that the opacity was not pursued in the development of visual techniques, although it offered a promising alternative. A new approach of using a combination of symbol size and interaction in representing uncertainty in the raster data model was another idea.

In the development stage the selection of appropriate data for the derivation of uncertainty measures was carried out. Remote sensing imagery provided suitable data that was uniform and free from processing error. The results of the classification process provided the main source of spatial data, as well as producing a file that contained information on the reliability of the performed classification. The satellite imagery determined the raster data model used in this study.

The development of visualisation techniques that provide immediate and accurate information about uncertainty was based on the findings of the exploration stage. Since the opacity and the combination of symbol size with the interaction offered an interesting alternative, two new techniques were designed. The first, called the *opacity* method, was based on the principle of the opacity variable, where low opacity represents low reliability and high opacity refers to high reliability of the classification. The second method, called the *squares* technique, uses square symbols that vary in size according to the confidence levels of their classification, and interaction that provides a tool for the accurate quantitative estimates of reliability and detection of local anomalies in data quality. Three other existing and documented techniques were developed for evaluation purposes. They included *colour saturation, three-dimensional reliability surface* and *blinking* methods.

The evaluation stage required the development of a customised assessment exercise to test the effectiveness of developed reliability displays. The WWW technology characterised by a broad accessibility, semi-platform independent format, and simplicity of use, provided a suitable environment for the development of a relevant questionnaire. A set of questions was developed to evaluate the effectiveness of the examined visual techniques, as well as to derive some patterns related to the respondents' opinions and their personal profile. An appropriate evaluation procedure was implemented to systematically process and analyse the collected information.

The analysis of 50 responses revealed that the majority of respondents were novices in the handling of spatial data. They represented a uniform level of skill and similar ability to identify patterns in spatial data.

The results provided valuable insights into the evaluation process by identifying the most effective methods of uncertainty communication and relating them to the respondents' opinions. The results revealed that the most effective methods were *squares* and *opacity* techniques, the *blinking* method was fairly effective, while *three-dimensional reliability surface* and *colour saturation* performed worst. However, the respondents' opinions were in opposition to these results. The *colour saturation* and *squares* methods were the most preferable, while the *opacity* method was consistently determined as their last choice. As a consequence, the method determined by the user is not necessarily the most effective one in visualising uncertainty. This urges the development of empirical tests, as users' perceptions of the effectiveness of a particular method could be very misleading.

Further analysis to investigate the findings for the whole sample could be performed in terms of the characteristics of respondents using the respondents' profile information. Breakdowns include possible differences between female and male responses, between novices and experts or by age, level of education, or main area of interest.

REFERENCES

Beard, M. and Mackaness, W., 1993, Visual access to data quality in geographic information systems. *Cartographica*, 30, pp. 37–45.

Brown, A. and van Elzakker, C., 1993, The use of colour in the cartographic representation of information quality generated by a GIS. In *Proceedings of the 16th Conference of the International Cartographic Conference*, Cologne, pp. 707–720.

Buttenfield, B., 1993, Representing data quality. *Cartographica*, 30, pp. 1–7.

Buttenfield, B. and Beard, M., 1994, Graphical and geographical components of data quality. In *Visualisation in Geographical Information Systems*, edited by Hernshaw, H. and Unwin, D., (Chichester: John Wiley & Sons), pp. 150–157.

Buttenfield, B. and Mackaness, W., 1991, Visualisation. In *GIS: Principles and Applications*, Vol.1, edited by Maguire, D., Goodchild, M. and Rhind, D., (London: Longman), pp. 427–443.

DiBiase, D., 1990, Visualization in the earth sciences. *Earth and Mineral Sciences, Bulletin of the College of Earth and Mineral Sciences, PSU*, 59, pp. 13-18.

DiBiase, D., MacEachren, A., Krygier, J. and Reeves, C., 1992, Animation and the role of map design in scientific visualisation. *Cartography and Geographic Information Systems*, 19, pp. 201–214 and 265-266.

Drecki, I., 1997, *Visualisation of Uncertainty in Spatial Data*, unpublished MSc thesis, (Auckland: The University of Auckland).

ERDAS Inc., 1994, *ERDAS Field Guide*, 3rd ed, (Atlanta: Earth Resources Data Analysis System Inc).

Evans, B., 1997, Dynamic display of spatial data–reliability: does it benefit the user? *Computers & Geosciences (special issue on Exploratory Cartographic Visualisation)*, 23, pp. 409-422.

Fisher, P., 1993, Visualising uncertainty in soil maps by animation. *Cartographica*, 30, pp. 20–27.

Fisher, P., 1994a, Visualisation of the reliability in classified remotely sensed images. *Photogrammetric Engineering & Remote Sensing*, 60, pp. 905–910.

Fisher, P., 1994b, Animation and sound for the visualization of uncertain spatial information. In *Visualisation in Geographical Information Systems,* edited by Hernshaw, H. and Unwin, D., (Chichester: John Wiley & Sons), pp. 181-185.

Glen, G., 1997, Customised implementation program of the assessment exercise. In *Visualisation of Uncertainty in Spatial Data,* Drecki, I., unpublished MSc thesis, (Auckland: The University of Auckland), pp. 164-165, (for the transcript contact: *igor@envbop.govt.nz*).

Goodchild, M., 1991, Issues of quality and uncertainty. In *Advances in Cartography,* edited by Muller, J., (Oxford: Elsevier), pp. 113–140.

Goodchild, M., Buttenfield, B. and Wood, J., 1994a, Introduction to visualizing data validity. In *Visualisation in Geographical Information Systems,* edited by Hernshaw, H. and Unwin, D., (Chichester: John Wiley & Sons), pp. 141-149.

Goodchild, M., Chih–Chang, L. and Leung, Y., 1994b, Visualizing fuzzy maps. In: *Visualisation in Geographical Information Systems,* edited by Hernshaw, H. and Unwin, D., (Chichester: John Wiley & Sons), pp. 158-167.

Hunter, G. and Goodchild, M., 1996, Communicating uncertainty in spatial databases. *Transactions in GIS,* 1, pp. 13–24.

Jiang, B., Kainz, W. and Ormeling, F., 1995, A modified HLS color system used in visualization of uncertainty. In *Proceedings of GeoInformatics '95, International Symposium on Remote Sensing, Geographic Information Systems & Global Positioning Systems in Sustainable Development and Environmental Monitoring,* Vol. 2, Hong Kong, (Hong Kong: Chung Chi College, Chinese University of Hong Kong), pp. 702-712.

Kraak, M., Muller, J. and Ormeling, F., 1995, GIS–cartography: visual decision support for spatio-temporal data handling. *International Journal of Geographical Information Systems,* 9, pp. 637–645.

MacEachren, A., 1992, Visualizing uncertain information. *Cartographic Perspectives,* 13, pp. 10-19.

MacEachren, A., 1994, *Some Truth with Maps: A Primer on Design and Symbolization,* (Washington D.C.: Association of American Geographers).

MacEachren, A., 1995, *How Maps Work: Representation, Visualisation, and Design,* (New York: The Guilford Press).

MacEachren, A., in collaboration with Buttenfield, B., Campbell, J., DiBiase, D. and Monmonier, M., 1992, Visualisation. In *Geography's Inner Worlds,* edited by Abler, R., Marcus, M. and Olson, J., (New Brunswick: Rutgers University Press), pp. 99–137.

MacEachren, A., Howard, D., von Wyss, M., Askov, D. and Taormino, T., 1993, Visualizing the health of Chesapeake Bay: an uncertain endeavor. In *GIS/LIS'93 Proceedings,* Vol.1, Bethesda, pp. 449-458.

McGranaghan, M., 1993, A cartographic view of spatial data quality. *Cartographica,* 30, pp. 8–19.

NIST (National Institute of Standards and Technology), 1992, *Federal Information Processing Standard, Publication 173 (Spatial Data Transfer Standard),* (Washington D.C.: U.S. Department of Commerce).

Pang, A., Wittenbrink, C. and Lodha, S., 1996, Approaches to Uncertainty Visualisation, (for more information contact Alex Pang: *pang@cse.ucsc.edu*).

Rejeski, D., 1993, GIS and risk: A three culture problem. In *Environmental Modelling with GIS*, edited by Goodchild, M., Parks, B. and Steyaert, L., (Oxford: Oxford University Press), pp. 318-331.

Romaniuk, A., 1997a, Customised implementation program for the interactive squares technique. In *Visualisation of Uncertainty in Spatial Data*, Drecki, I., unpublished MSc thesis, (Auckland: The University of Auckland), pp. 134–138, (for the transcript contact: *igor@envbop.govt.nz*).

Romaniuk, A., 1997b, Customised implementation program of the static squares technique. In *Visualisation of Uncertainty in Spatial Data*, Drecki, I., unpublished MSc thesis, (Auckland: The University of Auckland), pp. 139-144, (for the transcript contact: *igor@envbop.govt.nz*).

Siegel, S., 1956, *Nonparametric Statistics for the Behavioural Sciences*, (Tokyo: McGraw–Hill Book Company Inc).

Schweizer, D. and Goodchild, M., 1992, Data quality and choropleth maps: An experiment with the use of color. In *GIS/LIS'92 Proceedings*, San Jose, pp. 686-699.

Veregin, H., 1995, Developing and testing of an error propagation model for GIS overlay operations. *International Journal of Geographical Information Systems*, 9, pp. 595–619.

van der Wel, F., Hootsmans, R. and Ormeling, F., 1994, Visualisation of data quality. In *Visualisation in Modern Cartography*, edited by MacEachren, A. and Taylor, D., (Oxford: Elsevier), pp. 313-331.

CHAPTER ELEVEN

Uncertainty in Polygon Filling Using Particle System Model

Tinghua Ai and Zhongliang Cai

ABSTRACT

Visualization of uncertainty in polygon filling is an important aspect of metadata display related to data quality. In order to explore and visualize uncertainty polygon filling and further to take into account its impact on decision-making, the traditional static visualization method which is based on Bertin's visual system requires to be extended. This chapter presents a new approach to animate uncertainty polygon filling by means of particle system model. The algorithm is developed on the basis of pixel random movement including stages of birth, action and death. The description of each pixel in raster map involves position, movement speed, movement direction, colour brightness and life span. The randomization change of these variables enables the reader to feel that the attribute distribution is uncertainty within a polygon. Through Monte Carlo simulation method, the number of different colour pixels can be controlled consistently with the proportion of corresponding attribute class.

11.1 INTRODUCTION

Spatial objects have a geometric and an attribute description. Both can be uncertain because of fuzzy attribute distribution of complex objects or loss of enough cognitive knowledge. As for area object, such as soil class distribution, from a different analytical viewpoint we can say that one determinate attribute has an indeterminate spatial distribution (the distribution boundary is indefinite), or within a determinate polygon the attribute distribution is uncertain with a blend of multi-feature classes. If a soil of type x is mapped as being at a particular location, it is far from certain that an investigator sampling that location would actually find soil type x. Instead soil y might be found, and added as an inclusion in the area mapped as x (Fisher,1993).

Traditionally, we visualize area-value data using a choropleth map, proportional symbol map and dot map (Robinson *et al.*, 1984; Dykes, 1994; Fisher, 1996). A choropleth map describes continuous phenomena change and is usually applied to represent such a density theme. Proportional symbol maps and dot maps portray distribution number and are usually applied to represent count data. The above methods have the assumption that phenomena distributes evenly and independently. Obviously it is far from the real case. To reveal geographic phenomena distribution exactly, more and more attention is paid to uncertainty information visualization (Fisher 1989, 1991, 1993, 1996; MacEachren, 1992; Goodchild *et al.* 1994; Goodchild, 1995). A number of cartographic methods have been achieved with digital visualization technology. Traditional mapping methods, which apply single colour or single texture to fill a polygon, result in the error in representation of uncertain attribute distribution. In order to explore and visualize geographic phenomena uncertainty and to take into account its impact on decision-making, the conventional static visualization method, which is based on Bertin's visual system, requires to be extended. The data quality information associated with uncertainty, reliability and precision should be added to the map visualization. Based on the fact that casual dynamic symbol movement or animation enables one to feel the displaying is indeterminate (Ai, 1998), the noisy display or random animation can be implemented to visualize uncertainty phenomena. Buttenfield and Beard (1994) presented a fog visual metaphor approach.

Fisher (1993) developed an algorithm to visualize soil class distribution by means of different colour pixels stochastically appearing within a raster polygon. In his method the number of one kind of colour pixel represents the corresponding soil class proportion role. In order to analyze and compare the animation of uncertainty polygon filling, the initial algorithm of Fisher (1993) is listed as follows:

Procedure Random2 (Original _ Value, X,Y, classes2)
A. Take a random number, R1, from a uniform distribution, where $0 \leqslant R1 \leqslant 1$;
B. Find the entries in classes corresponding to Original _ Value;
C. Compare the accumulated frequency of each inclusion with R1 until accumulated frequency>R1;
D. Display a square at X,Y with colour for the corresponding soil types;
E. End procedure.

In the animation method above, each pixel of raster map is initially filled with one colour through calling the above procedure, and then draws two random numbers to decide a pixel to be filled with colour through calling the above procedure again. The latter process is continuously repeated until interrupted by the user.

The latter repeated process does not take into account the shared pixel location when randomly selecting pixels. Some pixels with one colour may be covered by another colour pixel. It means the sentence C that is derived from Monte Carlo simulation cannot exactly guarantee that the number of different colour pixels are proportional to the corresponding soil type distribution in display snap. With regard to the stochastic cognition conveyed by the algorithm, it is not strong. The random variable only involves the pixel position. Once the pixel with one colour stands out, the local scenery is fixed unless it is occupied by another colour pixel.

This chapter attempts to improve the algorithm to enhance the visual sense of random display. We consider the pixel as a movement element containing limited life span and extend random variables to wider descriptions. An approach called particle system model is applied in the improved algorithm. In the particle system model, circle or box casually move through stochastic process that is based on some dot. A set of variables is introduced to describe the pixel movement, including position, movement speed, movement direction, colour brightness and life span. The experiment shows the improved algorithm has a better visualization effect in revealing uncertainty properties.

The chapter moves on to the introduction of the particle system model in Section 11.2 and then presents an improved algorithm in Section 11.3. After the analysis of the experiment result and the comparison with the initial algorithm in Section 11.4, the chapter discusses the potential application of the particle system model in the conclusion, Section 11.5.

11.2 PARTICLE SYSTEM MODEL

Particle system modelling is proposed by Reeves (1985, 1983). Some simple geometric elements such as dot, box, circle, ball and so on serve as a micro particle in the model. Through random movement of a set of description variables, the particle system model could be applied in simulation of fuzzy object distribution or description of irregular dynamic scenery, for example, the animation of fireworks or irregular texture distribution. This model is well known in the field of computer graphics. In this chapter we introduce it in cartographic visualization.

The particle movement in the model includes three stages: birth, action and death. Each stage has the properties of randomization. When one particle is born, it obtains random life span and other associated description variables such as size, colour, brightness, movement speed, movement direction etc. In the action stage, the variables stochastically changing leads to the modification of particle state. Each of the variables can be computed with recursion random function, for example, movement speed variable $vi = vi\text{-}1 + a*(fi\ \text{-}fi\text{-}1)$, where a represents acceleration and fi represents time in state i. With time past, the life span of an action particle reduces correspondingly. When fi reaches 0, it means the particle is dead and the particle display vanishes from the environment.

When used in map visualization such as polygon filling, the establishment of random function for particle movement needs to consider the statistical rules of geographic phenomena distribution. It means some variables should be limited to satisfy some determinate probability distribution such as normal distribution. For this purpose Monte Carlo method could be introduced. Another consideration is that the particle description variables have to be divided into qualitative and quantitative groups. According to the visualizing theme, some variables act as common or fixed to convey non-quantitative information. For instance, the particle colour is usually implemented to represent attribute class and it is pre-determined. In polygon fill visualization, the raster pixel corresponds to the particle whose size is often fixed and its stochastically moving requires it to be controlled within the polygon extent.

We set up a tuple to describe particle as follows:

$$s(ti) = (di,\ si,\ pi,\ hi,\ bi,\ fi)$$

Where ti stands for time and $di,\ si,\ pi,\ hi,\ bi,\ fi$ represents movement direction, movement speed, position, colour hue, brightness and life span respectively. We use colour hue to describe such things as soil class which is not a random number. Except for hi, the other 5 variables are defined as random distribution variables. In a raster map, the movement direction *di* of a pixel particle is defined as uniform distribution [0,7]. Each direction meaning is portrayed in Figure 11.1.

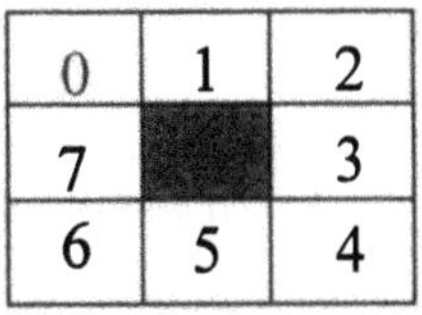

0	1	2
7		3
6	5	4

Figure 11.1 Eight neighbour movement directions of one pixel.

Movement speed describes how far the pixel moves in each step and is defined as a uniform distribution [0,S], where S is pre-determined by the user. Position *pi* is decided by random row and random column number within the polygon extent. Based on a hue-brightness-saturation colour system, *bi* can be defined as a random number varying from 0 to 255. Life span *fi* is a random number distributed in [0,F], where F is pre-determined by the user.

11.3 IMPROVED ANIMATION METHOD

Take the soil class distribution for example. Within a determinate polygon, the main inclusion and extra non-main inclusions are known, including the attribute class and associated proportion. We establish an algorithm to animate the soil inclusion distribution and to reveal the statistic distribution rule. The method is divided into three steps.

11.3.1 Raster polygon conversion

After a vector polygon is converted to the raster format, it is stored in RLC (running length code). Every row contains several column pairs. The corresponding data structure is defined as RASTER_POLY. In the next procedure a raster pixel serves as one particle if it belongs to a non-main class.

11.3.2 Particle initialization

Inclusion class structure is defined as follows:

```
typedef structure
{ int    class;   // soil class code
  float  rate;    // inclusion proportion
  int    colour;  // corresponding fill colour hue
} CLASS_TYPE
```

Array variable classes belongs to CLASS_TYPE and classes[0] describes the main soil class. In the next algorithm, the polygon is initially filled with main soil colour classes[0].colour as background. At first the number of live particles is assigned 0, N=0.

Algorithm 1
Init_Poly (CLASS_TYPE *classes*, RASTER_POLY *grid*)
Initially fill all grid pixels with colour classes[0].colour;
Initialize live particle number N to zero;
For (i=1; i ≤rows of grids; i++)
For(j=begin column in RLC i; j ≤end column in RLC i; j+ +) //for each pixel in polygon;
  { Take a random number , R1 ∈[0,1],from a uniform distribution;
  Initialize accumulator s to 0;
  For(k=0;k ≤element number of array classes; k++)
    { accumulator s=s+classes[k].rate;
    If R1>s break;
    }   // Find corresponding soil type according to random number R1.
  If k=0 continue;   // For main soil type, no particle generated.
  Else //For extra soil type, generate a particle and assign initial descriptions.
    { particle[N].p←[i,j];   // position assignment;
    particle[N].c←classes[k].colour;   // colour
    particle[N].d←random number R2 ∈[0,7];   // movement direction
    particle[N].s←random number R3 ∈[0,S];   // movement speed
    particle[N].b←random number R4 ∈[0,255]; // brightness
    particle[N].f←random number R5 ∈[0,F];   // life span
    Fill particle [i,j] with colour particle[N].c;
    N=N+1;   // for next particle;
    }
  }

The procedure produces a number of live particles with non-main inclusion colour. Under the control of accumulator s and circulation k, the algorithm guarantees that the number of different soil inclusion is proportional to its associated rate. When non-zero valued k is found, it means a stochastic process resulting in a non-main inclusion soil class. Correspondingly a set of initial random description variables are assigned to the generated particle. These variables will control particle movement and change display scenery in the next procedure.

11.3.3 Particle movement

Based on the born particles in algorithm 1, the polygon animation can continuously repeat through the particle birth, action and death. The life span of a particle decides the living length from birth to death. During life span, the live particle effects convey the feeling of polygon animation. The number of different kinds of particle remains the same in the whole procedure. The particle movement procedure is continuously repeated until interrupted by the user.

Algorithm 2
Particle_Movement (int N)
Repeat
{ For (i=0; i<number of particles N; i+ +)
If (particle[i].f<>0)   // Life span being greater than zero means it is active.
    { Compute the new position of particle[i] according to movement variables;
    If (new position is not occupied by other live particle)

         Move the particle particle[i] to the new position;
         particle[i]. f= particle [i]. f-1;
        }
Else   // the life span being zero means the particle is dead.
      { Make particle[i] disappear;   // Through changing colour to background.
       Repeat
          Take two random numbers, R6 $\in$[1, ROWS],
                      R7 $\in$[begin COLUNM, end COLUMN];
       Until position [R6,R7] is not occupied by other particle;
       particle [i].p $\leftarrow$[R6, R7];   // Generate a new particle
       particle [i].d $\leftarrow$random number R8 $\in$[0,7];
       particle [i].s $\leftarrow$random number R9 $\in$[0,S];
       particle [i].b $\leftarrow$random number R10 $\in$[0,255];
       particle [i].f$\leftarrow$random number R11 $\in$[0,F];
    }
Fill particle [R6, R7] with colour particle [i].c;
}
Until interrupted by user.

     Before moving a particle, algorithm 2 checks whether the computed position is occupied by another live particle or not. If so, the particle stays in its original position. This strategy guarantees N particles are not overlapping each other and exactly guarantees the proportion of one kind of particle is fixed in the whole animation procedure. The life span of a particle reduces one unit each step in the circular process. When the life span reaches zero, the displaying particle vanishes and immediately a new particle locating in a random position is generated. Like the process of particle birth in algorithm 1, a set of description variables are assigned to the new particle by means of randomization. But the colour hue variable *particle[i].h* inherits the hue of the previous dead particle. The inheritance retains the number of different kinds of particle to make the animation respect the distribution probability.

11.4 EXPERIMENT AND ANALYSIS

We designed several sets of inclusion tables with pairs of classes and rates to test the above algorithms. Each inclusion class corresponds to one distinguished colour hue. When test data Table 11.1 is used, the display result of running algorithm 1 is portrayed in Figure 11.2 in the colour plate section, and the display snap of running algorithm 2 is shown in Figure 11.3 in the colour plate section.

Table 11.1 Inclusion class and rate of test data.

Inclusion class	0 (main class)	1	2	3
Rate	75%	12%	8%	5%

     In the animation demonstration various colour pixels move at random and the scenery looks like a flashing picture. In order to obtain a suitable flashing scenery that satisfies the user's visual feeling, a delay time sentence can be inserted into algorithm 2. The delay time length depends on the polygon size and complexity.

When the running procedure of algorithm 2 is interrupted any time, we can find various colour pixels exactly respect the rule of soil inclusion distribution. With regard to the uncertainty feeling conveyed by algorithm 2, it is enhanced because of the introduction of particle movement and more random description. The improvement leads to a more effective result in contrast to Fisher's initial method.

11.5 CONCLUSION

Based on particle system model, the paper improves Fisher's initial animation algorithm of uncertainty visualization in polygon filling. Monte Carlo simulation is inherited in the improved method so the quantitative statistical rule of attribute distribution is retained. But the simulation exactness of quantitative distribution and conveyed uncertainty feeling are enhanced in the new method. Based on the fact that random movement leads to uncertainty sense, particle system model provides a good description for uncertainty phenomena through dividing it into different pixel elements. It is able to be applied not only in computer graphic simulations such as fireworks, coarse texture, but also in simulation of some uncertainty properties of geographic phenomena. The stochastic process of geographic phenomena can be revealed through random particle movement.

This model has several potential applications in other uncertainty visualization fields. In the display of uncertainty line object distribution, the particle can be limited within the line buffer and its position is controlled through Monte Carlo method so that the particle cloud is able to reveal like the normal distribution rule. This application belongs to uncertainty boundary visualization or ε-band error distribution visualization (Veregin, 1994). The particle movement can also be used to visualize fuzzy membership of geographic phenomena. The relationship between particle movement and membership degree could be established by introduction of some mathematics function and the description elements of particle can serve as different information transformation media. The selection of description variables of particle depends on the visualizing theme and usually qualitative and quantitative description has to be distinguished. The movement of three stages is decided by the phenomenon distribution and the uncertainty characteristics.

REFERENCES

Ai, T., 1998, Dynamic Symbol and Dynamic Map. *Journal of Wuhan Technical University of Surveying and Mapping*, 23(1), pp. 47-52.
Buttenfield, B. and Beard, M. K., 1994, Graphical and Geographical Components of Data Quality. In *Visualization in Geographic Information Systems*, edited by Hearnshaw, H. W. and Unwin, D. J. (Chichester: Wiley), pp.150-157.
Dykes, J. A., 1994, Visualizing Spatial Association in Area-value Data. In *Innovation in Geographic Information System*, edited by Worboys, M. F. (London: Taylor & Francis), pp. 149-159.

Fisher, P. F., 1996, Animation of Reliability in Computer-generated Dot Maps and Elevation Models. *Cartography and Geographic Information Systems*, 23(4), pp. 196-205.

Fisher, P. F., 1993, Visualizing Uncertainty in Soil Maps by Animation. *Cartographica*, 30(2&3), pp. 20-27.

Fisher, P. F., 1991, Modeling Soil Map-unit Inclusions by Monte Carlo Simulation. *International Journal of Geographic Information Systems*, 5(2), pp. 193-208.

Fisher, P. F., 1989, Knowledge-based Approaches to Determining and Correcting Areas of Unreliability in Geographic Databases. In *Accuracy of Spatial Databases*, edited by Goodchild, M. F. and Gopal, S. (London: Taylor & Francis), pp.45-54.

Goodchild, M. F., 1995, Attribute Accuracy. In *Elements of Spatial Data Quality*, edited by Guptill, S. C. and Morrison, J. L. (Oxford: Pergamon), pp. 59-79.

Goodchild, M. F., Buttenfield, B. and Wood, J., 1994, Introduction to Visualization Data Validity. In *Visualization in Geographic Information Systems*, edited by Hearnshaw, H. W. and Unwin, D.J. (Chichester: Wiley), pp.141-149.

MacEachren, A. M., 1992, Visualizing Uncertainty Information. *Cartographic Perspective*, 13(Fall), pp. 10-19.

McGranaghan, M., 1993, A Cartographic View of Spatial Data Quality. *Cartographica*, 30(2&3), pp. 8-19.

Reeves, W. T., 1983, Particles System----A Technique for Modeling A Class of Fuzzy Objects. *Computer Graphics*, 17(3), pp. 359-376.

Reeves, W. T., 1985, Approximate and Probabilistic Algorithms for Shading and Rendering Structured Particle System. *Computer Graphics*, 19(3), pp. 313-322.

Robinson, A. H., Sale, R. D. and Morrison, J. L., 1984, *Elements of Cartography*, 5th edn, (Chichester: Wiley).

Veregin, H., 1994, Integration of Simulation Modeling and Error Propagation for the Buffer Operation in GIS. *Photogrammetric Engineering and Remote Sensing*, 60(4), pp. 427-435.

CHAPTER TWELVE

A Policy-Maker Point of View on Uncertainties in Spatial Decisions

Bernard Cornélis and Sébastien Brunet

ABSTRACT

While public decisions concern society or some communities, they are formalised by the policy-makers. Policy-makers face widespread types of information, each one having their own uncertainties associated with them. The particularity of public decision-makers is to deal with those uncertainties. By contrast, scientists limit the uncertainties in their experiments by enunciating hypotheses. In any decisional process, decision taking is based on a decision analysis phase which relies on a documentation/information phase. It is then followed by the implementation and the evaluation phases. The foundation of the documentation information step is composed of three interconnected aspects: data, issue and context. Decision, being the act of making up one's mind, can be linked to space or influenced by space or not at all related to space. Public decisions are all spatially referenced if not spatially induced since they apply to a certain territory. The territorial scale of application of a decision intervenes not only in the choice of data and analysis, but also in the way the issue is expressed and on the degree of certainty acceptable for decision-makers at the different steps of the decision process. This contribution approaches the uncertainties not from a data perspective, but from the decision-making process point of view. After defining the concepts of spatial and public decisions, the different types of uncertainties and their consequences are presented using a generic model of decision processes.

12.1 INTRODUCTION

Geographic information systems have emerged, according to Laaribi (2000), from a technological development, away from the spatial decision-makers' desiderata. Furthermore, he states that this development did not benefit from an existing spatial decision theory. The same applies to spatial decision support systems, although some decision theories are usually used. Concerning uncertainties in spatial decisions, the issue is similar.

In the following pages, those questions are addressed from a public decision-maker's perspective. This essay first gives a definition of the spatial and public decisions concepts – thorough definitions of the "spatial decision" concept can be found in Cornélis (2000). It then introduces a generic model of decision-making processes. This model is therefore used to present the different types of uncertainties seen from a policy-maker's standpoint. Two examples illustrate the concepts before the conclusions.

12.2 A MATTER OF DEFINITION

12.2.1 What is a spatial decision?

With the development of spatial decision support systems (SDSS), one can wonder what is so special about spatial decisions compared to "regular" decisions. In the characterisation of spatial decision problems by Malczewski (1997), only the spatial variability of the outcomes or consequences of the decision alternatives is specific to spatial decisions. When presenting the concepts of SDSS, Densham (1991) avoids defining the notion of spatial decision by focusing on the technical aspects of SDSS compared to DSS and GIS. On the complexity of spatial decision making, Leung (1997) mentions different components of decision making without expressing the spatial specificity(s) of decision. This lack of concept definition could lie in the triviality that a spatial decision is a decision which uses spatial information.

Decision is the act of making up one's mind or the judgement on an issue under consideration (Berube *et al.*, 1995). Decisions can either be linked to space or not. An **aspatial** decision is a decision not related at all with space. For example, one choosing a number or a colour in a lottery is making an aspatial decision. To a certain extend selecting a beverage for lunch is also an aspatial decision. The spatial decisions - decisions related to space - can either be **spatially referenced** or **spatially induced**, according to whether the decision is taking place in space or over a certain area or whether the decision is influenced by the territory. For example, someone going from Brussels to Hong Kong is taking a spatially referenced decision since he has to choose between going through Frankfurt, Paris or Zürich. It is the same for one deciding whether or not to build a railway between two towns. The driver deciding where to park his car is taking a spatially induced decision since his decision will be influenced by his final destination –as close as possible being the most frequent criteria–, by the availability and the size of a parking spot, and by the geometrical characteristics of his vehicle.

This decision classification is influenced by the way the problems or decision questions are expressed. The categorisation of decisions is not easy to perform since decisions are not only context sensitive, but also inherit from decisions previously taken. For example, "To have a drink on a café's terrace or not?" is a spatially referenced decision. If it is raining, this question does not seem relevant even though it still leads to a spatially referenced decision, itself leading to a spatially induced behaviour. In this context, one might expect the question "To have a drink or not?" and take an aspatial decision. If the terrace is right next to ongoing work or to a heavy traffic road, someone thirsty feeling like being in a quiet place will take a spatially induced decision. When looking *a posteriori* at decision results to identify the type of decision taken, misclassification is possible. For example, the location of lighthouses along a coastline is most likely to be a spatially induced decision. Nevertheless, if the decision-maker just decided that the lighthouses would be located next to village V, hamlet H and dwelling D without taking care of the surroundings (substratum, relief,...), he took a spatially referenced decision.

12.2.2 What is a public decision?

Just as a decision can be characterised as spatial, it can also be defined to be public. In contrast to individual decisions, public decisions concern the collective future of a community. Although personal decisions may have an effect on an organised group of individuals, only the actions taken by politicians, judges and Administration will be considered, in this article, as public decisions. Those people answer questions such as "How do we want to live?", "What are our values? And how do we rate them?". These decisions are characterised by the political organisation they take place in, by their complexity and by the mandate given to the public servants.

12.2.2.1 Political organisation

The group of individuals representing a community and working for its management can be identified as a political organisation since its members are carrying a common project for the society. In terms of participation in the public decision-making process, three families of political organisation are usually identified : monarchy, oligarchy and democracy (Rodee *et al.*, 1957). Monarchy is the generic term designing a political system in which only one person is deciding. Tyranny and despotism are particular monarchies characterised by an absolute power, and by its arbitrary use in the former case. Oligarchic decisions are taken by a few selected persons or group of persons. Aristocracy, partycracy and technocracy are variants of oligarchy involving respectively a social class, political parties and scientific experts. In democracies the entire community is involved directly (direct democracy) or not (representative democracy) in the decision process. Anarchy is not part of this classification since it is characterised by the absence of political organisation (Leleux, 1997).

12.2.2.2 Complexity

Politics is concerned with the management of the society status. So the upholding of the existing situation and/or its adaptation to changes are the main tasks of decision-makers (Renwick, 1993). Since their decisions concern the organisation of life in the community, decision-makers should define what is the common interest out of the different visions expressed or not by the components of the community. Compared to other dynamic systems, the difficulty of handling this one lies not only in its evolution but mainly in the unexpected behaviours of the individuals composing it e.g. hidden agenda and cultural differences etc. The definition of the common interest is a continuous process involving the different stakeholders. It is the key point on which public decision-makers base their actions and define the priorities and guidelines (Chevallier and Loschak, 1974). For Roqueplo (1997) reality has as many facets as there are perceptions of it, so the common interest can be in partial or total opposition to personal ones and its definition is always biased. Thus, the claim to an objective definition of the common interest should be considered as a political story-line (Hajer, 1995). The complexity of public decisions is also reinforced by the multiple consequences of matters under consideration. This results in the difficulty of evaluating public decisions (Meny and Thoenig, 1989).

12.2.2.3 Mandate

Since the 16th century, the State is the institutional framework in which public decisions are elaborated. It has four essential elements : government, sovereignty, people and territory (Rodee *et al.*, 1957). The area of application of public decisions is the territory of the state defined by arbitrary borders. The policy-makers (government) are the only ones (sovereignty) recognised by the community (people) to take public decisions within the limits of the state (territory). The spatial characteristics of the territory (resources, shape, ...) are variables independent from the political organisation but which can be taken into account. The administrative division of the territory is the public decisions structure where skill level and space should match each other (Zwetkoff, 1995). So, to each problem, one administrative level (international, national, local,...) is more appropriate. Therefore, public decisions are all spatially referenced if not spatially induced since they apply to a certain territory. Added to the legitimacy of the political organisation, the area of application of public decisions defines the competencies or mandate given to the decision-makers.

Whatever the political organisation, the public decisions will mainly be justified through either some kind of violence or through three types of ideological arguments: tradition, charisma or rationality. For Weber (1971), the legitimacy evolution goes from a traditional power to a rational power, passing through a charismatic power. The traditional authority is inherited or originates from habits. The charismatic authority is based on the exceptional personality of a character who will be recognised as the leader. In the case of the rational authority, the power justification comes from the reasonable character of previously established rules.

Nevertheless, one has to recognise that most public decisions are taken today outside of the traditional structures of the state. For example, techno-economic actors such as multinational firms are changing our world. This kind of "sub-politics" (Beck, 1992) activity suggests and promotes specific values and visions of the future. Of course, this political production has neither the same relation to the space nor the same legitimacy as institutional public decisions.

12.3 A MODEL OF THE DECISION-MAKING PROCESS

Looking at the decision-making process is not an easy task and can turn out to be very puzzling. So to grasp its different facets, the decisional fountain model will be used. This generic model apprehends decision-making through three complementary aspects: data, issue and context and through the following phases:

documentation/information (DO), decision analysis (DA), decision taking (DT), decision implementation (DI) and decision evaluation (DE) (Cornélis, 1999).

12.3.1 The decision bases

Decisions rely on three inter-linked bases: issue, context and data.

12.3.1.1 Issue

The issue can be stated in terms of problem(s), of need(s) or of objective(s). These formulations are just ways to express the actual concern of the decision-maker(s), his point of focus of the moment. Each one of them can be reformulated as one another, which in turn might orient the way the issue will be dealt in its temporality, way of getting to a solution, and so on.

12.3.1.2 Context

The context refers to the attitude, the experience and the knowledge. There are four basic attitudes: proportional, egalitarian, maximalist and fluctuating (Figure 12.1). They can be combined into decisional strategies. The experience is a personal or collective form of knowledge. The knowledge relates to the hard and soft scientific knowledge as well as to background information not necessarily stated or well-defined.

12.3.1.3 Data

The data can either be raw, integrated or processed. The processed data for someone might be considered as raw data for another party, just as a decision turns to information once it has been taken.

12.3.2 The decision phases

The fields dealing with decision science are numerous: operational research, mathematics, psychology, management, political science, and so on. Each one of them developed some kind of decision culture, paying more attention to some of its aspects.

12.3.2.1 Documentation/Information (DO)

The documentation/information phase assembles the decision bases and all their relations. The connections between issue, context and data are manifold, to the point of presenting fuzzy boundaries. Data can generate a problem by showing unexpected values. The other way around an issue will identify data to fulfil needs. If an objective is expressed in terms of maximisation of benefits for example, it already includes attitude elements. Knowledge is nowadays extracted from databases. Ignoring the connections or one of the bases is like wearing blinkers.

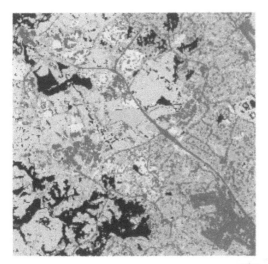

Plate 1 Classified SPOT satellite image of Albany Basin, North Shore City, New Zealand. Colour represents eight land cover classes.

Plate 2 Bivariate *opacity* technique where low opacity refers to low confidence in land cover classification and high opacity represents high confidence in the classification. Colour refers to land cover class.

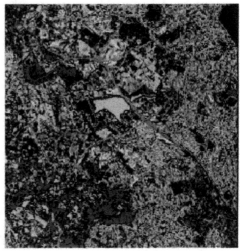

Plate 3 Bivariate *squares* technique where visual variable of size is used to visualise uncertainty. Larger squares represent higher confidence and smaller ones refer to lower confidence in the classification. Square colour refers to land cover class.
Note: this image covers 1/36 (1km x 1km) of the original classified scene.

Plate 4 The initial state of particles.

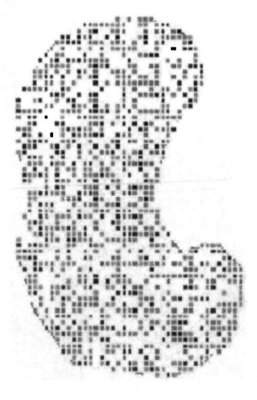

Plate 5 A snap display extracted from particle movement.

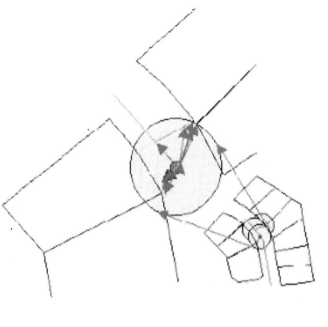

Plate 6 Creating directions and distances from the digitized themes.

Plate 7 "Old" and "new" digital cadastral database (DCDB).

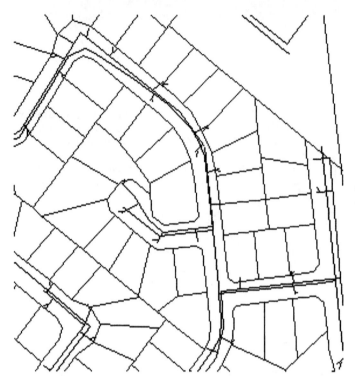

Plate 8 "New" DCDB with "old" sewer and water utilities.

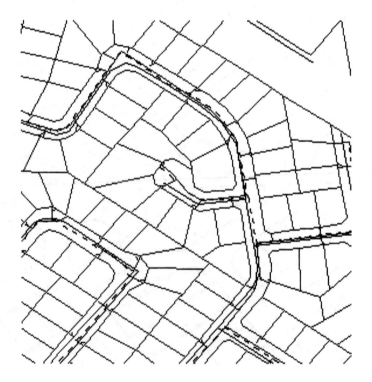

Plate 9 "New" DCDB with adjusted utility data.

Although this sub-division can be seen as arbitrary, it identifies the three principal components of the documentation/information phase. Out of all this, documentation and information are extracted and will serve for the rest of the decision process.

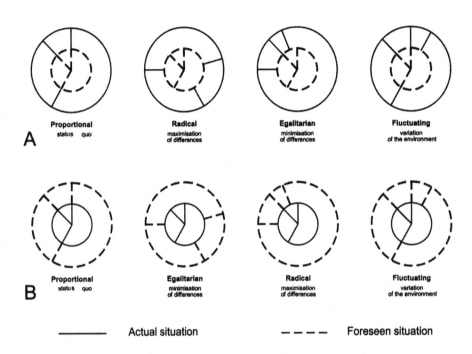

A

| Proportional | Radical | Egalitarian | Fluctuating |
| status quo | maximisation of differences | minimisation of differences | variation of the environment |

B

| Proportional | Egalitarian | Radical | Fluctuating |
| status quo | minimisation of differences | maximisation of differences | variation of the environment |

—————— Actual situation   – – – – Foreseen situation

Figure 12.1 The four basic decisional attitudes in the case of a decrease in resources (A) and of an increase of resources (B). The radical and egalitarian labels have been interchanged.

12.3.2.2 Decision analysis (DA)

The decision analysis step is the phase which processes and integrates the information based on the documentation, both originating from the bases. It is characterised by the methodology and tools used, by the information and documentation taken into account, as well as by the information it returns.

12.3.2.3 Decision taking (DT)

The decision-taking phase is the moment at which the choice is made. It is characterised by the person(s) taking the decision, their behaviour and by the formal procedure followed. Who is taking the decision? Is it one person, several persons with their own interests and objectives, a group of people sharing the same interests and objectives, or several groups of people with their own interests and objectives. The procedures followed can be despotic, negotiated, agreed, voted, freely taken, controlled, weighted, etc. At the moment of taking decisions, decision-makers can go for seven elementary behaviours or for a combination of them.   The two basic decision dispositions are either positive (selection) or negative (elimination).   The elementary decision behaviours can either be binary, plural, ordered, null, self-conflicting, conditional or nuanced (Cornélis, 1998; Cornélis and Viau, 2000).   The random decision could be added to the list, but then it does not conform to the definition of decision given in Berube *et al.* (1995) – one making up his mind.

12.3.2.4 Decision implementation (DI)

Decision implementation can impact the different phases of the decisional fountain – the fountain metaphor originates from the analogy of water splashing the fountain foundation.   It is characterised by the scales of decision: the stability in time (stable, evolving, definitive), the time of application, of prescription, the scale of application (international, national, regional, municipal or individual property levels).   This phase turns the decision into action, policy, knowledge, and so on. The decision confidence level and the diffusion methods are typical of this level. The implementation of a decision can be a time for legitimisation and justification of the decision taken.

12.3.2.5 Decision evaluation (DE)

The evaluation of decisions is tricky in itself.   It cannot be complete and will be based only on one or a few standpoint(s).   Just like when looking at a picture of a landscape, the decision evaluation will look at the projected image of the core of the fountain.   This point of view might present an emphasis on the issue, or on the data or on the context.   But just like the picture of the landscape, it will only be a partial view.   Even a zenithal view would not be complete since the implementation would be diffused over the bases.   The evaluation gets more difficult as times goes by and as the implications are numerous.

12.4 UNCERTAINTIES IN DECISIONS

While scientists limit the uncertainties in their experiments by enunciating hypotheses, public decision-makers have to deal with a wide range of uncertainties. To comprehend this aspect of public decisions, the decisional fountain model will be used.   It appears that uncertainties come along the entire decisional process and that they can be found at the different stages (Figure 12.2).

12.4.1 Uncertainties in the DO phase

Among the uncertainties at the documentation/information level, the uncertainties concerning data are the most investigated ones.   In the field of spatial sciences, it has been studied only for a few years (UCGIS, 1996).   The International Symposium on Spatial Data Quality in Hong-Kong (Shi *et al.*, 1999) is a good example of the ongoing research in the field.

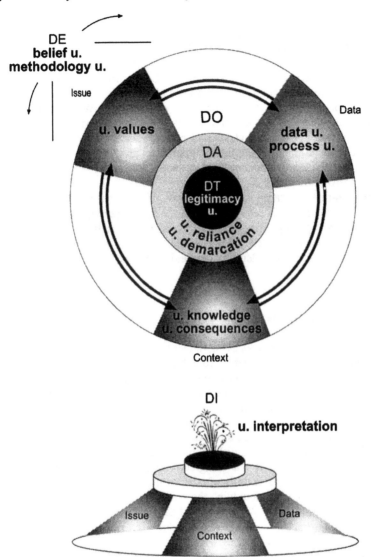

Figure 12.2 Uncertainties in the decisional fountain seen from a zenithal view at the top and from an oblique view at the bottom. DO stands for Documentation/Information, DA for Decision Analysis, DT for Decision Taking, DI for Decision Implementation, DE for Decision Evaluation, u. for uncertainties.

Two types of uncertainties can be identified: the ones linked to data itself (for example positional, attribute, conceptual) (Griffin, 1995; Leung and Yan, 1997, 1998; Fisher and Wood, 1998), and the ones linked to the processing of the data (for example visualisation, interpolation, modelling) (Hearnshaw and Unwin, 1994; Fisher, 1998; Heuvelink, 1998).

In the context base, uncertainties are also present. The absence or the lack of experience or of knowledge, whether scientific or not, creates uncertainties. Major environmental issues where lack of information is important are said to be situations of great uncertainty. Fuzzy analysis can be seen as a scientific method for dealing with an **uncertainty of knowledge** (Leung, 1997). Experience is based on past events but, time going by, not all the elements involved are remembered, introducing uncertainty in cases similar to the present case. When the outcomes of the options identified by the decision analysis are not known, the uncertainty is called by Hansson (1996) **uncertainty of consequences**. These uncertainties can be apprehended through probabilities, imprecise probabilities, possibilities and unknown possibilities (Bouchon-Meunier and Nguyen, 1996; Hansson, 1996; Turban and Aronson, 1998).

Decision-makers transpose the demands of the community into problems, objectives and needs. This reformulation work introduces uncertainties due to the interpretations and perceptions of the policy-makers themselves. The interpretation is based on their moral values and on the ones of the stakeholders. These values are subject to change, to misidentification and introduce the **uncertainty of values** (Hansson, 1996).

12.4.2 Uncertainties in the DA phase

The decision analysis techniques being numerous, there is an uncertainty of method choice. This uncertainty originates from the definition of the common interest. Where several decision analysis techniques are used, their results might be partially or totally opposed to each other. The same applies to experts' opinion. Moreover, the decision analysis phase simplifies reality. It turns out that some aspects of reality are left over which introduces some uncertainty in the analysis. All these uncertainties fall in what is called **uncertainty of reliance**. When talking about the experts, Hansson (1996) distinguishes three degrees of uncertainty of reliance: when experts are recognised but disagree; when the *bona fide* experts are not identified; and when the question is whether or not experts should be relied on. In the light of the decisional fountain, the reliance of data, context and issue is questioned during the decision analysis phase. The reliance of documentation/information can be expressed in terms of weights given to the different elements in the decision analysis.

Policy-makers are facing **uncertainty of demarcation** when the options are not well determined. This occurs when it is not known whether all available options have been identified (*unfinished list of options*) or when there is no agreement on the issue (*indeterminate decision horizon*). In the former case, there are three ways to overcome this uncertainty (Figure 12.3). The first one is to content oneself with the available list of options and to choose one of them (*closure*). The next one is to postpone the decision and search for other options (*active postponement*). The last one is half way between the two previous ones. *Semi-closure* selects and carries out one of the available options but at the same time searches for improved alternatives and plans for later reconsideration of the issue (Hansson, 1996). This shows the possible interaction between DA and DT.

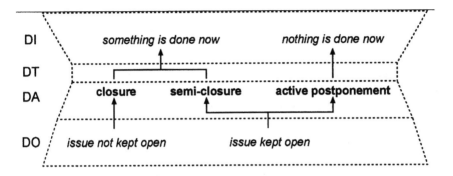

Figure 12.3 The three ways to overcome uncertainty of demarcation of an unfinished list of options (after Hansson, 1996) corresponding with the decisional fountain phases.

12.4.3 Uncertainties in the DT phase

In the decision-taking phase, the uncertainties are due to the decision-taking procedure and to the nature of human beings. The DT phase is usually defined by rules answering the questions : *Who?*, *In which way?* and *On which topic?*. These rules are influenced by the political organisation and exist to avoid **uncertainty of legitimacy**. This uncertainty leads to contestation. For example, in a decision procedure in which representatives of pressure groups participate in the decision-taking phase, one uncertainty might be who will be representing one tendency. The uncertainty for the spokesman of a group is, at the time of taking decision, to decide on a way of representing his party. Procedures themselves may carry uncertainty through particular or evolving rules. The mastering of these rules assures the power. Those participating in the elaboration of rules can opt for procedures which are in their favour. In front of the multitude of demands and of the lobbying activities, policy-makers can be unsure of the ones they should give priority to. Public decision is the result of a comparison between several options, which turns in to choosing between several types of uncertainties.

12.4.4 Uncertainties in the DI phase

Usually, policy-makers are not the ones carrying out the decisions. As a consequence, the decision is interpreted by the actors involved in its execution (Grawitz and Leca, 1985). This interpretation can focus on the content of decision (precision), on the way to carry it out (execution) and on the decision legitimacy (acceptability). These three readings belong to the **uncertainty of interpretation**. Moreover, the transmission of decisions might be distorted by the complexity of the system structure or by the different systems composing the organisation (information system, operation system, decision system, complex system and auxiliary system) (Pantazis and Donnay, 1996; Pantazis and Cornélis, 1997).

The *temporal uncertainties of interpretation* originate from the integration of time in the decision implementation phase. The decision stability and the time of application (of prescription) fluctuate over time as the context, data, issue and analysis techniques evolve. A succession of decisions may be necessary and influences the perception of the original decision. Time might prove predictions, prevision and forecasts correct and reduce some uncertainties introduced in a decision process. The other way around is also possible. Variations in a decision process might keep the uncertainty level stable.

12.4.5 Uncertainties in the DE phase

In the decision evaluation phase, the uncertainties are twofold: **belief uncertainty** and **methodology uncertainty**. The former happens when what is evaluated is not what has to be evaluated (Meny and Thoenig, 1989). For example, when the effects of a public decision under evaluation are the consequence of outside events. The latter occurs when it is uncertain that the technique(s) used for evaluation is (are) appropriate. Finally, in evaluation involving decision(s), all the uncertainties typical to the decision processes can be observed (uncertainties of knowledge, consequences, values, reliance, demarcation, legitimacy, interpretation).

12.5 ILLUSTRATION OF THE CONCEPTS

To illustrate the concepts exposed above, two cases will be presented: the common situation of parking a vehicle and the taking into account of drought at the European level. The first one is a very pragmatic spatial decision example which everybody can experience. The second is seen from the perspective of an institution having to manage its territory.

12.5.1 The case of parking a vehicle

Through this example, the decisional fountain concepts will be illustrated; the different uncertainties will be depicted in the second case study. Parking a vehicle is a one soul decision which, according to the previously given definition, can be categorised as spatially induced. For the driver, the issue is "to park the vehicle". The context tells whether this issue can be refined to "to park the vehicle as close as possible", "to park the vehicle as quickly as possible", "to park the vehicle for a fast departure", "to park the vehicle without manoeuvring", and so on. This refinement of the problem is based on the other objectives the driver has, such as "to load or unload the vehicle", "to be on time at a rendezvous", "to rob the bank", etc. or on his experience "in parking that vehicle", "in parking in-between obstacles", "in parking within the traffic". The context also includes the driver's knowledge of the local parking regulations and habits. The data cover characteristics of the vehicle and of the space in which the vehicle is evolving. The former is supposed to be already known by the driver. They include the geometry of the vehicle (its three dimensions, shape, etc.), and its thematic characteristics such as the licence plates. The latter are not as stable and vary in space and time. They comprehend the authorisation of parking, the presence of policemen nearby, the availability of a parking place, the size of the available parking space, the parking fee, and so on. All these elements form the DO phase.

The decisional analysis of the parking of a vehicle will be based on the issue, the context and the data. The analysis can either take into account the true characteristics of space, such as the availability of an empty parking place, or perceptions of the characteristics, such as the perceived size of the parking place compared to the mental size of the vehicle. The analysis can be performed by the driver or by a passenger or by a parking usher. No matter who is making it, it usually has to be executed relatively fast, otherwise the vehicle might already have passed by the opportunity. In itself, the analysis also includes a decision answering the question "Is this place suitable for the vehicle?". This decision turns into information for the decision-maker.

The final decision is usually taken by the driver when answering the question "Will I park here?". Doing so, he will have a binary decision behaviour. Where the parking place is indicated by a parking usher or a policeman, the driver is not the one making the spatial decision.

The decision implementation is executed by the driver. If the answer to the question "Will I park here" was negative, the implementation consists in going over the process of searching for a suitable parking place. If the answer is affirmative, the result can either be a parked vehicle, or an attempt at parking the vehicle.

The driver's first evaluation of the decision will verify that the result meets his original objectives and that his analysis was correctly performed. The driver might do another evaluation after some time. His objectives might be different from the ones he had at the time of parking the vehicle. If the result does not satisfy him, he can park the vehicle again or get over it. The decision evaluation can also be performed by other drivers, by pedestrians, or by policemen. In these cases, it is the implementation, the context or the data which will be the focus point of the evaluation.

12.5.2 The drought case at European level

Drought is a normal feature of climate occurring in virtually all climatic regimes. It is a complex and poorly understood phenomenon which affects more people than any other natural hazard and which differs from other natural hazard by its creepiness and by its spatio-temporal scales (Wilhite, 1993, 1996). Droughts can affect from local to worldwide social systems through their impact on the environment, on human health, on the agriculture, on transportation, on power generation, on commerce and industry, on urban areas, on water resources, on education, on government operations, etc. (Riebsame *et al.*, 1991). The spatio-temporal aspects and the complex interactions between environment and society make it difficult to evaluate the damage caused by a drought (Vogt and Somma, 2000a). In Europe, drought occurs just about every year in places ranging from Scandinavia to the Mediterranean (Bradford, 2000). The impacts of such drought can be estimated to several billions of Euros (Garrido and Gómez-Ramos, 2000). Scientists have proposed drought planning as a way to mitigate drought impacts

(Wilhite, 1991; Vogt and Somma, 2000b). Now, let's have a look at the decision uncertainties which can be associated with the drought question addressed from a policy-maker viewpoint at a European level.

The value uncertainty is the most determining uncertainty in the DO phase at a European level. This uncertainty can be expressed through the following questions: "Is drought really an issue?" – a link can be made to Wilhite's hydro-illogical cycle (Wilhite, 1990), "Should we take care of the population affected by drought or of the land's resources (agriculture, forest, animals,...)?". As Kerrey (1987) says *"The single most difficult task for a representative democracy is to engage in long-range planning. If we are planning for something which, moreover, no one believes (or wants to believe) will happen and which might require us to work or restrain now to avoid a possible disaster later, the task is even more difficult"*.

The uncertainty of consequences can be put into words such as "Will this decision have an impact on drought conditions?". It can either be ignored or it may slow down the decision process and in particular its DA and DI phases.

The uncertainty of knowledge can be formulated in questions such as "Isn't drought a created problem (Zimmer and García Rodrígo, 1995)?", "Isn't it specific to arid countries?", "Isn't it the same thing as desertification?". It can originate from semantic definition not unanimously accepted, from the ignorance of the people involved in the decision process, from cultural differences – in which case the language translation is playing its part, from the misunderstanding of phenomenon, and so on.

The data and process uncertainties arise from the scale of the data, from the way in which it was produced, from lack of data history, from the data handling, etc. These uncertainties are the most ignored ones, not only because of some trust in the data providers, but also because, at this level of decision, data is used to justify spatial discrimination in the DI. In any case, having some data is better than having none.

The uncertainty of reliance can be encountered when the question is "Which expert's opinion, which report conclusions, or which recommendations should be followed?". This uncertainty is typical but not limited to scenario analysis.

The uncertainty of demarcation is met when a question like "What could be done at European level to mitigate drought?" is raised. Apart from some relief funds given to drought-affected areas within or outside the Union, the European attitude is more a postponement type than an active postponement.

The uncertainty of legitimacy can explain this attitude. "Is it up to the European Union to take care of drought matters?". "Shouldn't it be addressed by the Member States, or by the regions?". The actual European position can be seen in the European Spatial Development Perspective (European Commission, 1999) as its program for combating drought falls under the Interreg framework.

The uncertainty of interpretation is quite frequent in the administrative field. In its ten priorities for economic and social cohesion, the European Commission (2001) identified six of them with an important territorial dimension. One can wonder if drought-prone areas fall in the "areas with severe geographical or natural handicaps" priority. The answer is no, only outermost regions, islands, mountain areas, peripheral areas and areas with very low population density are the actual focus of interest. So, let us take the example of mountain areas. What does it mean for the European Union? The Council Directive 75/268 (European Union, 1975a) defines them in this way:

"Mountain areas shall be made up of local government districts or parts thereof characterized by a considerable limitation of the possibilities for using the land and an appreciable increase in the cost of working it, due: - either to the existence, because of the altitude, of very difficult climatic conditions the effect of which is substantially to shorten the growing season, - or, at a lower altitude, to the presence, over the greater part of the district in question, of slopes too steep for the use of machinery or requiring the use of very expensive special equipment, - or to the combination of these two factors, where the handicap resulting from each taken separately is less acute, provided that this combination gives rise to a handicap equivalent to that caused by the situation referred to in the first two indents".

Since the definition is not unambiguous, a number of other legislative acts have been passed to clarify what is meant for each country concerned (European Union, 1975a, 1975b, 1975c, 1975d, 1976a, 1976b, 1981, 1986a, 1986b, 1986c, 1992, 1995a, 1995b, 1995c). This results in a case-based meaning of the concept.

The belief uncertainty could result from evaluating the drought mitigation measures in terms of money spent or in terms of ground water level after an unusually rainy period.

The methodology uncertainty can be formulated by questions such as "Should the people give their opinion on the decision and on its implementation?", "Should it only be a population survey?", "Or an experts evaluation?".

While to get to know the in-and-outs of a decision is a very complicated task, to have a complete picture of the uncertainties associated with a certain decision process is an even more titanic work.

12.6 CONCLUSIONS

Uncertainties in spatial decisions have been identified by screening public decisions through the decisional fountain model. Out of this analysis, it appears that the GIS community has only been busy with some types of uncertainties which politicians and decision-makers would like to sweep under the rug (Goodchild, 1998), having sufficient uncertainties to worry about. The integration of the decision process in time increases the complexity of decision making to the point of challenging the certainty of policy-makers. Going from a local level to an international level, the diversity of space and of demands increases complexity. A way out for policy-makers is to approximate elements comprised in the decision process. Another way is to focus only on a few aspects. Both increase uncertainties and facilitate decision taking. Finally, this contribution can be seen as an essay to fill the gap between the GIS community and the decision-makers underlined by Heywood *et al.* (1994). It strengthens Eagles's (1995) statements that "political scientists ought to be aware of the efforts of spatial analysts ".

REFERENCES

Beck, U., 1992, *Risk Society: Towards a New Modernity*, (London: Sage Publications).

Berube, M.S., Jost, D.A., Severynse, M., *et al,*. 1995, *Webster's II New College Dictionary*, (Boston: Houghton Mifflin Company).

Bouchon-Meunier, B. and Nguyen, H.T., 1996, *Les Incertitudes dans les Systèmes Intelligents*, Que sais-je?, 3110, (Paris: Presses Universitaires de France).

Bradford, R.B., 2000, Drought events in Europe. In *Drought and Drought Mitigation in Europe*, edited by Vogt, J.V. and Somma, F. (Dordrecht: Kluwer Academic Publishers), pp. 7–20.

Chevallier, J. and Loschak, D., 1974, *Introduction à la Science Administrative*, (Paris: Dalloz).

Cornélis, B., 1998, Managing decision in spatial decision support systems. In *Proceedings of the 1st AGILE Conference*, Enschede, http://www.uniroma1.it/DICEA/Agile/Paper6.html.

Cornélis, B., 1999, Errors and uncertainties in spatial decision support systems, In *Séminaire de Doctorat*, (Liège: Université de Liège), pp. 1–14.

Cornélis, B., 2000, Définition du concept de décision spatiale. In *Forum des Doctorats – Notes de Recherche n°12*, edited by Cornélis, B. and Halleux, J.M. (Liège: Société Géographique de Liège), np.

Cornélis, B. and Viau, A.A., 2000, Decision processes with regard to drought monitoring and mitigation. In *Drought and Drought Mitigation in Europe*, edited by Vogt, J.V. and Somma, F. (Dordrecht: Kluwer Academic Publishers), pp. 279–290.

Densham, P.J., 1991, Spatial decision support systems. In *Geographical Information Systems – Volume 1: Principles*, edited by Maguire, D.J., Goodchild, M.F. and Rhind D.W. (Harlow: Longman Scientific & Technical), pp. 403–412.

Eagles, M., 1995, *Spatial and Contextual Models in Political Research*, (London: Taylor & Francis).

European Commission, 1999, *ESDP – European Spatial Development Perspective*. (Luxembourg: Office for Official Publications of the European Communities).

European Commission, 2001, *Unity, Solidarity, Diversity for Europe, its People and its Territory – Second Report on Economic and Social Cohesion*. (Luxembourg: Office for Official Publications of the European Communities).

European Union, 1975a, Council Directive 75/268/EEC of 28 April 1975 on mountain and hill farming and farming in certain less-favoured areas. *Official Journal*, **L128**, pp. 0001–0007.

European Union, 1975b, Council Directive 75/271/EEC of 28 April 1975 concerning the Community list of less-favoured farming areas within the meaning of Directive No 75/268/EEC (France). *Official Journal*, **L 128**, pp. 0033–0067.

European Union, 1975c, Council Directive 75/270/EEC of 28 April 1975 concerning the Community list of less-favoured farming areas within the meaning of Directive No 75/268/EEC (Germany). *Official Journal*, **L 128**, p. 0010.

European Union, 1975d, Council Directive 75/273/EEC of 28 April 1975 concerning the Community list of less-favoured farming areas within the meaning of Directive No 75/268/EEC (Italy). *Official Journal*, **L 128**, pp. 0072–0225.

European Union, 1976a, Council Directive 76/401/EEC of 6 April 1976 amending Directive 75/271/EEC concerning the Community list of less-favoured farming areas within the meaning of Directive 75/268/EEC (France). *Official Journal*, **L 108**, pp. 0022–0038.

European Union, 1976b, Council Directive 76/631/EEC of 20 July 1976 concerning the Community list of less-favoured farming areas within the meaning of Directive 75/268/EEC (France - overseas departments). *Official Journal*, **L 223**, pp. 0007–0010.

European Union, 1981, Council Directive 81/645/EEC of 20 July 1981 concerning the Community list of less-favoured farming areas within the meaning of Directive 75/268/EEC (Greece). *Official Journal*, **L 238**, pp. 0001–0054.

European Union, 1986a, Council Directive 86/465/EEC of 14 July 1986 concerning the Community list of less-favoured farming areas within the meaning of Directive 75/268/EEC (Germany). *Official Journal*, **L 273**, pp. 0001–0103.

European Union, 1986b, Council Directive 86/467/EEC of 14 July 1986 concerning the Community list of less-favoured farming areas within the meaning of Directive 75/268/EEC (Portugal). *Official Journal*, **L 273**, pp. 0173–0180.

European Union, 1986c, Council Directive 86/466/EEC of 14 July 1986 concerning the Community list of less-favoured farming areas within the meaning of Directive 75/268/EEC (Spain). *Official Journal*, **L 273**, pp. 0104–0172.

European Union, 1992, Council Directive 92/92/EEC of 9 November 1992 amending Directive 86/465/EEC concerning the Community list of less-favoured farming areas within the meaning of Directive 75/268/EEC (Federal Republic of Germany) 'New Länder'. *Official Journal*, **L 338**, pp. 0001–0039.

European Union, 1995a, Council Directive 95/212/EC of 29 May 1995 concerning the Community list of less-favoured farming areas within the meaning of Directive 75/268/EEC (Austria). *Official Journal*, **L 137**, pp. 0001–0041.

European Union, 1995b, Council Directive 95/412/EC of 25 September 1995 concerning the list of less-favoured farming areas within the meaning of Directive 75/268/EEC (Finland). *Official Journal*, **L 241**, pp. 0023–0039.

European Union, 1995c, Council Directive 95/498/EC of 23 November 1995 concerning the Community list of less-favoured farming areas within the meaning of Directive 75/268/EEC (Sweden). *Official Journal*, **L 287**, pp. 0033–0052.

Fisher, P., 1998, Improved modeling of elevation error with geostatistics, *GeoInformatica*, **2** (3), pp. 215–233.

Fisher, P. and Wood, J., 1998, What is a mountain? Or the Englishman who went up a boolean geographical concept but realised it was fuzzy, *Geography*, **83** (3), pp. 247–256.

Garrido, A. and Gómez–Ramos, A., 2000, Socio-economic aspects of droughts. In *Drought and Drought Mitigation in Europe*, edited by Vogt, J.V. and Somma, F. (Dordrecht: Kluwer Academic Publishers), pp. 197–207.

Goodchild, M.F., 1998, Uncertainty: The Achilles Heel of GIS? In *Geo Info Systems*, November, pp.50–52.

Grawitz, M. and Leca, J., 1985, *Traité de Science Politique - Tome 4 les Politiques Publiques*, (Paris: Presses Universitaires de France).

Griffin, C., 1995, Data quality issues affecting GIS use for environmental problem-solving. In *Seminar Proceedings of the National Conference on*

Environmental Problem-solving with Geographic Information Systems, Cincinnati, (Washington D.C.: US Environmental protection agency), pp. 15–30.

Hajer, M., 1995, *The Politics of Environmental Discourse*, (Oxford: Oxford University Press).

Hansson, S. O., 1996, Decision Making Under Great Uncertainty. *Philosophy of the Social Sciences*, **26** (3), pp. 369–386.

Hearnshaw, H.M. and Unwin, D.J., 1994, *Visualization in Geographical Information Systems*, (Chichester: John Wiley & Sons).

Heuvelink, G.B.M., 1998, *Error Propagation in Environmental Modelling with GIS*, (London: Taylor & Francis).

Heywood, I., Oliver, J. and Tomlinson, S., 1994, Building an exploratory multi-criteria modelling environment for spatial decision support. In *Proceedings of the 5th European Conference and Exhibition on Geographical Information Systems – Volume 1*, Paris, edited by Harts, J.J., Ottens, H.F.L. and Scholten, H.J., (Utrecht: EGIS foundation), pp. 632–641.

Kerrey, R., 1987, Foreword. In *Planning for Drought – Toward a Reduction of Societal Vulnerability*, edited by Wilhite, D.A., Easterling, W.E. and Wood, D.A. (Boulder: Westview Press), pp. xi–xii.

Laaribi, A., 2000, *SIG et Analyse Multicritère*, (Paris: Hermès).

Leleux, C., 1997, *La Démocratie Moderne*, (Paris: Les Éditions du Cerf).

Leung, Y., 1997, *Intelligent Spatial Decision Support Systems*, (Berlin: Springer).

Leung, Y. and Yan, J., 1997, Point-in-polygon analysis under certainty and uncertainty. *GeoInformatica*, **1**, pp. 93–114.

Leung, Y. and Yan, J., 1998, A locational error model for spatial features, *International Journal of Geographical Information Science*, **12** (6), pp. 607–620.

Malczewski, J., 1997, Spatial decision support systems. In *NCGIA Core Curriculum in GIScience*, http://www.ncgia.ucsb.edu/giscc/units/u127/u127.html, posted October 6, 1998.

Meny, Y. and Thoenig, J.-C., 1989, *Politiques Publiques*, (Paris: Presses Universitaires de France).

Pantazis, D. N. and Cornélis, B., 1997, Designing and implementing a GIS in an international context. *Transactions in GIS*, **1** (4), pp. 301–320.

Pantazis, D. N. and Donnay, J.-P., 1996, *La Conception de SIG : Méthode et Formalisme*, (Paris: Hermès).

Renwick, A., 1993, *Basic Political Concepts*, (London: Thornes).

Riebsame, W.E., Changnon Jr, S.A. and Karl, T.R., 1991, *Drought and Natural Resources Management in the United States – Impacts and Implications of the 1987–89 Drought*, (Boulder: Westview Press).

Rodee, C.C., Anderson, T.J. and Christol, C.Q., 1957, *Introduction to Political Science*, (New York: McGraw Hill).

Roqueplo, P., 1997, *Entre Savoir et Décision, l'Expertise Scientifique*, (Paris: Éditions INRA).

Shi, W., Goodchild, M.F. and Fisher, P.F., 1999, *Proceedings of the International Symposium on Spatial Data Quality '99*, (Hong Kong: The Hong Kong Polytechnic University).

Turban, E. and Aronson, J.E., 1998, *Decision Support Systems and Intelligent Systems*, (London: Prentice Hall International).

UCGIS, 1996, Research priorities for geographic information science. *Cartography and Geographic Information Systems*, **23** (3), pp. 115–127.

Vogt, J.V. and Somma, F., 2000a, *Drought and Drought Mitigation in Europe*, (Dordrecht: Kluwer Academic Publishers).

Vogt, J.V. and Somma, F., 2000b, Introduction. In *Drought and Drought Mitigation in Europe*, edited by Vogt, J.V. and Somma, F. (Dordrecht: Kluwer Academic Publishers), pp. 3–5.

Weber, M., 1971, *Économie et Société*, (Paris: Éditions Plon).

Wilhite, D.A., 1990, *Drought Planning: a Process for State Government*. (Lincoln: IDIC Technical Report Series 90-1).

Wilhite, D.A., 1991, Drought planning: a process for state government. *Water Resources Bulletin*, **27**, pp. 29–38.

Wilhite, D.A., 1993, The enigma of drought. In *Drought Assessment, Management, and Planning: Theory and Case Studies*, edited by Wilhite, D.A. (Boston: Kluwer Academic Publishers), pp. 3–15.

Wilhite, D.A., 1996, A methodology for drought preparedness. *Natural Hazards*, **13**, pp. 229–252.

Zimmer, D.E. and García Rodrígo, C., 1995, Dürre hausgemacht. *Zeit–Magazin*, **21** July, pp. 6–12.

Zwetkoff, C., 1995, *Science Administrative*, (Liège: Université de Liège).

CHAPTER THIRTEEN

The Registration of Quality in a GIS

Henri J.G.L. Aalders

ABSTRACT

Quality information of geographical information has been a topic of research in many organisations. Reported documents are available from the ICA, the CEN, ISO, etc. Quality of geographical information is an important element for potential users to decide whether the offered data set does fit the intended use.

Quality of geographical information refers to sets of data that may consist of one single occurrence of an attribute, one occurrence of an entity, one attribute type or several types of attributes, types of entities, selection of occurrences, a whole data set, a combination of several data sets, etc.

According the mentioned reports, many different elements can be used to describe the quality of a geographical data set. And so many possibilities for quality description exist referring to many different types of geographical data sets resulting in large amounts of data that will become inaccessible for users.

In this chapter a proposal for the design of quality registration for geographical data using user-defined data sets in a relational database is given. This may be done through occurrences of quality entities especially describing the quality of a data set, having each data occurrence referring to a specific quality entity or the quality definition by entity type.

13.1 INTRODUCTION

Since the application of digital techniques in geoscience, geographical source data are often merged using various sources in order to meet the requirements of user's applications. The end result is that data collectors, who originally collect data about a feature, have lost the degree of control over the data sets which traditional data collectors for analogue maps enjoyed. As a result, collectors of data are lacking the description of the quality of the spatial data by the producer, so that it appears to meet the demands of the potential customers.

From another perspective, the data user must select between competing data sets, while the criteria users may apply to make this decision are unknown. Such decisions depend on the quality, availability, ease of accessibility and the cost of obtaining the data.

Traditionally cartographers, who have a reputation for excellent maps, have mastered the use of these criteria, often unconsciously. For today's graphically sensitive world, the user must still impart to visualisation indications of the data quality. In the digital world, two aspects become important for the conscientious user of digital spatial data, i.e.: what spatial data quality information is required and which methods are available to impart to a visualisation the quality information of a digital set of feature data.

This chapter discusses quality elements that can be used in practical applications. Both the quality of geometric data and the quality of non-geometric attributes, following the rules that are applied in information technology, are considered.

Designing a system for quality registration for geographical data using a user-defined data set in a relational database system allows an indication of quality to be given to each entity or attribute instance in the database or to combinations of entities and/or attributes by defining a set's membership for instances or entities or attributes. In this way, the quality information becomes an integral part of the data set (Tobler, 1989).

Despite giving the quality information on instances, as indicated above, the metadata database containing quality information about the data set also has to be developed for judging the data set in the situation of data exploration. One has to distinguish between these two different aspects using quality. Different quality information has to be stored with the data set in order to enable the two different functions the quality information fulfils in the exploration and use of the data set.

13.2 HISTORY OF QUALITY IN GIS

In the analogue era, concern for the quality of spatial data and concern for the accuracy of the planimetric position of an object relative to its position on the earth were used as synonyms. And certain institutions, over a long period of time, earned a reputation and prestige for producing quality products based on their spatial characteristics, including their visualisation in printed map form. For instance, on topographic maps this concern concentrated on elevation of the land surface above a precise datum. Only rarely was a similar degree of attention paid to other attributes of the data.

In contrast, today's technology allows every user, surfing the Internet, to collect data sets from a variety of sources, and download them to the terminal. The visualisation of each layer of these data sets can be made to look precise and accurate regardless of the original quality of spatial data (Naisbitt, 1994).

The user may often find that data from different sources will not perfectly agree, i.e. the visualisation of the points, lines and areas representing the same earth object do not match exactly, or semantic attributes of the same object type instance may contradict each other. In the absence of quality information, or any a priori knowledge, the user has no logical way to resolve these differences.

This points to a need for spatial data quality information. To provide quality information about spatial data, three areas of general agreement are necessary. These are (1) the definition of the elements of spatial data quality; (2) the derivation of easily understood measures or parameters of spatial data quality that may accompany a data set; and (3) the methods for exchanging data quality information, understanding it, and representing or rendering the specified data quality in a visualisation.

Quality information as part of a data set has been a topic for research in GIS for a long period. It appeared at first during standardisation processes developing the SDTS standard, where it was felt necessary to provide users with quality data to be transferred from the provider to the user, both for data exploration and data use. The standard introduced the following parameters: *source, resolution, accuracy (positional and thematic) completeness and logical consistency* (see Table 13.1).

Table 13.1 Quality indicators and quality parameters in different standards.

SDTS *(1992)*	ICA *(1996)*	CEN/TC287 *(1997)*	ISO 2001
Source	Source	Source	Overview (source,
		(Potential) usage	Purpose, usage)
Resolution	Resolution		Resolution
Metric accuracy	Metric accuracy	Metric accuracy	Accuracy
Thematic accuracy	Thematic accuracy	Thematic accuracy	Thematic accuracy
Completeness	Completeness	Completeness	Completeness
Logical consistency	Logical consistency	Logical consistency	Logical consistency
	Semantic accuracy	Meta quality	
	Temporal accuracy	Temporal accuracy	Temporal accuracy
		Homogeneity	
			Testing and conformance

Research from the ICA working group resulted in the publication of a book about accuracy for spatial objects. In this research, two parameters were added compared to the SDTS standard i.e.: *temporal and semantic accuracy* (Guptill and Morisson, 1995). The temporal accuracy referred to the accuracy related to temporal aspects, while the semantic accuracy referred to the quality of the definition of the entities and attributes in the metadatabase.

The developments for standards by the CEN resulted in a list of quality parameters that could be used for transferring data quality information between users. The CEN working group on data model design added two more parameters as listed above. One was *usage*, describing the use of data sets prior to the actual intended use, in order to allow the new user to understand the previous applications of the data set to judge whether his application may become successful too. The other quality parameter introduced by the CEN working group was *homogeneity* to indicate how much the quality parameters are valid for the whole data set or whether they are valid for only parts of the data set.

The ISO working group on quality principles added another parameter: *testing and conformance* in order to describe also the methods used that lead to the specific quality parameter. Though this was already indicated by the CEN working group, it was stated as being a process description and not a data quality parameter. According to the Vienna Agreement between ISO and CEN and the arrangements between the two Technical Committees of ISO and CEN, the descriptions relating to processes were left to the ISO working group.

13.3 DATA QUALITY DEFINITION

In general a statement of data quality represents for the user the fitness of that data set for a potential application (*'fitness for use'*). This is a very general, but also very often used, interpretation of data quality. In this presentation, this definition is used for quality assessment of spatial data. A more precise statement, adhering to the ISO 8402 Standard, describes quality as being the totality of characteristics of a product or a service that fulfils both the stated and implied needs of a user (ISO 1993).

One can distinguish between an internal and an external quality. Internal quality measures adherence to data collection specifications (compare semantic quality (Guptill and Morrison, 1995)), while the external quality indicates how well these specifications fulfil the user needs (i.e. fitness for use).

13.4 QUALITY IN GIS USING IT METHODS

In this chapter GIS databases are seen as databases developed for general IT, using their theory and methods. The only difference between the general IT database systems and the GIS database systems are the presence of spatial data in a GIS database, requiring specific developments related to the spatial aspects in a GIS data set, that do not appear in other IT applications. This also accounts for the quality parameters which appear in GIS data sets.

Quality descriptions in a GIS database may be attached to entities in the form of attributes. It is assumed that quality-attributes describe the quality of one or more instances of entities and/or attributes that are contained in a data set.

Quality is stored in metadata sets describing the quality parameters for each data set since data on quality is metadata. For this purpose, one can apply set theory and quality attributes following the concept of metadata sets. Quality attributes are valid for all occurrences of entities or attributes in a data set that are reported in a specific metadata set, i.e., quality metadata sets describe the quality of homogeneous entities or entity-sets. In practice, data sets that have a quality reference are usually not fully homogeneous and an indicator is required to mark the homogeneity of a data set.

13.5 UNIVERSE OF DISCOURSE

To define a data set describing real world objects by position, theme, and time - i.e. to make these objects intelligible and representable in a database - a universe of discourse (UoD) is defined to model the potentially infinite characteristics of the objects in the real world.

The use of the UoD involves the selection and abstraction of objects from the real world into an ideal concept. To model the quality of a data set, it is also necessary to precisely define the process that allows one to derive the data set from the real world. So the process of building a database containing data about geographical objects is decomposed into two steps:

- *conceptualisation*, which contains both the product specifications of what should be considered in the real world and the abstraction of the selected objects; and
- *mensuration*, which specifies the measuring methods and the measurement requirements for the capturing and storing of the data itself.

To fill the conceptual gaps between the real world and the UoD, and between the UoD and the data set, quality indicators are defined for the specifications and abstraction of real world objects. The nominal ground forms the basis for specification, against which the quantitative contents of data sets can be tested.

These specifications are usually defined by either the users or producers of geographical data sets and should depend directly on their intended use. The accompanying quality model consists of a set of quality indicators that will allow users or producers to define the relation between the specifications and the geographical data sets. So, the nominal ground can be considered as an abstract description (specifications) of the user-intended database (Aalders, 1998).

Although one can speak of the quality of the modelling process (see also semantic accuracy (Guptill and Morrison, 1995)), one should realise that the intended use is defined in the data set contents specification. Since quality is defined as a combination of characteristics for both the expressed and the intended use, it is obvious that any quality model for an existing data set should refer to the quality of both the data modelling and data capturing process(es).

13.6 META-QUALITY

The existence of the quality of each quality statement in a data set should also be noted. In other words, to each quality indicator a meta-quality statement should be attached, expressing the reliability of the quality indicator (i.e. the quality of quality parameter, also described by meta quality). This information comprises:

- a measure of the confidence of the quality information, indicating the level of confidence in the data set;
- a measure of the reliability of the quality information, reflecting how well the information represents the whole data set;
- a description of the methodology used to derive the quality information, to indicate how this result was obtained;
- a measure of the abstraction to account for the differences between reality and the UoD (i.e. *the semantic quality*). This measure is not relevant for factual statements concerning quality that are independent of the content of the geographical data set.

13.7 STRUCTURING OF QUALITY ELEMENTS

In 1982 in the United States of America, a National Committee on Digital Cartographic Data Standards (NCDCDS) was established under the auspices of the American Congress of Surveying and Mapping (ASCM). The reports of this committee (Moellering 1985, 1987) contain sections on data quality which specify five components.

The ICA Commission on Spatial Data Quality, instituted in 1991, added two additional components (Guptill and Morrison, 1995). In the CEN/TC 287 Quality model (CEN, 1995) even more quality components were considered and the ISO/TC 211 added testing and conformance (see Table 13.1). Below all elements are discussed.

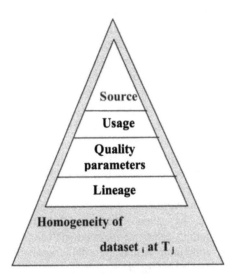

Figure 13.1 The pyramid of quality elements.

Quality indicators in a quality model are of four main types (Figure 13.1).

- **Source of the data set.** This indicator lists the name and the organisation responsible for the data set, as well as the purpose, date, method of data capture, source documents, if appropriate, and creator of the data set's original production, including their address.

   Source information can be considered as part of the metadata of the data set as can all quality information. Source information can also be seen as part of the "overall" data set's lineage. In this context, source information is considered as the minimum quality information that should always be available for a data set or a data (sub)set.

   The attributes of the source information may tell the present user in general terms about the possible validity of the data. This impression of the data may be created from a consideration of the time interval that has elapsed since the original creation of the data set, an awareness of the reliability by the producer, and who should be considered to be ultimately liable for errors in the data set. The original intended purpose for the data set should also be included.

- **Usage.** Any previous use of a data set by other users for various applications may be a good indication of the fitness of the data for the present use and the general reliability of the data set in different circumstances. For each usage a separate statement should be given indicating the organisation that has used the data set, the type of usage and its perceived fitness and any possible constraints or limitations that were imposed or discovered during that use. Usage of a data set is considered to be important from the point of view of new users. If a data set has been used for applications similar to the one envisaged by the current user, it clearly adds to the potential user's confidence. Also previous combinations of the actual data set with other data sets can be an indication of the quality of the data set. However, information about usage, as with all descriptive quality information, is only useful if it is honest, and errors, constraints, and limitations are also recorded.

  Also the provider's vision on potential application(s) of the data set may be given;

- **Quality parameters.** Quantitative descriptions of measurable aspects of the performance of occurrences in a data set for an intended use can be considered as a quality parameter. These parameters (in Guptill and Morrison (1995) referred to as elements) are described in the next section.

- **Lineage.** The description of the processing history which each occurrence of an entity has undergone since its original creation. For each process, a statement has to be given describing the processing the entity has undergone (including details of the methods used and possibly references to documents containing actual algorithms applied), who performed the process, when it was performed, and why.

  Processes that have been performed on each occurrence of an entity in the data set may refer to the type of transformations, selections, generalisations, conversions, updates, consistency checks, validations, etc. For this purpose meta entities may be defined and related to occurrences of entities in the data set.

- **Homogeneity.** Describes how well the quality information is applicable to all occurrences of entities in the data set.

Quality indicators may refer to a data set that consists of an entity, a set of entities, an occurrence of an entity or a set of such occurrences, or to occurrences of an attribute or a set of occurrences of attributes. One may notice that the amount of data describing each of the quality indicators increases from source to usage and on to quality parameters and detailed lineage. This is graphically represented by the position of the quality indicator in the pyramid of Figure 13.1.

Occurrences of specific entities may be assumed to follow the quality statements of the other entities in the same meta class, thereby enabling the user to obtain their quality properties. This involves a complex logical structure within the data set.

Any quality indicator must be reported with reliability, which is an indication of the quality of the quality indicator. The reliability can either be qualitative, such as how the quality indicator was determined, or quantitative, such as a confidence level and interval of the quality indicator.

13.8 QUALITY PARAMETER SET

A quality parameter is a specific instance of a particular quality parameter type and represents its performance characteristic. Each quality parameter may have one or more 'metrics' which is a quantitative statement of the quality parameter. Metrics may be statistical (expressed in real numbers, with or without units) or Boolean (indicating conformance) and in some instances may be descriptions. All metrics should be dated to indicate their temporal validity. The descriptions of the algorithms, used to determine the metrics, could be described as part of a meta quality methodology component.

The quality parameter set may consist of different quality parameter metrics for each of the following elements of spatial data quality: accuracy, completeness and logical consistency.

Accuracy is the probability of correctly assigning a value. It describes the stochastic errors of observations and is a measure to indicate the probability that an assigned value is accepted. There are different types of accuracy to be distinguished in a geographical data set.

Thematic or attribute accuracy. This refers to the accuracy of continuous (also called scalar or quantitative) and discrete (also called qualitative or nominal) values associated with a feature or relationship in the data set. Accuracy values are given in the same unit and reference system as in the measured values.

Metrics for scalar attributes may involve the same procedures as those for positional accuracy. For discrete attributes possible metrics include the probability of a correct classification or misclassification, and the probability of correctly assigning alternative values. The accuracy metrics are different for the quantitative and the qualitative thematic attributes. Also, one has to distinguish between *single-valued quantitative thematic attributes* and *multiple-valued quantitative thematic attributes*. The single-valued quantitative attribute is a measure to indicate how far the assigned value is from the best estimation value (the mean of all observations). The accuracy metrics for this may be expressed by root mean square error (RMSE) or standard deviation (STD), bias, systematic error, range (minimum and maximum error), histogram of deviation from the mean value or by confidence interval and confidence level.

The multiple-valued quantitative thematic attributes can be expressed by a list of accuracy values for single valued quantitative thematic attributes, the correlation matrix or the eigen value of the correlation matrix, the correlation function or by the range (minimum and maximum error).

The accuracy of a *single-valued qualitative thematic attribute* is the probability that a correct value has been assigned to this attribute, including the probability of correctly assigning alternative values. In general, the metrics for a single valued qualitative thematic attribute is a vector $\{i \text{ or } \omega_i, P(X=\omega_i)\}$ where ω_i are the possible values and X is the assigned value. This corresponds to the density distribution of a quantitative attribute. Reductions of the description is possible by only the maximum probability ω_i , $P(X=\omega_i)$, the m (m is predefined) largest probabilities $\{\omega_1, P(X=\omega_i) ; \ldots ; \omega_k, P(X=\omega_k)\}$, all values where assignment has the probability larger than a threshold p, $\{\omega_i , \omega_j , \omega_k\}$ with $P(X=\omega_i) \geq p$; $P(X=\omega_j) \geq p$; $P(X=\omega_k) \geq p$ (this corresponds to the confidence interval for quantitative valued thematic attributes) or by m values $\{\omega_i, P(X=\omega_i) ; \omega_j, P(X=\omega_j) ; \omega_k, P(X=\omega_k) ; \ldots \}$ where $\sum P(X=\omega_i) \geq p$ (for i = 1, . . . , m) and where p is a predefined confidence or significance level.

The accuracy of a *multiple- valued qualitative thematic attribute* that might be correlated is usually indicated by a correlation matrix or its eigen value, lists of maximum probabilities: ω_i , $P(X=\omega_i)$ or error ellipse.

Positional accuracy. According to the NCDCS report, the quality report section on positional accuracy must include the relationship of the data coordinates to latitude and longitude (or to a standard reference datum). It includes measures of the horizontal and vertical accuracy of the features in the data set, and it must also consider the effects on the quality of all transformations performed on the data as well as the result of any positional accuracy testing performed on the data (Moellering, 1985,1987).

Generalisation error may lead to a wrong structure of defining positional accuracy: this type of error refers to the ability for abstraction. One can say that accuracy is made of four components: these are the measurement accuracy; ability of abstraction (this is a component that accounts for the amount of generalisation in the terrain during measurement by defining the crisp boundary to be surveyed); fitting the result of the measurements in the existing reference system; and internal accuracy of the reference system. Usually the positional accuracy corresponds to the first three components.

For positional accuracy one can speak of *absolute accuracy,* i.e. the accuracy measured (in both the horizontal (x, y) and vertical (z) dimensions) against the spatial reference system; *relative accuracy,* i.e. the accuracy of the positions of occurrences of entities relative to each other (in both the horizontal (x, y) and vertical (z) dimensions), and the results of testing procedures.

Positional accuracy refers to geometric primitives of types point, line, surface, grid-point, and pixel. The accuracy metrics for points can be described as the metrics for a couple or triple of reals (e.g. X,Y (Z)). Therefore its accuracy metric is the same as the accuracy of a quantitative attribute (see thematic attribute accuracy). The accuracy metrics for line are given by an ordered set of points (including a starting and end point) and by the connections between these points expressed by a function to define the interpolation between the points (shape). For the accuracy of the points see the positional accuracy of a point; the accuracy of the length of the line is defined as a quantitative attribute. The accuracy of the connections between points in a line can be described by the accuracy of the distance between two lines, such as maximum lateral distance or Hausdorff distance. On a data set the parameter is defined by standard deviation, range, quadratic means, or bias. The accuracy of the connections between points in a line can be also described by the maximum curvature radius difference between two lines (on a data set the parameter is defined by standard deviation, range, covariance function) or by required point density regarding the tolerance and interpolation type.

A surface may be defined by an ordered set of its boundary lines, so the accuracy of the surface is defined by the accuracy of its boundaries (see positional accuracy for lines). The calculated area value of a surface is a quantitative attribute (for its accuracy see quantitative thematic accuracy).

Grid-points and pixels are defined by a couple of integers representing the row and column numbers in a grid. (For their accuracy refer to thematic qualitative attribute accuracy).

The accuracy of *description* being errors in logic e.g.: two partially overlapping objects are defined to be disjoint or errors in semantics e.g.: incorrect spelling or incorrect meaning.

Temporal accuracy. This describes the correctness of time and updating of a data (sub)set (currentness) by such metrics as the moment of last update (in case of creation, modification, deletion or unchanged use), rate of change of entities per unit of time, trigger value (indicating the number of changes before a new version of the data set is issued), temporal lapse (giving the average time period between the change in the real world and the updating of the database) or temporal validity (indicating data to be out of date, valid or not yet valid).

In Moellering (1985, 1987) temporal accuracy is not treated separately, but as an integral part of each of the five quality aspects mentioned there.

The temporal element of spatial data quality during the first part of the twenty-first century requires and will probably receive the most attention. It is postulated that by sometime early in the twenty-first century positions and attributes will be routinely recorded in digital files to a sufficient level of accuracy to satisfy most users needs.

An important consideration being recognised is that the metrics and quality parameters of the temporal element may affect, to a very large extent, the overall size of most data files. The position and description of one feature and its attributes contributes a fixed amount of information in a data file. It is conceivable that, as that feature evolves and the historical character of the feature is archived, time stamps of changes, processes applied and uses will contribute significantly to an increase in the size of the accompanying quality information file.

Completeness. This indicates the estimation of errors of omission and commission which can be expressed by percentages of missing or over-complete data in the data set relative to the specification; *currency* estimates of change of occurrences of entities between the last update of the data set and the actual time of use and density of occurrences relative to one another.

Completeness is given by two ratios. These are *missing data* and *over-completeness*. Missing data (also called omission), express the ratio of the number of occurrences that are not in the data set according to the definition by the nominal ground divided by the number of occurrences that should be in the data set according to the nominal ground. Over-completeness (also called commission) is expressed by the ratio of the number of occurrences that should not be in the data set according to the definition of the nominal ground divided by the number of occurrences in the data set.

Completeness can only be evaluated on a data set with occurrences of object types, attribute types or relationship types. Missing data on a set of attribute instances correspond to the absence of some values, i.e. for some attributes instances the values are set as a NULL value.

Over-completeness on attribute values may be defined when the domain value is set.

The metrics that can be used for the definition of completeness may be (including the definition of the data set for which the completeness is defined) a percentage of missing data, percentage of over-completeness in the number of instances or in length or area, the number of errors, being the sum of missing data and over-completeness or type of evaluation including the date of evaluation i.e.: guessed, propagated (e.g. from different sources) or checked. And if checked, the type of checking should be indicated by checking on the full set or sampling; and the reference to a testing procedure such as Acceptable Quality (AQ) or Nominal Limiting Quality (NLQ);

Logical consistency. This describes how well constraints are met by a data set defined by the relationships encoded in the data structure (Veregin, 1989). For example, it describes the number of features, relationships, and attributes that have been correctly encoded in accordance with the integrity constraints of the feature's data specification. As such, tests should be both topological and visual.

Consistency includes static and dynamic consistency. Static consistency describes the validation of data and constraints of data relations, while dynamic consistency describes the validation of the processes. In the context of data transfer standards for geo-information, only static consistency will be considered.

Validation is a test of a constraint defined in the object types specifications in the nominal ground. It includes domain validation (on object types, attribute values, geometry and their sets) and relational validation (on relations) as described in the conceptual schema.

Among these are attribute database consistency tests, metric and incidence tests, topological and order-related tests, as well as scene consistency checks. A metric for topological consistency in one dimension could be reported by indicating the percentages of junctions that are not formed when they should be, or in two dimensions by indicating the percentage of incorrectly formed polygons.

The results of redundancy checking the correctly formed relations, file structure adherence testing for file structure conformance, and validity checks on all attribute values are examples of metrics to describe logical consistency. These metrics may indicate whether or not such a check has successfully been performed.

The metrics for constraint may be the result of the validation process (e.g. the number of violations of a topological rule), the type of validation process e.g. by check / assessment / calculation or the result and the date of the validation process.

Semantic accuracy. Salgé (Guptill and Morrison, 1995) defines semantic accuracy as "the quality with which geographical objects are described in accordance with the selected model. Related to the meanings of 'things' of the universe of discourse (reality), semantic accuracy refers to the pertinence of the meaning of the geographical object rather than to the geometrical representation". He believes that "semantic accuracy is an element of the evaluation of 'fitness for use' that users have to perform before using spatial data".

As an illustration, one aspect of semantic accuracy could be termed *textual fidelity*. One measure of textual fidelity could indicate the accuracy of spelling, perhaps by the percentage of wrong spellings. Another measure could be of the use of exonyms or alternative spellings. A third metric may indicate the consistency of abbreviations.

13.9 STORING QUALITY IN GIS

In order to obtain a GIS including quality reports in a normalised manner as required by IT in a relational database the data set is introduced. Members of a data set may be entity, attribute or relationship types in any mutual combination or their occurrences. The quality reports appear in occurrences of a meta quality entity containing all quality parameters defined before. For each occurrence in the set the appropriate quality values are mentioned. The occurrences in the data set use secondary keys to link to the primary keys in the quality metadata set. In this way, all combinations of entity, attribute or relationship types and their occurrences can be member of any occurrence in a quality metadata set.

13.10 CONCLUSIONS

The above model for data quality contains a set of parameter descriptions that can be used. This description contains the parameters as indicated in different research projects, such as the development of SDTS (SDTS, 1992), the ICA publication on Elements for Spatial Data Quality, as well as the developments of standards for transferring data quality information between users by CEN and ISO (CEN/TC287, 1995; ISO, 1993). In addition, an extensive list of description for metrics, which can be used for quality description, is given. This list may not be complete, as previous researches have been indicating.

Furthermore, it will be difficult to supply all parameters by the original data collector and it will be even more difficult and expensive to maintain all possible quality parameters in a data set. Each provider and user has to decide which parameters should be determined and stored in the database, to create a usable, workable and maintainable database from a technical and cost-effectiveness point of view.

Further application research is necessary to provide a system for data sets that contains a minimal yet sufficient set of quality parameters in practical applications.

The basic assumption in the above-described quality model is to design an object-oriented database with homogeneous data (sub)sets, and then to consider these as new entities. This makes it possible to describe the different aspects of quality in terms of quality elements and their metrics and apply this quality information to all members of a data subset occurrence. However, this additional data quality information will make the database structure extremely complex and will extend the database size to a multiple of its original size. As yet we have no experience in implementing such a model, completely and correctly, from either a theoretical or a technological point of view. Experience with attempts to implement such a model is needed, and should help users decide whether this model is feasible, and/or useful.

REFERENCES

Aalders, H.J.G.L., 1998, *The Universe of Discourse in GIS*, (Lisbon: GISPlaNET, CD-ROM, 9-11 September 1998).

CEN/TC 287, 1994, WG 2, Data Description: Quality, *Working Paper N15*, August 1994.

CEN/TC 287, 1995, PT05, Draft Quality Model for Geographical Information, *Working Paper D3*, January 1995.

CEN/TC 287, ENv 126561, 1998, Geographic Information - Data Description - Quality, October 1998. NEN Delft, The Netherlands NVN-ENV 12656 November 1998 ICS 07.040; 35.240.70.

Guptill, S. and Morrison, J., 1995, *The Elements of Spatial Data Quality*, (Oxford: Elsevier).

ISO, 1993, 3534-1:1993, Statistics – *Vocabulary and Symbols Part 1: Probability and General Statistical Terms*, (Geneva, Switzerland: International Organisation for Standardisation).

ISO, 1993, *Guide To Expression of Uncertainty in Measurement*. Guide established by a working group supported by seven organisations, including BIPM and ISO.

ISO 2001, CD 19114 *Geographic Information - Quality Principles*. ISO/TC 211: URL: http://www.statkart.no/isotc211/dokreg12.htm, N1049

Moellering, H., 1985, Digital Cartographic Data Standards: An Interim Proposed Standard, *Report #6, ACSM* 1985, National Committee for Digital Cartographic Data Standards, Ohio State University, Numerical Cartography Laboratory, USA.

Moellering, H., 1987, A Draft Proposed Standard for Digital Cartographic Data, *Report #8, ACSM* 1987, National Committee for Digital Cartographic Data Standards, Ohio State University, Numerical Cartography Laboratory, USA.

Naisbitt, J., 1994. *Global Paradox*, (New York: Avon Books).

SDTS 1992, *The American Cartographer*, Vol. 15 No.1, Jan 1988, pp. 21–137.

Tobler, W., 1989, Frame Independent Spatial Analysis. In *Accuracy of Spatial Databases*, (New York: Taylor & Francis), pp. 115–122.

Veregin, H., 1989, Error Modeling for the Map Overlay Operation. In *Accuracy of Spatial Databases*, (New York: Taylor & Francis), pp. 3–18.

Introduction to Part III: Applications Studies Concerning Data Quality

Peter F. Fisher

In this part of the book chapters focus on two important issues of spatial data quality: data quality as spatial metadata and use of the quality information. The first is the place where the issues of spatial data quality have come to the attention of a wide community of users, while the second is really fundamental to everything that comes before, because it is crucial that the quality research described elsewhere in this volume should become a standard part of the application of geographical information technology.

Data quality has a special status in spatial data standards. Data quality under its five headings (i.e. lineage, positional accuracy, attribute accuracy, consistency, and completeness) was specified as a necessary part of the original draft of the Spatial Data Transfer Standard (Digital Cartographic Data Standards Task Force, 1988) which has had such a huge influence on standards work ever since. It is therefore appropriate that the first two chapters in this part look at the results of surveys among National Mapping Agencies (NMAs).

Dassonville and colleagues (in chapter 14) discuss the influence of the ISO 9000 standard for total quality management on the activities of NMAs. They find that the influence has been large, and has determined much of the current activity in NMAs and that all who responded either had or were planning an ISO 9000 type quality regime. More importantly, they report on the work on data quality and metadata under CEN-TC 287.

Chapter 15 by Jakobsson examines in some detail the work on data quality in one of the European NMAs, the National Land Survey of Finland, and compares it with activities in the Norwegian and French NMAs. Specific measures and tests of quality are discussed, and compared across the agencies.

Qiu and Hunter, and Gan and Shi (Chapters 16 and 17, respectively) present the results of their work developing prototype systems for browsing and maintaining quality metadata. Qiu and Hunter report a system which provides access to quality information describing information from the dataset level to the individual object. Gan and Shi demonstrate a system which not only enables users to view the quality information at a similar set of levels, but also includes automatic incorporate of lineage information and update of quality measures when features are revised or updated with information of higher quality. Qui and Hunter work with ESRI's ArcView while Gan and Shi use Intergraph Microstation. Both pairs of authors have integrated the GIS component with database management by Microsoft's Access.

The last three chapters in Part III are all concerned with using data quality in analysis. This is a topic of great importance. Standards have principally been generated within the data production industry, reflecting the concerns of traditional cartographic production. The consequences of a particular value of a data quality measure on an analysis is, however, not clear. These three chapters show how quality issues may be propagated into spatial analyses in three particular contexts. The three are drawn from issues across the possible applications of GIS – from the analysis of infrastructure, of crime and of forestry.

In chapter 18 Merritt and Masters address the specific problems of topology (relativity) maintenance between a cadastral database and a utility infrastructure such as water of phone lines. The work discusses enforcing the maintenance of topological relations. Clearly the utility lines need to connect to buildings, not end in thin air, and to be located in the street thoroughfare. This is related to the update problem of new and old cadastral mappings.

In chapter 19 Law and Fisher examine one of the fundamental problems of spatial analysis using Monte Carlo simulation to explore the consequences of data aggregation on the assessment of crime prevention measures. They look at the consequence of what has been termed cross-area estimation between spatial aggregation units, and the influence of that problem on the evaluation of the effectiveness of crime prevention measures. The level of uncertainty they report may seem high to some, but within the policy context was not considered important.

Finally, Weir looks at the costs and benefits of data quality in estimating forest yields. He uses Monte Carlo simulation to estimate the alternative areas of forest stands. He examines a monitoring application where snap shots of the conditions are taken at different times. Given that the surveys at those different times are of different quality, he reports on the pattern of quality which is most desirable over time, showing that you can improve on a poor quality dataset – the output is not as bad as the worst input – but if you have a low quality dataset late in the analysis it can have dramatic consequences on the quality of the outputs.

REFERENCES

Digital Cartographic Data Standards Task Force, 1988. The proposed standard for digital cartographic data. *The American Cartographer*, 15 (1) pp. 21-137.

CHAPTER FOURTEEN

Quality Management, Data Quality and Users, Metadata for Geographical Information

Laure Dassonville, François Vauglin, Antti Jakobsson and Claude Luzet

ABSTRACT

CERCO (Comité Européen des Responsables de la Cartographie Officielle) gathers participants from European National Mapping Agencies. It has created several working groups for exchanging information and for discussing geographical information issues. One of them, the Working Group on Quality, was created in September 1997 and is working on three main subjects: Quality management and ISO 9000; Data quality issues; and Standards. This chapter presents elements which have been given to the questions raised up to now and emphasises common and different points of view of the National Map Agencies. The first subject consists of analysing the opportunity to implement ISO 9000 for National Map Agencies. A questionnaire has been written and this chapter presents the results. The second subject aims to share experience about geographical data quality during the entire life cycle of a product such as cadastre, topographic products or orthophotography. A lot of topics are raised: such as methodologies for identifying users requirements or processes to evaluate the fitness for use. The point is to identify relevant questions among all these issues. The third subject concerns experience in applying the two CEN/TC 287 standards: Metadata and Quality. Metadata standards give very useful common structures for geographical dataset description. This group decided to focus on a subset of the most comprehensive standards at a level named "core-metadata", presented in this chapter.

14.1 INTRODUCTION

CERCO (Comité Européen des Responsables de la Cartographie Officielle) is a forum for the heads of 35 European National Mapping Agencies. It was created in 1980 with three missions: mutual information, discussion, and cooperation between European National Mapping Agencies. The chairship of CERCO rotates every two years among the heads of NMAs. CERCO includes several active working groups focusing on geographical information issues of legal and economic affairs, geodesy, updating digital geographical databases, and more recently on quality. Another working group, created in 1991, is considering the concept of a European network for geographical information. That working group gave birth to MEGRIN (Multipurpose European Ground Related Information Network) in 1995 which aims to stimulate the geographical information market and develop common activities of its members. Following preliminary discussions the Working Group on Quality was created at the CERCO general assembly in September 1997 with the following program of action: ISO 9000, Data quality issues, Quality-measurement algorithms, CEN/TC 287 results implementation, and subcontractors. The Working Group first met in November 1997 and split into three sub-working groups: Quality management and ISO 9000; Data quality; and Standards (CERCO WGQ,1997).

14.2 QUALITY MANAGEMENT SYSTEMS AND ISO 9000

14.2.1 Quality management concepts

The quality of a product or a service is defined as "the totality of characteristics of an entity that bear on its ability to satisfy stated and implied needs" (ISO 8402, 1994). Quality is not an intrinsic characteristic of a product, but is related to an overall perception of user satisfaction. Quality is a result which is observed when using the product. Thus, it is clear that assessing and checking the quality of a product or service only at the final stage of the production line is not enough, and that such an approach is incorporated within the quality of the production processes. By the time a final product is available, quality is a concern of the whole organisation. Therefore the concept of Total Quality was introduced during the 80's, and may be defined as " a management approach of an organisation, centred on quality, based on the participation of all its members and aiming at long-term success through customer satisfaction" (ISO, 1994). That approach aims at continuously increasing value to customers by designing and continuously improving organisational processes and systems. A Quality Management System (QMS) consists of all parts of an organisation that deal with the quality of processes and products. A QMS involves defining all responsibilities, interfaces, main processes and procedures required to achieve the quality objectives of the organisation with respect to the quality policy.

The ISO 9000 standards published in 1987 were reviewed in 1994 (all ISO documents are available on the ISO official web site www.iso.ch). They provide a normative framework for defining a QMS in a consistent way, paying special attention to continuing improvement and user feedback through the consideration of user satisfaction as a core concept. There are several standards depending on different scopes: design, production, inspection and delivery (9001); production, inspection and delivery (9002); and final inspection (9003). These three models are to be used in contractual situations, whereas the 9004, which is the widest quality management standard, concerns a voluntary approach from an organisation (see www.iso.ch on ISO 9001 to 9004). These standards provide quality requirements, explaining the elements of a QMS, without specifying how they are achieved. A QMS consists of quality controls throughout the production process, and should give customers trust in the product, before they use it. ISO 9000 standards, however, have been primarily designed for manufacturing goods, and their application to geographical information is still a matter for discussion.

14.2.2 Action lines of the ISO 9000 sub-working group

The CERCO ISO 9000 sub-working group had to analyse reasons for implementing a QMS within a NMA. It had to develop arguments for use of ISO 9000 and certification in NMAs, to analyse opportunities and risks, and, eventually, to identify its applicability to geographical information. Consideration of ISO 9000 has been extended to Quality Management Systems generally, and a questionnaire survey was organised so that each of the collaborating European NMAs could give its own opinion. The issues addressed by the survey are what kind of QMS is needed, why and when a QMS is required, and how it has been implemented.

14.2.3 Questionnaire

14.2.3.1 Questionnaire structure

The sub-working group decided to write a document entitled "Good reasons for implementing a Quality Management System" with three objectives: identifying key issues, common knowledge, and experiences to be discussed. For that purpose, a questionnaire was prepared. The questionnaire structure is based on four main sections (CERCO SWG A,1998) presented in Figure 14.1.

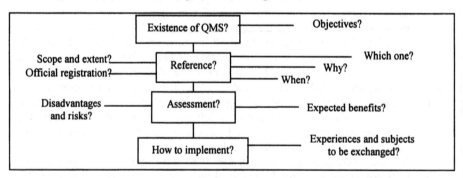

Figure 14.1 Structure of the questionnaire about a Quality Management System.

14.2.3.2 Questionnaire results

The questionnaire was sent in December 1998 to 32 European NMAs, as well as to the OEEPE (Organisation Européene d'Etudes Photogrammétriques Expérimentales / European Organisation for Experimental Photogrammetric Research), EUROGI (European Umbrella Organisation for Geographic Information) and MEGRIN. Among the European map agencies, 70% of questionnaires were returned, showing a high degree of interest in quality management (CERCO SWG A, 1999). Fifty percent of NMAs that answered have already implemented a QMS, and all the others have the intention of doing it in the future. This means that geographical information providers are concerned to improve quality processes and organisation as in other industrial fields.

ISO 9000 standards are the reference adopted by 80% of the NMAs responding. Half of them selected the ISO 9001 and their QMS covers their whole organisation. ISO 9001 seems to be very attractive because of its chapter about design and specification. Thirty percent of NMAs with no QMS say that they intend to adopt ISO 9000 standards; the extent and the scope of these QMS concern different departments, including mapmaking, metrology, cadastre, digital geographic information production, data capture and storage. The picture of a QMS is seen at a detailed level and expected benefits are not clearly set up on the global organisation. NMAs which have already implemented a QMS cope with the fact that a QMS covers all the organisation, even if it is cut and structured into small independent QMS. The choice of the ISO reference is justified by the fact that it is an existing international model, used in industry, answers customer demand, has external requirements, and support and training are available. The number of NMAs implementing a QMS has doubled since 1995, following the trend among wider industries concerning ISO 9000 certification. About 70% of the NMAs that answered the questionnaire intend to apply for certification, and three of them have already obtained it. Certification is seen as a way of improving reliability, official recognition, improving subcontracting policy and maintaining the level of quality inside the organisation.

The advantages and benefits identified by NMAs already working with a QMS may be arranged into four areas.

- They expect to get better management, which includes a strong intention of increasing benefits and productivity, by relating the amount of product inspection and cost investments in production control to the level of user satisfaction. It includes also an explicit and feasible approach to solving organisational problems by clearly defining responsibilities and interfaces, giving transparent rules applicable by every employee, and introducing continuing improvement as a regular part of day-to-day leadership. It promotes a philosophy of measurement, calibration and accountability wherever it is required, when any risks may occur; costs are saved when quality inspection is positioned thoughtfully in areas where there is a risk of low-quality. From that organisational point of view, a QMS seems to be a good tool for reducing overlapping work, for protecting know-how and for reducing training costs when staff change.

- They anticipate improving fulfilment of customers needs. A QMS gives a framework for having a consistent and improved approach to customers, and it provides a model for customers and partners, which perhaps sets the need of a mutual understanding (and therefore a common language for explaining needs and translating them into specification, but also for answering customers' claims).

- Improving the quality of products or services is the third main reason for implementing a QMS. This means as much defining the product closest to the users' requirements (specification) as reducing non-conformance and tracking what happened during production. It implies developing a high level of control on processes, by describing and harmonising them, observing them, simplifying and optimising them. The knowledge of processes gives the possibility of error detection and management, and avoiding repeating the same error. It also provides control and inspection to customers, giving more information on the products.

- The last main reason mentioned is compulsion; in some countries the NMA must satisfy the trading rules or laws of their country by implementing a QMS.

NMAs who do not yet have a QMS focused on customer demand (reliability, trust and credibility) and interfaces, and co-operation with subcontractors and other organisations, promotion of their public image, and improvement of products and processes, are nonetheless interested in having compatibility with other NMAs, showing that they need a common understanding.

Disadvantages and risks foreseen in the long term include the increase of production costs and bureaucracy, with more paperwork but without actual results and consequent decrease in worker satisfaction. These are the risks of building a very complex system, with many documents, which are difficult to maintain, and with strong procedures which slow work down and bore the production staff. The difficulties of convincing top management have been mentioned, showing the contrast between the large amount of effort and time required for implementing a QMS, and the short time period before top staff want to see benefits.

All NMAs expressed their interest in exchanging their experience in QMS. The favourite subjects are the application of ISO 9000 to geographical information, the analysis of procedures and processes and also the quality management organisation and documentation. Three other subjects have been mentioned: education, training and subcontractors; users satisfaction and non-quality costs; and quality management terminology. The interest in the applicability of ISO 9000 standards is justified because it is crucially important to find new ways for improving production processes and quality of products.

14.2.4 Conclusion: what future for QMS in geographical information field?

The questionnaire was a valuable exercise and the participation rate shows that the subject is very relevant for NMAs. ISO 9000 standards seem to be just one step in the direction of quality management. That concept combines a better care of customers and also a continuous improvement of the whole organisation. Quality management appears as a new way for improving the quality of the product. In the future, NMAs will exchange their experience interpreting ISO 9000 chapters and will connect that standard to practical issues of geographical information. A seminar has already been planned for this purpose. The next step will be to analyse the methods for implementing a QMS, to compare the difficulties and the solutions encountered. A general method for implementing a QMS for geographical information may arise.

14.3 DATA QUALITY ISSUES

14.3.1 Fundamental concepts and current need of NMAs

As defined in Section 14.2.1, quality is well defined for products or services. According to several working groups on geographical data quality, the definition given by ISO (ISO 8402, 1994) can be divided into an "external" component and an "internal" component when applied to geographical data.

- External quality is the appropriateness of the specifications to the users' needs: is our product well defined and does it correspond to an actual need among the community of users?
- Internal quality is relates to whether specifications are respected: have we produced what was intended in the specifications?

A sub-working group, focusing on geographical data quality involving both internal and external components, has been created within the CERCO Quality Working Group. The actual aim of this sub-working group (called SWG B) is to share experience about geographical data quality during the entire life cycle of a product. Products can be as different as cadastre, topographic products or orthophotography.

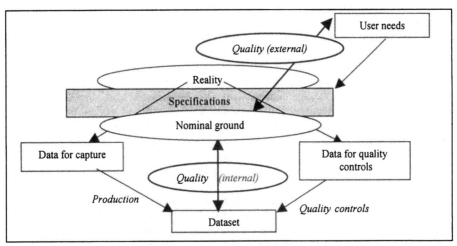

Figure 14.2 Concepts on data quality.

The life cycle of geographical products is defined here as being concerned with all production processes from identification of users needs to delivery. The starting point of SWG B is examination of the fundamental concepts of quality and is summarised in Figure 14.2. This view was developed by expert groups from CEN ("Comité Européen de Normalisation" / European Committee for Standardisation) as well as from the NMAs. Quality is good if both internal and external quality are good; quality may be poor if one or both components are poor.

14.3.2 Action lines of the sub-working group on data quality

14.3.2.1 Life cycle of products

As a group of experts on data quality from European NMAs, it is important for SWG B to adhere to the way production is organised. Hence the structure of production needs to be taken into account when designing new methods or controls.

For these reasons, SWG B decided to start by studying the life cycle of products before identifying issues and action lines within that framework. After comparing NMAs organisation of production, five major steps have been identified:

- identify requirements of users;
- specify products meeting those user requirements;
- produce data meeting the specifications of step 2;
- check that the data have been correctly produced;
- deliver data checked to users and to prepare to test if requirements have been correctly identified in step 1.

These five steps are organised as a continuous cycle: step 1 follows step 5. This enables control over all steps of production. At each step, there is a check on whether the results of earlier steps were satisfactory. This cycle acts like a feedback loop ensuring a convergence of user needs and satisfaction. The life cycle of geographical products is presented in Figure 14.3.

The dashed arrows of Figure 14.3 present very straightforward questions that must be raised at each step; but the corresponding answers are never simple (e.g. describing a methodology or an update strategy for a geographical databases is complex). Identifying user requirements is a tricky topic involving technology and marketing. The need for methodology in that field is obvious. Similarly, it is necessary to specify geographical databases and control and update production. Checking also raises an important question: how to handle data that are not bad but that do not meet the specific requirements of the specifications? Some feedback from users must be associated with the delivery strategy so as to enable some retroactive correction aiming at a closer representation of users needs in the specifications.

All these questions become tougher when NMAs are subcontracting their production, which is becoming more and more frequent. In that case, there must be a policy of clear documentation and contract handling. The emerging geographical information market sets requirements for NMAs to show the quality of their products and even more to meet customer expectations that data can be used for varied purposes. NMAs share a concern for a common reference such as a shared standard for geographical data quality, which will help sharing methods and experiences between organisations. To formulate some answers, SWG B defined two action lines: data quality issues in NMAs and quality measurement algorithm exchange. A questionnaire has being prepared according to those guidelines.

14.3.2.2 Data quality questionnaire

The purpose of the questionnaire to be sent to all European NMAs is to better know what NMAs are doing about data quality and their data quality needs. This questionnaire will be about photogrammetric, topographic and cadastral data quality.

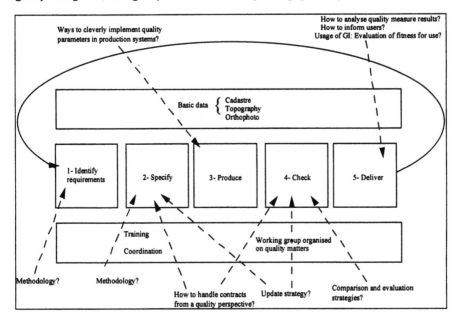

Figure 14.3 Data quality issues and life cycle of products.

The current state of the questionnaire is presented here (CERCO SWG B,1999). The structure of the questionnaire is derived from the life cycle of geographical products as presented in Figure 14.3. Other similar initiatives will also be useful (Östman,1997).

- **Identification of user requirements.** Who are the users? What are they willing to buy? What are the requirements for producing the databases? What are the legal requirements? What are the market requirements?
- **Specify products.** Are there written product specifications? What are they? Have they been written with a specific method? Do they include an abstract definition of geographical features? This will help to identify whether a clear conceptual model (nominal ground) is defined and how. Do specifications include criteria on data quality? How are the requirements specified for subcontractors?
- **Data production.** Is the work subcontracted and why (political decision, costs, lack of resources, acceleration of production)? According to CEN/TC 287 definitions, what quality components are taken into account during production? What sources are used for production (existing maps, field campaigns, aerial or satellite imagery, etc.)? What methods are used for capture? What is the update strategy (continuous, periodic, irregular, no updates) and how is it defined? Is there an audit of the production line? These questions provide a good representation of the production methods and of the way they have been defined.
- **Quality check.** What methods are used to assess data quality? Are existing standards involved, like ISO 2859 or ISO 3951? What is the sampling procedure (simple random, stratified, multi-stage, etc.) and what

quality components as defined by CEN/TC 287 are checked (semantic, positional and temporal accuracy, completeness, logical consistency, usage, lineage)? Are there acceptance or rejection criteria for datasets and how have they been defined? The answers to these questions describe in detail the quality control policy of NMAs.

- **Data deliver.** Does the quality of data affect the price? How is delivery handled (resellers, internet, direct sell, service agreement, etc.)? How are after sales and feedback handled? Is there a warranty for quality? What documents are given to the users (especially concerning quality)? These questions should identify whether and how delivery and after sales are used to increase the quality within the loop described in Figure 14.3 and whether and how fitness for use is assessed.

Other questions will illuminate other important issues like the training policy and the different data types produced. Many questions could be divided into two parts: those on methodology explaining the principles and the reasons that are involved, and those questions on technical choices exploring the impact of technique over quality.

14.3.3 NMAs and data quality

With the tremendous expansion in the amount of digital geographical data and of quality concepts among the whole industry, NMAs need to interact and exchange information on geographical data quality. Simply drawing up such a questionnaire among experts on geographical data quality from 35 European NMAs has greatly increased communication at a European level. The answers will probably improve data quality at lower costs across Europe, demonstrating the advantages of a European structure like CERCO.

14.4 STANDARDS

14.4.1 Workplan of the standards sub-working group

Among all the subjects related to quality of geographical information, the standards issue is central to NMA concerns. Testing the implementation of CEN/TC 287 is essential for defining how standards may be applied and how their interpretation may be tuned. In the current context, the sub-working group has limited its mission to two standards: metadatata (ENV 12656, 1998), which contains a set of elements for describing datasets, and quality (ENV 12657, 1998), which defines elements and criteria for assessing dataset quality. The group defined several actions focused on metadata, especially core metadata and transferring practice from one metadata service to another.

The concern over core metadata consists of defining a set of the most used metadata for creating a common core-metadata so that NMAs could communicate and exchange meta information more easily. The action line on transferring practice for metadata information is introduced here because even if NMAs use metadata based upon the CEN standard, the data model may be slightly different, which creates many problems for transferring metadata; it means that one NMA having a national metadata service is not able to communicate with another, and therefore it is not possible to update easily their part of the European metadata service.

14.4.2 The metadata context in European organisations

This activity collaborates with other European initiatives with similar objectives including MEGRIN, the "Laclef" and ESMI projects. MEGRIN is a consortium of 19 NMAs and is tasked with stimulating the creation of pan-European datasets from source data owned by its members. Among other projects, it aims to inform users about the availability of European geographical data via its free Internet service GDDD, created in 1994. GDDD (Geographical Data Description Directory) is an innovative metadata service created and maintained by MEGRIN. The service describes the main geographical information products from NMAs of 22 European countries and is a pioneer initiative in implementation of the emerging CEN/TC 287 standards on metadata.

Within a new EU-funded INFO2000 project named LaClef, which aims at making public sector information more available and accessible to the general public, the GDDD will be greatly upgraded by MEGRIN and its partners, and will include the implementation of a prototype for an electronic commerce service. The ESMI (European Spatial Metadata Infrastructure) project started in January 1997, and is another project in the European Union's INFO2000 program. It aims to develop the mechanisms needed to link European metadata services and is concerned with the co-ordination of distributed metadata. MEGRIN is member of the ESMI consortium. For more information, see http://www.megrin.org.

14.4.3 Metadata and core-metadata

Metadata standards give very useful common structures for dataset descriptions. Metadata is data about data, providing reliable search and access for geographically referenced datasets to geographical information users. It provides information on sources, data quality, spatial extent, general content, production processes and responsibilities. The CEN/TC 287 metadata standard is composed of more than one hundred metadata elements, and the working group decided to concentrate its effort on a subset of the more comprehensive standards at a level named the core-metadata. This approach is based on the main lessons learned from operating the existing metadata services. One lesson was that many potential metadata/data providers would not implement standards because they are too complicated. The other lesson was that most users do not look for detailed metadata, and are satisfied with information that can be provided by a set of some 10 to 50 elements. Desk research and workshops have shown that a core-metadata set of some 40 to 50 elements are extremely similar in the main standards, and that mapping between the different structures is an exercise that can easily be automated. The core-metadata level is usually sufficient for data discovery, high level evaluation of fitness for purpose, and eventually access to the data and/or the data provider.

As MEGRIN defined it, core-metadata is the minimum information necessary to provide the users with an effective service and access to European datasets. The users can get answers to three main questions: Is there any dataset on this area (extent)? Is it relevant for a specific need (content, quality, usable data format)? What is the organisation contact and what are the data distribution conditions?

The currently discussed core-metadata structure is composed of only about 40 metadata elements, divided into 9 sections:

- dataset identification;
- dataset overview (summary and sample);
- dataset quality elements;
- administrative metadata (organisation and contact);
- distribution (restrictions on use, formats, price);
- extent;
- metadata reference (last update date);
- spatial reference system; and
- classification (keywords for search, giving an idea of the content of the dataset).

The list has been created by MEGRIN on the basis of the CEN ENV 12657 standard together with its previous experience in setting up the GDDD for 35 countries and in ESMI, but also from analysing other metadata standards. From the CEN standard, it kept those fields that are necessary to get a minimal running metadata service (mandatory fields). Other additional fields may be found: conditional fields (mandatory under certain criteria) and optional fields (to improve the implementation of search and inquiry features, such as the child-parent relationship between a dataset series and its component units – map series, for instance).

This list was proposed and discussed during the MEGRIN metadata workshop in October 1998, with representatives of 15 European NMAs. This list is not yet finalised and some elements are still under discussion. It does not contain the list of classes, attributes and links to the dataset because they are not required for the first prototype metadata service, but the list will be reviewed with experience from the LaClef prototype.

When testing the CEN standard against the requirements of the intended services, it was found that most needs were covered by the CEN elements. For practical implementation purposes, it has been seen as useful to sometimes rely on the more detailed provisional ISO standard for structuring multiple information within one element (e.g. within bounding coordinates, or contact addresses). However, one domain is not sufficiently covered by both the standards and the requirements for e-commerce. This has not been covered within the previous studies, and it is the topic at the heart of today's LaClef investigations.

14.4.4 Questionnaire about metadata

To identify the existence and use of metadata and core-metadata within NMAs, a questionnaire was sent to CERCO members in November 1998 (CERCO SWG C,

1998), after the metadata workshop (see Figure 14.4). It contained definitions of metadata and core metadata and also a proposed list of core metadata, especially for digital products, aerial photos and paper maps. A question about controlled lists has been included, to establish whether NMAs have developed tools for fully consistent semantics. Questions about language have been set because the idea is to have a multilingual user-interface as well as multilingual metadata content.

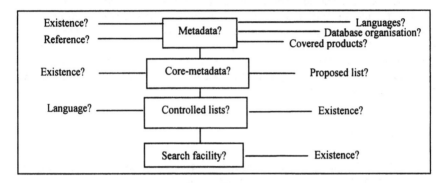

Figure 14.4 Metadata questionnaire.

The questionnaire was sent to 30 NMAs and 18 countries completed the questionnaire (MEGRIN, 1998). The importance of CEN seems to be clear with 50% who have chosen the CEN standard, but the fact that 43% either have a national reference or have no standard at all emphasises the necessity for common core-metadata. Various countries expressed the opinion that information should be supplied in English, but, additionally, 47% of organisations may supply their metadata in their national language only: this shows the need to ensure that the metadata service supports national languages. Fifty percent of respondents have metadata in digital format: a serious issue remains with the 34% who have no metadata or metadata in paper format. Only 24% of respondent NMAs use controlled lists of keywords to facilitate searching and data entry: there is an actual need for providing organisations with appropriate controlled lists.

The results of the questionnaire show that even if metadata still have a low priority within NMAs, there is a need for a common core-metadata structure and an improvement in definitions, abbreviations and glossary; the use of such a structure will also imply update tools and maintenance procedures, solving the problem of data transfer format. The NMAs, however, seem to set a generally low priority for metadata development, and a broad spectrum of experience and maturity was reflected, from minimal paper-based metadata to full Internet service.

14.4.5 Next steps: practical solutions for implementing metadata

The European geographical information market is still nationally focused and it is difficult to know what datasets are available, with what quality and distribution conditions, and also concern exists as to data consistency across borders. Metadata services appear therefore to be a solution to these problems and the demand includes easy access, where users can discover the available data, purchase and be supplied at the same time (Wandinger, 1999).

The starting point of the CEN standard investigation is already in progress with the definition of a common core-metadata structure, using all the experience obtained through the GDDD and the European ESMI project. The actions of the CERCO standards sub-working group have been transferred to the LaClef project to avoid overlaps. The sub-working group completed its objectives and was closed down at the end in April 1999. The work done raised the awareness of metadata in Europe, expressed as the need to access and discover datasets through a common core-metadata structure. The European project LaClef will continue by implementing the core-metadata prototype. A Metadata Across Europe workshop was planned by LaClef in September 1999 and was to be an opportunity to review the core-metadata prototype.

14.5 CONCLUSION

The CERCO Working Group on Quality was created to discuss and exchange information about geographical information quality. The three main actions that NMAs investigated are linked and complementary. The first issue involves better care of customers and users. Methods for a better understanding of their needs are a challenge. How to translate demands and needs into product specifications? How to assess customer satisfaction and fitness for use? Another important issue is considering entire life cycles of products. Quality is not obtained at the end of the production line, but begins with the need's translation, the specifications draft, the production control and only ends with the quality inspection. The quality management concept adds value by requiring proof about what happened during the production processes. The last main issue concerns accessibility of European geographical information. How to discover what data are available? Is it possible to transfer and to use them? A core-metadata is the current solution tested by several NMAs. The participation of NMAs in this working group shows the need for sharing concepts, quality vocabulary, methodologies and practice.

The Working Group has already produced questionnaires and reports during the last 18 months. Also, it organised several meetings and workshops for discussing results. Up to now, its working method has been based on three steps: to identify key issues, status and need within NMAs; to discuss the issues and produce reports; and to produce documents expressing a common view (core-metadata structure, for instance). As a result of this collaborative approach among NMAs, common concepts and practices are being identified and formalised. They give guidelines and tested solutions for solving some difficult quality aspects related to geographical information. The next step for the Working Group is to focus on the production of more guidelines. Then, the way NMAs use and implement these guidelines inside their own organisation will be a matter for future research.

14.6 REFERENCES

All information and documents about OEEPE, EUROGI, MEGRIN, CEN and ISO can be found on their official web sites: www.oeepe.org - www.eurogi.org - www.megrin.org – www.cenorm.be - www.iso.ch – MEGRIN and CERCO have recently merged into one body called Eurogeographics. Its official web site is www.eurogeographics.org

CERCO Working Group on Quality, 1997, Report to the CERCO general assembly (www.eurogeographics.org)

CERCO Working Group on Quality – SWG A, 1998, Questionnaire on good reasons for implementing a quality management system. (www.eurogeographics.org)

CERCO Working Group on Quality – SWG C, 1998, Metadata questionnaire. (www.eurogeographics.org)

CERCO Working Group on Quality – SWG B, 1999, Questionnaire on data quality – first draft. (www.eurogeographics.org)

CERCO Working Group on Quality – SWG A, 1999, Good reasons for implementing a Quality Management System in European NMAs – draft 1.0. (www.eurogeographics.org)

BI67, 1997, Qualité d'une Base de Données Géographiques: Concepts et Terminologie. In *Bulletin d'Information de l'IGN* n°67, Edited by B. David and P. Fasquel. Vol. 67.

ENV 12656, 1998, Geographic information – Data description – Quality. European Standard - Draft for Public Enquiry, Comité Européen de Normalisation. (1996) (www.cenorm.be)

ENV 12657, 1998, Geographic information – Data description – Metadata. European Standard, Comité Européen de Normalisation. (www.cenorm.be)

ISO 8402, 1994, Quality management and quality assurance - Vocabulary. (www.iso.ch)

ISO 9001, 1994, Quality systems - Model for quality assurance in design, development, production, installation and servicing. (www.iso.ch)

MEGRIN, 1998, MEGRIN Metadata Workshop – Summary and proposed core metadata. (www.megrin.org)

Östman A., 1997, The specification and evaluation of spatial data quality. In *Proceedings of the 18th ICA International Cartographic Conference*, Sweden, (Gävle: Swedish Cartographic Society), pp. 836–847.

Wandinger M., 1999, Upgrading the GDDD towards the ideal pan-European GI marketing platform. (www.megrin.org)

CHAPTER FIFTEEN

Data Quality and Quality Management –Examples of Quality Evaluation Procedures and Quality Management in European National Mapping Agencies

Antti Jakobsson

ABSTRACT

National Mapping Agencies (NMAs) in Europe have recognised the importance of quality but issues like process management, especially related to quality evaluation procedures, data management issues such as data specification, selecting quality elements, reporting quality results and competence of personnel remain high priorities. This chapter will give some examples of quality evaluation procedures that are used to assess the quality of topographic data sets in European NMAs. It is based on the author's experiences at the National Land Survey of Finland (NLS) since 1992, when the Topographic Database System (TDB) was introduced (Jakobsson, 1995, 1999). The system covers all the topographic databases that we compile at the NLS at scales 1:10,000, 1:20,000, 1:50,000, 1:100,000, 1:250,000, 1:500,000, 1:1000,000, 1:2000,000 and 1:4500,000. We have developed generalisation processes in order to use the Topographic Database in other databases so that the quality of the data will be maintained. European references are derived from the numerous symposia and working groups that the author has attended since the 1990's.

15.1 INTRODUCTION

National Mapping Agencies in Europe have recognised that data quality and quality management are important issues and they are trying to introduce quality evaluation procedures for data production. This chapter will explain some of the results.

The principles of spatial data quality are first discussed, using concepts accepted in the forthcoming family of International Standard (ISO 19100, etc) for geographical information. Completion of these ISO standards is now approaching. The data quality of geographical information is described as the difference between the universe of discourse and the data set, where the data set is defined as an identifiable collection of related data and the universe of discourse as the view of the real world that includes everything of interest. The universe of discourse is described by a product specification, against which the quality content of a data set is evaluated.

The quality evaluation procedures discussed in this chapter are based on the quality evaluation model that is defined in ISO 19114 Quality Evaluation Procedures (ISO/TC 211, 2001b). Examples of the quality evaluation procedures used for topographic data sets are given using the experience of the NLS and other European mapping agencies.

15.2 QUALITY OF GEOGRAPHICAL INFORMATION

The quality of geographical information is defined in the ISO/DIS 19113 Quality Principles (ISO/TC 211, 2001a; Godwin, 1997). In Figure 15.1, the quality of the data set is the difference between the universe of discourse and the data set. This has two different perspectives: the data producer's perspective and the user's perspective. If the user requirements and the product specification are the same, then the quality is also the same.

REAL WORLD

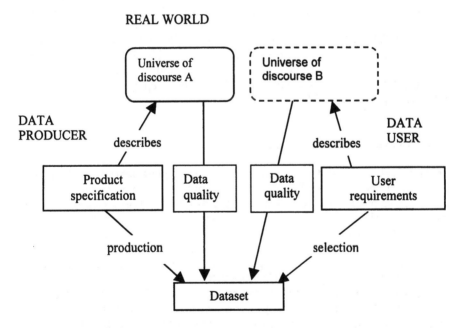

Figure 15.1 The framework of data quality concepts (ISO TC211, 2001a).

A data set is defined as an identifiable collection of related data. The universe of discourse is a view of the real or hypothetical world that includes everything of interest. It is described by a product specification, against which the quality content of a data set is evaluated.

15.2.1 Data quality and quality management

Quality is defined as the totality of characteristics of a product that bear on its ability to satisfy stated and implied needs (ISO 8402, 1994). In the new ISO/DIS 9000:2000 standard (ISO, 2000) the definition of quality is: "Ability of a set of inherent characteristics of a product, system or process to fulfil requirements of customers and other interested parties". This indicates that data quality and quality management are very closely related. Data quality is part of the organisation's total quality management.

Quality management can be based on the ISO 9000, self-assessments using quality award criteria e.g. the Excellence Model of the European Foundation for Quality Management (EFQM) (www.efqm.org), the Malcolm Baldrige National Quality Award (www.quality.nist.gov) in the USA, practical experiences within the company or a combination of these. The ISO 9000 and quality award criteria require the quality system to be documented.

The ISO 9000 Standards series has evolved from the need to control the quality of production. There was a clear need to move from product inspection to an assessment of the ability to produce products that meet the specifications. Recent developments in the series have moved towards process management control. An organisation can also apply for a certification showing that its quality management system meets the requirements of the ISO 9000.

There has been considerable criticism of the ISO 9000 because developing and updating the quality system can be expensive. There is a clear difference between Europe and the United States in the popularity of the ISO 9000. Self-assessment has been very popular in the United States, applying the criteria given in the quality awards model.

15.2.2 Characteristics of data quality

Geographical information or data consists of a geographical location (e.g. coordinates) and attributes linked to it. Geographical location is commonly very stable but its precise determination can be difficult. Positional accuracy is often synonymous with data quality. In traditional cartography, mapmakers decide if the quality of the available data is sufficient to produce a multi-purpose map product. At best, the quality information included: (1) a quality statement furnished by the producing organisation; (2) the reputation of the producer; and (3) experiences resulting from using the product (Guptill, 1998).

Geographical information is a representation of a real phenomena. In order to collect data into a database, an object or a feature has to be defined using a data specification. Data compilation is carried out according to the data specification. The universe of discourse is seen as the truth when defining the quality of the data and anything that is not defined by the data specification does not exist in the universe of discourse. The data specification is thus one of the most important parts of the data quality because errors made in modelling are often difficult or impossible to correct. The quality information is reported against the data specification. For example, a building may have restrictions that are not collected i.e. small details or very small areas. It would show up as an error if objects were collected that do not meet the criteria, even thought they are buildings in reality.

Time is one of the special features of geographical information. It is often true that by the time the data has been compiled the information has already changed in reality. Data compilation processes can last several years, which can cause reliability problems. For the data producer, data compilation is virtually a continuous process.

Data compilation from several different sources often requires very good data quality. Positional accuracy, for example, can be the same in two data sets but the result combined may still not be correct i.e. there might be a topological error (a building overlapping a road). Furthermore, combining data sets that have no quality information can be very difficult or impossible.

Equipment used for data compilation does not guarantee results. The user of the data must be very careful in interpreting the quality results. A particular quality measure can mean little if there is no indication of how the quality results have been achieved. For example, positional accuracy can mean root mean square error (RMSE) or precision or something else. Producers may state the accuracy of the scanner but not the accuracy of the original map. It is often the case that the accuracy of the data set is not checked in the field because of the costs, and a customer has to rely on an estimated value.

15.3 DEFINING QUALITY

The forthcoming ISO and European standards describe data quality. Whether these definitions are the correct ones is not discussed here. A data producer can choose at least two different strategies. One is to have a defined data quality as we have at the NLS in our data quality model, and test the data sets against that level and the other is to only report the quality level that has been achieved. In the first strategy, a customer can make decisions even if the data has not yet been collected. That is, of course, the case in data revision.

15.3.1 Data quality definition in the European standard for geographical information

The European Standardisation Committee (CEN) has produced a European pre-standard that defines the data quality elements as follows.

- **Lineage** is a description of the history of the geographical data set in terms of source material, dates, processing applied and responsible organisations.
- **Usage** means that a geographical data set can have a set of records describing its previous usage.
- **Quality parameters** describe the performance of the geographical data set compared with its nominal ground. Quality parameters are presented in Table 15.1.

- **Homogeneity**, which is a textual and qualitative description of the expected or tested uniformity of quality parameters in a geographical data set.

Users are allowed to define their own quality parameters, indicators and measures.

Table 15.1 Quality parameters in the European standard (ENV 12656, 1998)

Quality parameter Quality indicators	Description
Positional accuracy	Describing the accuracy of geographical position within a geographical data set
Relative horizontal accuracy Relative vertical accuracy	RMSE and vertical bias or vertical threshold
Semantic accuracy Accuracy of classification Agreement for an attribute	Describing the accuracy of the semantic aspects of geographical data
Temporal accuracy	Describing the accuracy of the temporal aspects of geographical data
Accuracy in time measurement Last-update Rate of change Temporal lapse Temporal validity	
Completeness	Describing the presence and absence of entity instances, relationship instances and attribute instances
Omission Commission	
Logical consistency	The degree of conformance of a geographical data set with respect to the internal structure given in its specification.

15.3.2 Data quality definition in the International Standard for geographical information

The International Organisation for Standardisation (ISO) has produced a draft International Standard for quality principles (ISO/TC 211, 2001a). It defines the following data quality elements and sub-elements which are presented in Table 15.2.

The main difference between the CEN standard and ISO is that the latter allows users to define their own quality elements. A separate standard will also be produced for quality evaluation procedures not covered by CEN. The wording and some of the quality sub-elements are also different, but the basic concept of quality is the same.

Table 15.2 Data quality elements and sub-elements (ISO TC211, 2001a)

Data quality element Data quality subelement	Description
Completeness	Presence and absence of features, their attributes and relationships
Commission	Excess data present in a data set
Omission	Data absent from a data set
Logical consistency	Degree of adherence to logical rules of data structure, attribution and relationships
Conceptual consistency	Adherence to rules of the conceptual schema
Domain consistency	Adherence of values to the value domains
Format consistency	Degree to which data is stored in accordance with the physical structure of the data set
Topological consistency	Correctness of the explicitly encoded topological characteristics of a data set
Positional accuracy	Accuracy of the position of features
Absolute or external accuracy	Closeness of reported coordinate values to values accepted as or being true
Relative or internal accuracy	Closeness of the relative positions of features in a data set to their respective relative positions accepted as or being true
Gridded data position accuracy	Closeness of gridded data position values to values accepted as or being true
Temporal accuracy	Accuracy of the temporal attributes and temporal relationships of features
Accuracy of a time measurement	Correctness of the temporal references of an item (reporting of error in time measurement)
Temporal consistency	Correctness of ordered events or sequences, if reported
Temporal validity	Validity of data with respect to time
Thematic accuracy	Accuracy of quantitative attributes and the correctness of non-quantitative attributes and of the classifications of features and their relationships
Classification correctness	Comparison of the classes assigned to features or their attributes to a universe of discourse (e.g. ground truth or reference data set)
Non-quantitative attribute correctness	Correctness of non-quantitative attributes
Quantitative attribute accuracy	Accuracy of quantitative attributes

15.4 QUALITY EVALUATION PROCEDURES

15.4.1 Quality evaluation in the International Standard for geographical information

The ISO/DIS 19114 quality evaluation procedures (ISO/TC 211, 2001b) describes a process for evaluating and reporting data quality results. First, an evaluator selects applicable data quality elements and sub-elements. Next, the data quality scope (e.g. a certain area in the data set) and a data quality measure are defined. A the evaluator determines conformance comparing the data quality result. Quality evaluation methods are classified to direct and indirect methods. The direct evaluation method is further subdivided into internal and external. The internal method uses only data from the data set. Full inspection and sampling are the means to accomplishing the direct evaluation. The standard gives examples of sampling methods (e.g. simple random sampling, stratified sampling, multistage sampling and non-random sampling). ISO 2859 and ISO 3951 standards may also be applied. Indirect evaluation methods are based on estimates and may be derived from knowledge of the data set's lineage.

15.5 EXPERIENCES IN NATIONAL MAPPING AGENCIES

15.5.1 National Land Survey of Finland (NLS)

15.5.1.1 Quality management

The National Land Survey of Finland has developed quality management since the 1990's. It has been based on the ISO 9000 standard. The new process-focused organisation was adapted in 1999. The quality management is based on three perspectives: Organisation, processes and data sets. The organisation perspective has two different levels: NLS-wide and a unit level. Both NLS and different units have quality manuals defining the responsibilities and resources. Processes are defined with a process map. We have four key processes. Each of them has a process owner and a process team. A process team is responsible for process manuals. The manuals are updated over the intranet and every person in the organisation has access to them.

The Chief Director carries the main responsibility for quality. The management team will audit the main processes each year, and the surveying counsellors are responsible for the quality of the main processes. There are no quality management staff as such and the operative management carries the responsibility.

15.5.1.2 The data quality definitions in the NLS

In the National Land Survey of Finland, definition of data quality is based on several documents (Figure 15.2).

The Process Manual is based on ISO 9000 and defines the responsibilities and organisation for topographic data production. The Topographic Data Model defines the features in the real world and the Data Compilation Guide gives instructions on data collection. The Data Quality Model defines the quality elements and sets the requirements for quality. The missing parts are: **Quality reports** (should give information about the quality achieved) and **Product guides** (give information about a product, such as formats and copyright).

In the Data Quality Model, the quality factors are (Jakobsson, 1994):

- **lineage,** which means the source of data compilation, the data compilation date, the currency of the source material, the history of the data compilation and the name of those who were in charge of the compilation;

- **completeness,** which means how completely each data object was compiled into the data set;
- **currency,** which means how well the data set meets the required up-to-dateness;
- **accuracy of geographical location,** defined using positional accuracy, geometric accuracy and topological accuracy. (Positional accuracy is defined as the root mean square error (RMSE). Geometric accuracy means how well the geometric description (line, point or area) defines the real object. Topological accuracy means how well each object is compatible with other objects);
- **thematic accuracy,** defined by object class accuracy, attribute accuracy and temporal accuracy. (Object class accuracy defines how well each object class has been compiled. Attribute accuracy defines the quality of each attribute class, and temporal accuracy means the accuracy of the time measurement of the objects.); logical consistency, which means whether the objects meet the logical requirement defined in the data specification.

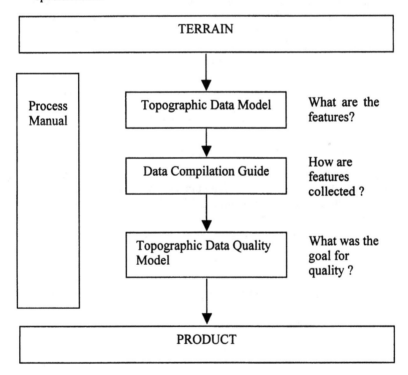

Figure 15.2 Quality documents in the Topographic Data System (TDS).

Quality requirements are defined for positional accuracy, completeness and thematic accuracy. The requirements for logical consistency are defined in the test procedures and all those should pass the tests. The positional accuracy requirement for a road is 3 metres (RMSE) in quality level A and the requirement for completeness for a road is that only 4 errors per hundred unit are allowed (acceptable quality level). For object class accuracy only 4 errors in classification of the road type is allowed. There can be a different requirement for each object class and attribute class (NLS, 1995).

15.5.1.3 Quality requirements and quality evaluation

From the point of view of the user, it is important how well the producer is able to guarantee the quality he has reported. The customer, of course, can go to the field himself and check some of the details but this is not usually possible for technical, cost or other reasons. On the other hand, it is important for the producer that the quality information is accurate, otherwise compensation for damages may be payable. Statistical tests are a good way of ensuring that the requirements are met at moderate cost. For example, the Ordnance Survey of Great Britain uses statistical methods for their test procedures. The NLS uses a statistical test for completeness and thematic accuracy.

In practice, we have found that customers tend to evaluate different quality requirements in the following order: coverage, currency and completeness, accuracy (positional, attribute), logical completeness, lineage and usage. As the requirement at the top of the list is met the next one becomes more important.

The quality evaluation procedures that were developed when the TDS was introduced in 1992 are documented below. The Topographic Database will now cover nearly the whole of Finland (some northern parts of Finland are still not covered). In fall 2000 the Topographic Database was exported to a Smallworld GIS and new quality evaluation procedures are now under development.

15.5.1.4 Testing of completeness and thematic accuracy at the NLS 1993-1998

Testing for completeness and thematic accuracy is carried out by applying the principles of standard SFS 4010 (Sample test procedures and tables; attribute inspection), which corresponds to standard ISO 2859 (1995). The standard defines the sample procedures for attribute inspection.

All the features on which data is collected are checked from the data source used if the quality requirement for the feature type is one nonconformity per 100 units (Acceptable Quality Level, AQL=1). An inspection based on sampling is made when the AQL of the feature type is 4 or 15. The inspection level is general inspection level I, the single sample programme for normal inspection (NLS, 1995).

15.5.1.5 Test for completeness

A lot must consist of map databases produced as far as possible at the same time and using the same methods. A lot is the minimum unit for which quality is evaluated. From the lot, an area of as many 1 km x 1 km squares is sampled as is needed to ensure a sufficient number of features with AQL=4.

The sampling is done using the default weights of the features. The weighting can be changed if necessary. In weighting, a default value of 1 is given to those features which have a significant presence in the lot or for which no AQL for completeness has been set. Features whose AQL is 4 or 15 are given a weight of 2 or 3. Weight 3 is given to features of which there are few. Otherwise, a weight of 2 is used.

Should the required sample size not be achieved, the program selects from the sampled squares those with the greatest number of stereoplotted features fulfilling the required sample size. At the same time, features with AQL=15 are tested according to the inspection level achieved. All features in the sampled squares are checked in the field. A feature does not conform if it is absent in the data set (omission) or if the feature in the 1 km x 1 km square does not exist in the field (commission).

15.5.1.6 Test for thematic accuracy

The test for thematic accuracy is made on the same material as the test for completeness. The number of errors permitted and the inspection level achieved are given on the test form if the completeness AQL is not 4. The quality supervisor inspects each item of attribute data on the basis of the source material. Attribute data are erroneous if they differ from the source material or are missing.

15.5.1.7 Processing nonconforming units

If the number of nonconformities is greater than permitted in the test of feature types, all features in the lot must be checked. The method is chosen on the basis of the source of information used, and is carried out on all features in the lot.

15.5.1.8 Experiences of tests performed

The results of tests performed provide information about the functionality of the data compilation processes and also about the quality of Topographic Database (TDB) data. Data digitised from graphic basic maps may also contain those errors made during the previous mapping process. Not all of the features are checked in the field during the data compilation process.

About 15 % of the annual production was tested during the year 1996. There were 33 tests made by our District Surveying Offices. Two or three 1 km x 1 km test squares can be checked in the field during one working day on average. The test results show that the biggest problems in completeness concern buildings, water less than 2 m wide, light-traffic routes and agricultural land. Not all of the features can be seen in a stereoplotter because of trees or shadows, for example. Time spent on field checking is minimised and not all the unseen features can be checked in the field. Results of tests for thematic accuracy have been mostly good, though there have been some errors with the usage information of buildings and classification of agricultural land.

The results of tests performed have come up to expectations in general. As a result of the tests, the quality requirements for some feature types have been changed and the instructions for data compilation have been adjusted. Quality tests provide information for maintenance of the TDS and its quality management. They also help the District Survey Offices to improve the quality and efficiency of their work (Pätynen *et al.*, 1997).

15.5.1.9 Testing positional accuracy and logical consistency

In 1996 and 1997 the NLS carried out a positional accuracy test in order to evaluate the accuracy of the TDB. The quality requirements set in the Data Quality Model were already tested in 1994. The test concerned covered about 1000 object instances (Jakobsson, 1994).

In 1997, 11 test areas where chosen for the positional accuracy test. The test was made using differential GPS. Only the most accurate feature classes, which have good identifiabilty in the field, were tested. They comprised buildings and roads that should have positional accuracy of 3 m according to the Data Quality Model. Five hundred buildings and 200 crossroads were measured and the result showed that the average positional accuracy of these feature classes was 2.2-2.3 m (Tätilä 1997).

During 1993–1998, the NLS carried out logical consistency tests. The number of errors per 1:10 000 map sheet decreased from a high 2.5 m to less than 1 m error in 1998. This included every kind of error possible in a data set.

15.5.2 Norwegian Mapping Authority

The Norwegian approach to quality evaluation is based on the Norwegian national standard for geographical information, called SOSI (Systematic Organisation of Spatial Information). The national standard includes data definitions for standardised geometry and topology, data quality, coordinate systems and metadata. All national mapping programs are based on the SOSI standard, which is also supported by all major GIS and mapping software providers (Sunde, 1998).

Quality control can be carried out easily using three different software packages. KVAKK (KvalitesKontroll av Kartdata – Quality control of map data) is a tool for automatic structure, coding and syntax control for data in the SOSI format. SOSIVIS is a free downloadable SOSI viewer and it can be used for interactive and visual content, structure, coding and syntax control. A full version called FYSAK is used internally at the mapping authority (Sunde, 1998).

15.5.3 Institut Géographique National (IGN), France

The Institut Géographique National (IGN) started to implement total quality control in 1987. In 1995 a decision was taken to start ISO 9000 implementation beginning at the topographic database 1:50 000 to 1:500 000 (BDCARTO) production. The organisation has published a quality chart that defines its quality policy and objectives. It has a quality director, with a separate quality unit for managing and coordinating the quality approach.

There is a separate control unit (MODV) that acts as a customer and has the right not to accept a product if it does not meet the specifications. The unit performs the quality evaluation of the BDCARTO, the topographic database 1:5000 to 1:25 000 (BDTOPO) and the road database (GEOROUTE). The MODV has (had) a staff of two and it takes about two weeks to inspect a one BDCARTO sheet.

The tools the IGN uses are:

- **a counting set** for statistics, detection of cross errors (e.g. missing attribute), verification of logical consistency and comparison of toponomy;
- **a comparison tool** with a reference data set, which calculates positional accuracy and semantic differences;
- **a visual check** with a raster (scanned data).

Using these tools, the control report is compiled and a decision is made on whether the data set should be accepted or not. The reference data set should be near to the nominal ground (the universe of discourse), so a surfaced road in BDCARTO is checked against a surfaced road in BDTOPO, for example.

In producing BDTOPO, quality is evaluated using two different types of measurements: the punctual **positional accuracy** of the data set entities compared with the nominal ground entities, the **exhaustivity** and **semantic accuracy** of the data set compared with nominal ground.

The measurement of geometric quality is carried out only to the feature categories represented by a point and they cover 19 point categories. The reference for horizontal accuracy is buildings and for vertical accuracy spot elevations. Two quality parameters are used in measuring semantic accuracy. Exhaustivity means the presence or absence of the data set entities compared with the nominal ground (rate of deficit, rate of excess), and semantic accuracy is the conformity of the data set entity classification compared with the nominal ground entities. Measurements are obtained by exhaustive sampling of part of the BDTOPO sheet. Any errors detected are noted on tracing paper that is kept for the next update of the database. Measures of deficits and excess items are accounted by entity class in an exhaustivity board. Measures of confusions are accounted by types of confusions between entity class or attribute values of an entity class in four different confusion matrices; practicability of roads, buildings, orography, hydrography and vegetation, and public, commercial and industrial activities (IGN, 1998).

All the results are registered and reported by the BDTOPO sheet. The results for all the year's measurements are registered, yielding a general report on the year and providing a mean for users of the database. Quality specifications are to be set in the future, when more measurements have been carried out.

15.6 CONCLUSIONS

In December 1996 the International Cartographic Association (ICA) and its Commission on Spatial Data Quality sent a worldwide questionnaire to producers of spatial data. The questionnaire went to 288 national mapping agencies and altogether 56 responded, most of them in Europe (59%). Table 15.3 gives a brief summary of results (Östman, 1997).

The questionnaire showed no quality estimate is made or only subjectively estimated in many instances. This situation needs to improve. As the examples show, national mapping agencies in Europe understand the importance of quality, but so far have not really invested in quality evaluation. The reasons are, of course, that the market has not demanded more and producers have got funding from government budgets. Now that the geographical information market is evolving and data sets are coming into everyday use, however, the role of the customer is but so far have not really invested in quality evaluation. The reasons are, of course, that the market has not demanded more and producers have got funding from government budgets. Now that the geographical information market is evolving and data sets are coming into everyday use, however, the role of the customer is changing. Customers are no longer professionals who know how the information is gathered and know what to expect. National mapping agencies therefore have to take quality seriously.

Table 15.3 Results of the ICA questionnaire: percentage of answers indicating a subjective evaluation or no evaluation as the quality assurance routine (Östman, 1997).

Quality parameters	Percentage
Positional accuracy	43
Thematic accuracy	48
Temporal accuracy	68
Completeness	56
Logical consistency	32

The European National Mapping Agencies have co-operated to improve the knowledge of quality management and data quality, especially through the work of EuroGeographics (formerly CERCO). The quality working group was created in 1997, and it has produced several documents related to these issues. In 1999 the working group conducted a questionnaire survey about data quality. The key issues are process management, data set specifications, standards, quality reports and training of personnel.

Quality management and quality information together constitute a significant competitive advantage for producers. Further better quality often saves costs too and raises process output.

REFERENCES

ENV 12656, 1998, Geographic Information - Data Description – Quality, Adopted European Prestandard.(www.cenorm.be).

Godwin, L.S., 1997, Coordinating Standards on Data Quality: An Important Ingredient for Cartographers. In *Proceedings of the 18th ICA International Cartographic Conference,* Sweden, (Gävle: Swedish Cartographic Society), pp. 533–540.

Guptill, S.C., 1998, Building a Geospatial Data Framework, Finding the "Best Available" Data. In *Data Quality in Geographic Information From Error to Uncertainty*, edited by Jeansoulin, R. and Goodchild, M. F., (Paris: Hermes), pp. 31–36.

IGN, 1998, Measuring the Geometric and Semantic Quality of the BD TOPO, Institut Géographique National, France, unpublished.

ISO 3951, 1989, *Sampling Procedures and Charts for Inspection by Variables for Percent Nonconforming,* The International Organisation for Standardisation (www.iso.ch).

ISO 2859-0:1995, 1995, *Sampling Procedures for Inspection by Attributes -- Part 0: Introduction to the ISO 2859 Attribute Sampling System*, The International Organisation for Standardisation (www.iso.ch).

ISO 9000: 2000, 2000, *Quality Management Systems -- Fundamentals and Vocabulary*, The International Organisation for Standardisation (www.iso.ch).

ISO TC211, 2001a, ISO/DIS 19113, *Geographic Information – Quality Principles, Text for DIS 19113*, The International Organisation for Standardisation (www.iso.ch).

ISO TC211, 2001b, ISO/DIS 19114, *Geographic Information – Quality Evaluation Procedures*, International Standard Committee Draft (www.iso.ch).

Jakobsson, A., 1994, Quality Procedures of the National Topographic Data System in Finland, *GIM, International Journal for Surveying, Mapping and Applied GIS*, October 1994, pp. 39–41.

Jakobsson, A., 1995, The Topographic Data System –A New Way to Compile and Update Topographic Information. In *Proceedings of the 17th International Cartographic Conference* Vol. 1, (Barcelona: Institute Cartogràphic de Catalunya), pp. 993–1000.

Jakobsson, A., 1999, Quality Evaluation of Topographic Datasets –Experiences in European National Mapping Agencies. In *Proceedings of the International Symposium on Spatial Data Quality*, (Hong Kong: Polytechnic University). pp.154–162.

NLS, 1995, *Topographic Data Quality Model*, (Helsinki, National Land Survey of Finland).

Östman, A. 1997, The specification and Evaluation of Spatial Data Quality. In *Proceedings of the 18th ICA International Cartographic Conference*, Sweden, (Gävle: Swedish Cartographic Society), pp. 836–847.

Pätynen, V., Kemppainen, I., Ronkainen R., 1997, Testing for Completeness and Thematic Accuracy of the National Topographic Data System in Finland. In *Proceedings of the 18th ICA International Cartographic Conference*, Sweden, (Gävle: Swedish Cartographic Society), pp. 1360– 1367.

Sunde, K.O., 1998, *Presentation of Data Standards, Specifications and Quality Measurement Tools Used for Data Quality Control*, unpublished.

Tätilä, P., 1997, Report on Positional Accuracy Tests, in Finnish, National Land Survey of Finland, unpublished.

CHAPTER SIXTEEN

A GIS with the Capacity for Managing Data Quality Information

Jibo Qiu and Gary J. Hunter

ABSTRACT

While spatial data quality statements are now routinely provided to clients by data producers, it is argued that their usefulness still remains low. We believe this is due to two problems, (1) that the current method of reporting data quality is poorly structured and remains physically separated from the actual data, and (2) that current geographical information system (GIS) packages continue to lack the tools to effectively manage (that is, to store, manipulate, query, update and display) any embedded data quality information.

Accordingly, this chapter proposes a model for managing data quality information in which a multi-level structure is adopted according to the hierarchy of objects in an associated GIS database. The model has been implemented using the Microsoft Access and ESRI ArcView software, and the integration of metadata with spatial data is achieved by using a relational database and Dynamic Data Exchange between the metadata database and the GIS. In this way we can store, access and report data quality information at different levels in the database.

16.1 INTRODUCTION

Metadata specifications have now been established by numerous government agencies around the world and in 1994, for example, the U.S. Federal Geographical Data Committee (FGDC) developed a metadata content standard to fit within the framework of that country's National Spatial Data Infrastructure (NSDI). Within the metadata standard, data quality information was included as an essential component, and in Australia/New Zealand data quality information has also been deemed to be one of the core elements of spatial metadata by the Australia New Zealand Land Information Council (ANZLIC).

However, we believe that data quality information embedded in metadata is still rarely applied by GIS users, and there are few cases where data quality information is included in databases and presented to GIS users in formats suitable for their intended needs. We consider this is due to the fact that data quality models proposed by current standards have inherent limitations such as documenting data quality information at global rather than local levels. In addition, many current GIS packages still lack the tools to manage (that is, to record, manipulate, query, update and display) data quality information. Some of these deficiencies have already been recognised by software developers, and recently we have seen the ESRI ArcInfo 8 product enhanced to include functions for simple management of metadata in the ArcCatalog module (ESRI, 2000).

In this module, metadata is stored in XML files and documented at more detailed levels such as feature classes and individual features. Furthermore, user-friendly interfaces are presented for editing and browsing metadata at these different levels of detail. Finally, basic metadata such as coordinate projection information is updated automatically when coordinate projection parameters are changed. However, these functions remain limited to a small subset of metadata types. As such, we believe there is still a demand for further research and development in the area of effectively managing data quality information.

Accordingly, this chapter proposes a GIS with the capacity for data quality management. It first presents a design strategy for such a system which consists of three basic aspects: the model for the system; the software design; and the style or appearance of the system. Since developing a GIS with the capacity for data quality management requires integration of related disciplines such as database technology, object-oriented approaches, human-computer interaction and cognitive ergonomics, these aspects will be given consideration in the design of the overall system. The discussion of the design strategy then leads to the formulation and development of a prototype GIS for managing data quality information.

16.2 SYSTEM DESIGN STRATEGY

The development of a GIS with the capacity for data quality management consists of three key design aspects, viz.:
- the data quality model—using a multilevel structure;
- the analysis and design of the system software—using an object-oriented approach; and
- the style of the system—using interactive processing.

A multilevel data quality model is presented and incorporated in the proposed system, and it serves as the basis for the design of a database to store data quality information. As an important part of the system development, software that provides the functionality required for data quality management needs to be constructed.

An object-oriented approach in both the analysis and design of the system software is adopted, which aims to provide improved software functionality. Furthermore, data quality information management involves a series of processes such as accessing and updating metadata, and the processes need to be carried out in an interactive manner. In this section each of the aspects is discussed in association with related disciplines.

16.2.1 The data quality model—a multilevel structure

A hierarchy is a structure within which all elements have only one-to-many relationships with each another (apart from terminal nodes). According to standard terminology every element, except at the highest level, has a "parent" which is the element immediately above it. The descendants of an element are called "children". The higher level elements tend to be more abstract than those at lower levels. This hierarchy also exists in GIS databases, which are collections of objects representing geographical phenomena and their relationships to each other. Moving from the general to specific classes, data can be held at four levels: the data set level, data layer level, feature class level and individual feature level (Figure 16.1).

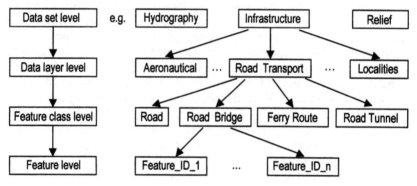

Figure 16.1 The hierarchy of objects in a GIS database.

This hierarchy is also recognised in the design of many systems as identified by Fisher (1993).

Thus, in a GIS database a hierarchical set of object relations at different levels of granularity can be defined as follows:

- The data set is the highest level and is a collection of logically related geographical information about a particular theme. For example, the TOPO-250K data product prepared by the Australian Surveying and Land Information Group (AUSLIG, 1992) contains three data sets representing hydrography, infrastructure and relief. Each dataset contains one or more data layers and thus a one to many (1:n) relationship exists between the data sets and data layers.

- At the second level, a data layer is a grouping of features representing compatible spatial objects that depict a specific theme. For example, a polygon layer of waterbodies contains features such as lakes, swamps, land subject to inundation and watercourses sufficiently wide enough to be shown as polygons—which are all related to the specific theme of hydrography.

- Next, if a certain description of characteristics is common to a group of features, the group constitutes a feature class. A feature class is an abstract object, since it provides a specification with no implementation detail. In most cases, it is user-defined.

- Finally, the greatest degree of granularity occurs at the feature level which contains detailed information for individual features and their relationships in each data layer. Each feature may be represented by points, lines and polygons in a vector data structure or pixels in a raster data structure. Attached to each feature are one or more attributes.

According to this hierarchical structure of objects in a GIS, a multi-level approach is also proposed for modelling and storing data quality information. The method assumes that data quality information is available and can be associated with geographical objects at different levels. The quality information at each node (or level) is shared by its children, and data at each node not only inherit quality information from their parent node but may also have their own distinct quality information.

It is recognised here that the descriptions of different elements of spatial data quality vary considerably, with some referring to individual features, others referring to objects at the feature class level, and some referring to objects at the data layer or data set levels. So it is essential to be clear about what the data quality descriptions attach to, since there are several different levels of granularity in the GIS data that quality measures may be associated with. For example, attribute accuracy can be either described at a data layer level by the proportion of points (pixels) correctly classified (the overall accuracy), or at a feature class level by the "error of omission" and "error of commission" (the accuracy of the individual categories).

The multi-level structure also has the advantage of facilitating storage of data quality information. Objects at lower levels will inherit the attributes of their parents, and this inheritance permits more efficient storage and maintenance of data quality information, since much of the data quality information for objects at a given level is identical. For example, the lineage details for individual features can be stored at the parent level.

The access and reporting of data quality information also benefits from the multi-level structure. Considering data granularity, users may want to examine data quality at different levels. Some may require a very detailed quality report of a given feature class or individual feature, while other users may only need a brief summary of quality information for a data layer or data set. Thus, a data quality model that adopts a multilevel structure provides the basis for designing a database to store data quality information.

Conceptually, the model ensures the integration of both data and data quality information physically and contextually, since modelling data quality is an integral part of the database design process. From an operational viewpoint, incorporating the data quality model into the database design will facilitate practical procedures for managing (such as querying and updating) data quality information in the GIS, since it allows quality information to be stored in a way that can be easily accessed and linked to the actual data it describes.

16.2.2 The approach—object-oriented analysis and design

Object-oriented (OO) concepts first emerged in the design of programming languages such as Smalltalk, Eiffel and C++. Since then, the term 'object-orientation' has been widely used in computer science and information technology. Nowadays, object-orientation has become a paradigm adopted in the whole procedures of software development, and the paradigm of object-orientation can be seen as the synergistic embodiment of three essential aspects (Henderson-Sellers, 1997):

- encapsulation;
- classification;
- inheritance and polymorphism.

The OO concept has been adopted in the software development in the proposed system. As Henderson-Seller (1997, p. 3) remarked, "[the] OO paradigm is a way of looking at the whole of the software life cycle: it's a way of modeling business problems in a more 'natural' way. It is also a whole new way of thinking about software development." In this research, the OO ideas are used in handling data quality information, and the concepts of encapsulation, classification and inheritance are employed in the analysis and design of the software for data quality management.

As data quality information is collected and stored as metadata in a database, a database table can be regarded as a group of objects subject to the same type of quality information. Each row in the table represents an object and each column represents an attribute of the objects. The functions for managing this type of quality information, such as accessing the information stored in the table, are also defined and encapsulated with the objects in that class. The idea is depicted in Figure 16.2 which shows a database table for storing quality information.

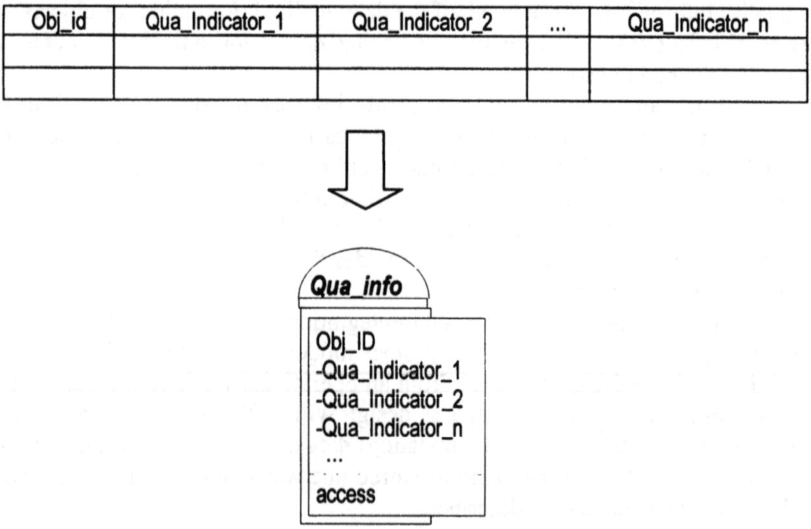

Obj_id	Qua_Indicator_1	Qua_Indicator_2	...	Qua_Indicator_n

Qua_info

Obj_ID
-Qua_indicator_1
-Qua_Indicator_2
-Qua_Indicator_n
...
access

Figure 16.2 Defining a class from a database table.

Accordingly, an object class is defined and represented by the icon (following the MOSES system of notation). For example, when a table containing positional accuracy information relating to cartographic displacement for each feature category is created, it can be regarded as an object class of "quality information on cartographic displacement for feature categories". The class specifies a set of attributes including the ID of the feature category and quality indicators, such as situation of displacement and typical displacement. It also determines the functions to access such information. Thus, tables in a database are regarded as object classes.

The classes can be further generalised into a superclass (parent class) if there exists an "is-kind-of" relationship between the subclass and superclass. For example, classes of quality information on parameters such as positional accuracy and logical consistency reported at the feature category level can be generalised into an object class of "quality information at the level of feature category", since both positional accuracy and logical consistency are quality information reported at feature category level. Hence, a superclass is created within an inheritance structure (see Figure 16.3).

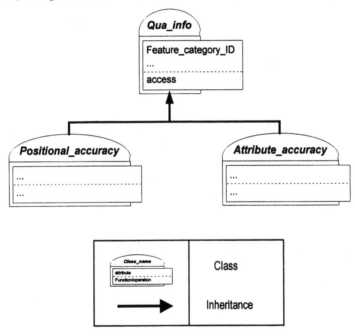

Figure 16.3 Hierarchical inheritance between superclass and subclass.

The inheritance allows the objects in the sub-class to inherit the operations (or methods) from their parent. This enables writing generic code that can be shared by many classes. For example, in Figure 16.3 the operation used to access quality information at feature category level is written in a code that can be shared by the sub-classes, which means quality information on positional accuracy and logical consistency can be accessed using the same code. Meanwhile, inheritance also allows any difference from this generic function to be easily defined in subclasses. For example, the function for visualising data quality information is subject to the type of quality parameters and indicators provided. Therefore, the operation to visualize quality information with the same type as that in the parent class needs to be defined for each subclass, and the method for that function will override that inherited from the parent class.

Polymorphism permits sending a message to an object to request an operation without knowing its class, and the appropriate method for that class will be invoked on the object. This is useful in managing different types of data quality information, since many of the functions are polymorphic which means that the one function may apply to many different classes by different implementation methods. For example, the function to "access" quality information may invoke different methods on classes of "quality information at feature class level" and "quality information at feature level". When the message is sent to an object to access quality information at a certain level, it only invokes the appropriate method for accessing the information at that level. Thus, even though we do not know the class that the object belongs to, we can send it the same message and the appropriate method is invoked depending on the class of the object.

Clearly, the benefits of an OO approach will be fully reflected in the analysis and design of a prototype system with the capability for data quality management. Unlike the conventional method for building software systems with large portions of the code to deal with closely related individual functions (Deitel and Deitel, 1994), OO features such as encapsulation, inheritance and polymorphism enable us to write code which is reusable and also flexible enough to be modified. Although the system software will be written in non-OO programming languages, the OO concepts can be mapped into the construct of the programming languages. As Henderson-Sellers (1997) remarked, it is possible to undertake an OO-based analysis and design, yet implement them in a non-OO programming language, in which OO analysis and design techniques will still provide a valuable precursor to implementation in a traditional language environment by providing a better design with a higher degree of encapsulation and information hiding.

16.2.3 The style—interactive processing

Because the processing of data quality management for the proposed system is operated through a human-computer interface (HCI), the design of the user-interface is considered a key issue. User interfaces are probably the most important aspect of HCI as they provide the ability for a user to complete a task or series of tasks and are thus the part of the system that the user sees and interacts with, as far as the user is concerned (Medyckyj-Scott, 1991). Therefore, the concept of user interfaces should be considered to extend beyond the simplistic representation of tools or channels within a system. Many previous studies have emphasised that designing a user-friendly interface for software is of major importance in system development (Nyerges *et al.*, 1995; Markham and Rix, 1995), and rules and guidelines in interface design have already been established in previous research (Markham and Rix, 1995) so that the users are provided with an environment that features ease of use and cognitive ergonomics.

Ergonomically, the interface should offer a simple means for carrying out actions and operations, such as the provision of push-buttons, option-buttons, scroll-bars and simple data entry fields in the dialog box. It should allow a user to operate the system in direct-manipulation style in both action and feeling. Thus, graphic user interfaces are suggested here such as forms and menus which have been conceived as being user-friendly.

With forms, the system decides the nature of a predetermined sequence of inputs and displays simultaneous requests for a complete set of inputs (Faulkner, 1998). A user completes all input requests and then submits to the system by clicking on an acceptance button. Figure 16.4 is an example of the form designed for querying quality information at the data set level. A formatted screen is used to collect the requests by filling in the form. Users construct a query statement by selecting a relevant data set and data quality parameters from the pick lists to perform information access in accordance with their requirements.

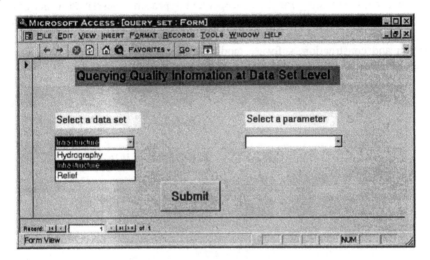

Figure 16.4 Queries constructed by the use of form-filling.

Similarly, menus are a common component of computer interfaces and are used frequently as an access mechanism. With menus, a user makes a choice among a number of options offered by the system. Menus may be structured hierarchically with the choice of an option resulting in the display of a new menu or form, which in turn provides a logical access path. This is termed "hierarchical menu selection" by Medyckyj-Scott (1991). While graphic user interfaces such as forms and menus are adopted as a design type, cognitive ergonomics pays attention to the actual design of the interfaces. Menu items should be cognitively organised and the emphasis is placed upon:

- concrete literal conceptual models to encourage greater attention to aspects such as direct graphics manipulation and cognitive ergonomics;
- system and user mental models, to focus attention on the human side of the interaction.

Accordingly, models of procedural sequences and how the users apply and think about the system ought to be reflected in the design of menus and forms so that the system can provide its user with a positive cognitive ergonomic structure, in both physical and mental forms. For example, a conceptual model of how the quality information is organised and stored in the system database can be provided to the users via the design of a menu. Figure 16.5 is an example of such a menu that implies the quality information is recorded and accessed at four distinct levels. Moreover, it is important that the hierarchical organisation of the menus provides a conceptual model of how options and functions within a system are grouped. Thus, the model should be carefully considered so that it will facilitate users' learning of an appropriate mental model of how the system operates.

16.3 A PROTOTYPE SYSTEM FOR DATA QUALITY MANAGEMENT

A prototype system has been developed during this research. Its main purpose is to conduct real data testing for the concepts and approaches described in the design strategy of the proposed system. The development of the prototype progresses through the setup and implementation of a system environment, the database design and construction, software module programming, the design of a system user-interface, and the implementation of the prototype using real data.

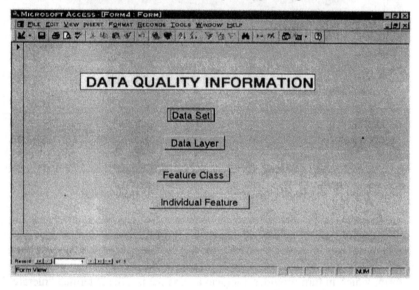

Figure 16.5 A menu interface for querying quality information at different levels.

16.3.1 Data resources

The data used for the evaluation of the system come from the TOPO-250K sample data produced by Australian Surveying and Land Information Group (AUSLIG). The data set is a digital version of the 1:250,000 Australian national topographic map series. The TOPO-250K data set uses a feature-based data model within which phenomena in the real world are represented by vector features. For example, a road is a feature and the characteristics of the feature are described by attributes such as the name of the road. A feature instance is the unique occurrence of a feature defined by its location and the values of its attributes. There are approximately 40 feature classes in the TOPO-250K data, with features having up to seven attributes (AUSLIG, 1992). The locational component of the data is provided by spatial objects in the form of points, nodes, chains and polygons, with each object being linked to attribute information that describe the feature.

The sample data are provided in ArcView format with the TOPO-250K Demonstration Package which consists of the Demonstration Data User Guide and data files for the Coffs Harbour map tile in the State of New South Wales. These files include three data sets: Hydrography, Infrastructure and Relief and each data set contains a number of layers. There is one data quality statement file for each data set that contains information about each tile.

Apart from the data quality information stored in the files, there is also a data quality table within which each record is linked to a feature instance in the TOPO-250K data. This table holds data quality information for each feature with respect to feature and attribute reliability, planimetric accuracy and elevation accuracy. The data quality information prepared by AUSLIG for the TOPO-250K product is an excellent example of detailed quality reporting and can be accessed online at http://www.auslig.gov.au.

16.3.2 System environment

An integrated system environment is required which can handle both spatial data and their metadata. This requirement is for two reasons. Firstly, spatial data have distinct characteristics such as geographical position, attributes, spatial relationships and currency, which make these data uniquely difficult to handle using conventional database systems. GIS software is therefore required for the storage and management of spatial data. In addition, quality information for data extending over a geographical region also has spatial characteristics, and should be regarded as spatial metadata in order to permit the local variation of data quality to be revealed (Wong and Wu, 1996). GIS software is thus needed as the development platform.

Secondly, data quality information can be expressed in different forms such as numerical values, text reports and graphical displays. The capability and efficiency of the storage and access of various forms of data quality information in a database are therefore crucial to the system's success. Hence, software selected for data quality management should be able to deal with different types of data and other limitations should not be imposed on them such as the size and type of the data. However, current GIS packages which can provide tools for handling spatial data are not necessarily ideal for managing large volumes and different forms of data quality metadata. As such, database rules are required to be developed to provide the linkage between the data and the data quality information in the various stages of the quality management process. Thus, it is suggested in this research that metadata can be stored using commercial database software, since it can provide structured data storage and fast data access, and also allows data sharing and publishing. In doing so, the data quality information can be linked to the actual data in GIS as most contemporary database packages can be interfaced with GIS software. Therefore, for the project investigations, Microsoft Access and

ESRI's ArcView have been chosen as the database management software and GIS software respectively for reasons of system availability, applicability of the outcome and data availability.

The actual spatial data and their attributes are stored in an ArcView project file, while the data quality information is recorded in a set of tables in MS Access. The integration of the two software systems is achieved by a client/server mechanism called Dynamic Data Exchange (DDE). It enables two applications to talk to each other by continuously and automatically exchanging data. When the applications interact, a conversation is established. One application is the server and the other application is the client. In this implementation, ArcView is a client to MS Access. Four conversations have been established in the implementation to access data quality information at the four levels. The conversations are carried out by coding programs in ESRI's Avenue scripting language, and they include the procedures for creating the ArcView DDE Client, initiating conversation, sending requests and presenting what is being received. In doing so, data quality information is associated with the actual data it describes through the conversation.

16.3.3 Spatial database design

Data supplied in ArcInfo export format are imported to ArcView. The data attribute tables, which are supplied as part of the spatial data, are also loaded to ArcView and stored in a project file. Objects in the database are represented at four levels: the data set, data layer, feature class and individual feature levels. A data set corresponds to a view listed in the ArcView project window. Each view is further made up of a set of themes that are collections of geographical features such as roads and rivers. Thus, each theme actually represents a data layer.

All the themes in a view are listed to the left of the map in the view's Table of Contents, which also shows the symbols used to draw the features in each theme. The check box next to each theme indicates whether it is currently displayed on the map or not. Associated with each theme is an attribute table, and each row in the table records attribute information of an individual feature in the layer, which is identified by its feature ID. The column labelled "feat_code" in the attribute table also shows the class that the feature belongs to. Thus, GIS objects at different levels are represented in ArcView by the project, view, theme or records in an attribute table respectively.

16.3.4 Data quality database design

According to the metadata model proposed in Section 16.2.1, the available quality information is organised using a multi-level structure and associated with the object it describes. Since the quality information for TOPO-250K sample data is available at four levels, data quality information is also structured in the same four levels. The model is then incorporated into the design of the metadata database which is a relational one for the prototype system. Quality information supplied in a text format is converted to a series of tables.

Other information such as the names of and the relationships between the data sets, data layers and feature classes is also stored in the database tables. To prevent modification anomalies, normalisation is carried out when defining the tables. Consequently, at the highest level (the data set level), the quality information is held in the following tables:

- THEME (Theme_ID, Themename, Description);
- QUA_THEME (Theme_ID, lineage, position, attribute, logical consistency, completeness).

At the data layer level, information for each data layer and its relationship with each data set is stored in separate tables as follows:

- LAYER (Layer_ID, Layername, Description,);
- LAYER_THEME (Layer_ID, Theme_ID);
- QUA_LAYER (Layer_ID, position, attribute, completeness, logical consistency).

At the feature class level, the following tables store the relationships between each feature class and each data layer, and the quality information of each feature class:

- FEATURE_CLASS (FClass_ID, Feature_code, FClassname, Layer_ID, view_ID);
- Carto_displa_3 (FClass_ID, test_id, position,);
- Attribute_3 (FClass_ID, test_id, attribute,);
- Logcon_3 (FClass_ID, test_id, logical consistency);
- Comple_3 (FClass_ID, test_id, completeness).

Finally, at the individual feature level, quality information for individual features is stored in the following table:

- QUA_FEATURE (Feature_ID, quainfo [pre-existing table in the TOPO-250K database containing quality information at the feature level]).

Each database table is assumed to include fields permitting linkage to attribute tables within the same database or to the GIS database within which the TOPO-250K data are stored. Underlined fields are primary keys for each table. The data quality database is then created using MS Access.

16.3.5 Software development

In Section 16.2.2, an object-oriented approach was suggested in the analysis and design of the software for data quality management. Accordingly, the OO approach is employed in developing the software for the prototype system. Figure 16.6 is an object model of the system. From the software architecture perspective, the classes represented in the figure need to be grouped and regarded as a series of modules that are formulated as programs. This is fully supported by OO design principles, since its modular structure causes no problems with coding as subroutines in procedural languages, and will enforce modularity into the procedural language (Henderson-Sellers, 1997).

Consequently, four modules were formed for querying, reporting and updating data quality information, and browsing data contents. The first module

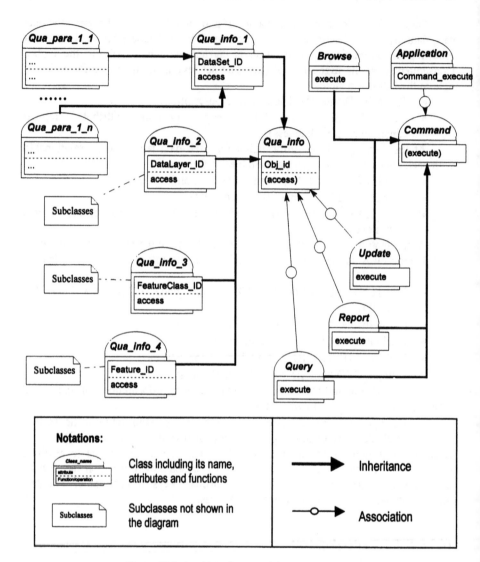

Figure 16.6 An object diagram of the prototype system.

was developed for querying data quality information. It provides the functions for a user to construct and submit a query to the database, and the results of the request are collected, passed back and presented to the user. The function of generating data quality reports was developed in the second module. It allows a user to retrieve data quality information associated with a data layer of interest from the database and presents the information in the form of a report. The report can be a brief summary which only contains quality information at global level or else a detailed statement including quality information at both the global and local levels. The module design leaves this choice to the user.

The third module was developed for updating data quality information. As data go through routine revision or a series of GIS operations, the metadata database needs to be updated. The module design provides for the user to access the corresponding tables in the database and modify their quality information. It further allows a user to create or delete tables when a new object is created or an old one is deleted from the GIS database. A detailed discussion regarding dynamic updating of data quality information and the function of the module has been presented elsewhere by Qiu and Hunter (2000). Each of these modules will in turn call submodules for accessing quality information at different levels either from the ArcView project window or the designed metadata database user-interface. Apart from those modules that provide functions for data quality management, the module for browsing data contents provides additional functionality for a user to inspect the data that are available in the database.

The functions for data quality management are thus accomplished by invoking the corresponding modules described above. Each module can call the submodules and communicate with each other under control modules, while a user-interface may be regarded as being the highest level of control. The modules are developed under the combination of Avenue script files, embedded SQL and Access Basic software.

Thus, the software is based on an OO analysis and design which provides a number of benefits. Firstly, there are different tasks involved in managing data quality information, and separating these tasks into different modules will simplify any modifications needed to accommodate changes or extensions to the system. Changes in underlying databases, such as adding a new type of data quality information to the database, only require modification of a single program module without affecting the rest of the system. Secondly, additional new functions such as visualising quality information can be added to the system by designing new modules. Thirdly, the actual software coding of the system modules is more tractable when small independent pieces are developed rather than a single large program.

16.3.6 A user-friendly interface

It is essential that sound programming methods and techniques be followed to evaluate the described system. These should also include the design of a user-friendly interface for the software. In an integrated environment, the programs developed will act as mechanisms, while the design interface will bridge the system and user. The prototype system can be accessed from within the ArcView interface. From the pull-down 'Project' menu located in the menu bar (Figure 16.7a), there is an option for opening the metadata database. In choosing this option the main menu named 'Metadata Database' is invoked in Figure 16.7b. It contains two main components that deal with browsing data content and data quality management—which in turn permits querying, reporting and updating of quality information (Figure 16.7c). The choice of any option in the menu in Figure 16.7c will invoke a submenu for completing the selected process. This forms a hierarchical menu selection as described in Section 16.2.2.

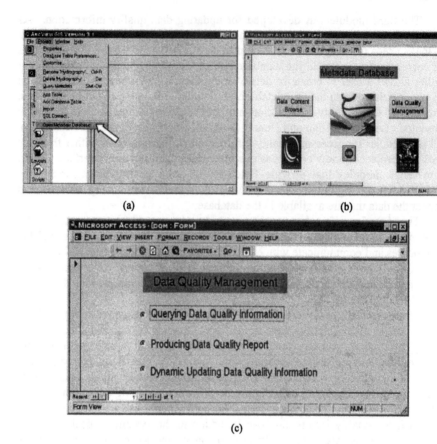

(a)   (b)

(c)

Figure 16.7 User interfaces in the metadata database.

Apart from the initial option of "Open Metadata Database" in the pull-down menu bar which links to the main menu of the metadata database, data quality information can also be queried in a GIS environment. The system allows the user to select any object in the map to query its quality information. In ArcView, the project window interface lists the name of all the views. While the project window is active, there is a 'Quality' button in the toolbar designed for querying data quality information at data set level. By selecting a view which also represents a data set and invoking it, a view window will be opened with all the data layers (known as themes in ArcView) listed to the left of the displayed layers. When the view window is invoked, a further 'Quality' button is available in the toolbar for accessing the quality information of each data layer. Quality information at the feature class level can further be accessed when a theme's attribute table is open and active, and finally quality information at the individual feature level can be accessed by selecting the feature in the displayed layer and again clicking the 'Quality' button.

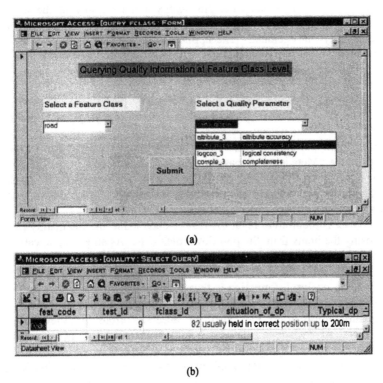

(a)

(b)

Figure 16.8 Querying data quality information in the metadata database environment.

From the cognitive ergonomics viewpoint, the hierarchical organisation of the menu system helps a user to progress through the necessary steps associated with the different data quality management functions. Each interface is organised in its operational sequence with a minimum of menu items. The provision of push-buttons, option-buttons and scroll-bars offers a simple approach to actions and operations. This will be reflected in both action and feeling since it allows users to simply click a mouse button and see what is happening, instead of imagining what has been happening as a consequence of the action they have just taken. Moreover, the menu button is cognitively organised to provide a conceptual model which is consistent with the multi-level structure of the metadata model. Furthermore, all the names of objects in the database become available to the user via the pull-down pick lists designed at different levels (Figure 16.8). Thus, a user can select information of interest without keying-in its name, which saves time and standardises the setting of variables.

16.3.7 System implementation

The prototype system has been implemented and demonstrated using real data. It is intended to further evaluate the practicality and flexibility of the multi-level structure data quality model as the basis of a GIS to provide the functions needed for data quality management. The applicability of the proposed system is also demonstrated. The implementation of the prototype is carried out in the experimental environment described in Section 16.3.2, which is a combined system environment using MS Access and ArcView.

The system is demonstrated and evaluated by performing the tasks of data quality management, that is, browsing data content, querying and reporting data quality.

The quality information stored in the metadata database can be accessed either from the "Metadata Database" interface in MS Access or via the Arc/View project window interface. The prototype system enables users to construct a metadata query interactively as previously mentioned. Figure 16.8a shows the menu for querying quality information at the feature class level.

The user is required to input the feature class name and the quality parameter by choosing the items from the pull-down pick lists. As an example, cartographic displacement is selected as the quality information required for the feature class "road". The constructed query is then submitted to the database system and the corresponding quality information is thus displayed in a table format as shown in Figure 16.8b.

Quality information can also be accessed from ArcView in a GIS environment. At each level, there is a button designed for invoking the query process which in turn initiates a conversation (DDE) between ArcView and MS Access. Using the same example of querying data quality information for feature class "road", the quality information at the feature class level is accessed when the theme's attribute table is open and active.

The query process is invoked by clicking the quality button "Q" on the left of the top toolbar in Figure 16.9a, which leads to the result of the query being displayed on screen with the selected feature class road highlighted in the yellow color in the ArcView project window.

The information relating to cartographic displacement is displayed in tabular format in a MS Access window at the bottom of the screen (Figure 16.9b).

In GIS, a data layer is the unit that is often used in spatial data representation, manipulation and analysis. It is sometimes required to produce a data quality report for a particular data layer so that users can examine the quality information when they assess the fitness-for-use of the data for a given task.

The prototype system provides the ability to generate data quality reports for the data layers, and Figure 16.10 illustrates the menu to assist the user to produce a data quality report from the available metadata stored in the database. A user can select the data layer of interest and construct a quality report by retrieving quality information reported at the necessary levels.

A user can either produce a brief summary by reporting the quality information at global level, or else a detailed report can be generated by further including the quality information at feature class level and feature level.

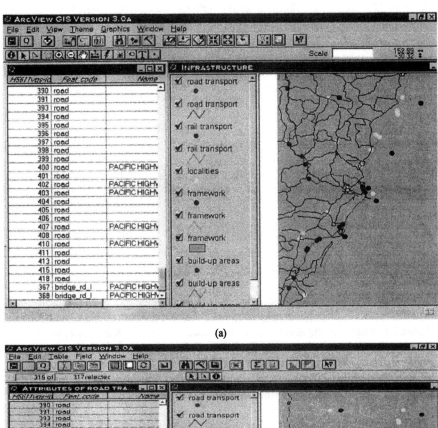

(a)

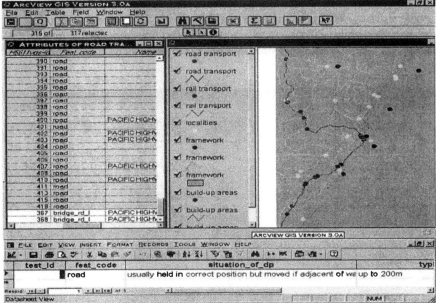

(b)

Figure 16.9 Querying data quality information in a GIS environment.

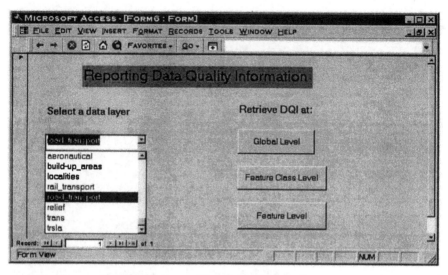

Figure 16.10 A menu for producing a data quality report.

Finally, there is often a need for a user to examine the data available for a particular task and/or geographical area (Medyckyj-Scott, 1991), and this is achieved through the capability to browse data contents in the prototype system. Selecting the option of "Data Content Browse" from the menu in Figure 16.7b will invoke the form in Figure 16.11. It allows the users to browse the data contents and provides information including what data sets are available, what the data layers are in each data set and what the feature classes are in each data layer.

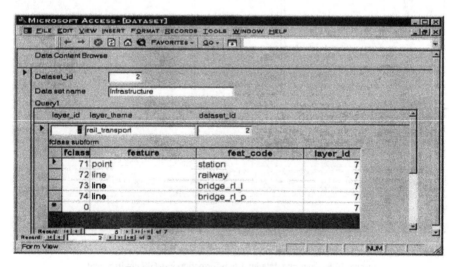

Figure 16.11 A form for browsing data content.

16.4 CONCLUSIONS

This chapter has proposed a GIS with the capacity for data quality management. It has shown how to incorporate the various concepts and procedures into a coherent body and develop them into a systematic approach for data quality management. The discussion has targeted three basic aspects of the proposed system:

- the metadata model incorporated in the system design;
- the object-orientated analysis and design of the system; and
- the interactive style of the system.

Firstly, to develop a GIS with the capability of data quality management, the proposed data quality model has been suggested as the basis of such a system. The model is incorporated into the design of a metadata database to ensure the integration of data and their quality information. Secondly, data quality management involves a range of tasks such as accessing, updating and reporting data quality information. The concept of object-orientation has been introduced and an OO approach is suggested for the design and analysis of the proposed system. This allows features such as encapsulation, classification, and inheritance to be fully utilised in the development of the system, in order to provide higher quality software which is more flexible, maintainable and has improved reusability.

Thirdly, many of the processes of data quality management are interactive operations which are the actions between users and computers, and users and information. Thus, the human-computer interface has also been fused into the design strategy for such a system. The importance of providing a user-friendly interface has been discussed so that formal rules and guidelines can be followed and incorporated into the design of the user interfaces for the proposed system in order to provide an environment that is easy to use and has cognitive ergonomics.

A prototype system has been presented to demonstrate the concepts and developments described in the design strategy. The selection of the system hardware and software was mainly based on the consideration of availability, applicability and data sources. The procedure of designing the prototype system consisted of database design, software development, interface design and implementation tests. A relational metadata database based on the proposed data quality model has been created in MS Access. The data quality information stored in the database is then linked to the spatial data in an ArcView GIS via DDE and primary keys defined in the databases. The development of the software as the database engine has been sub-divided into program modules based on the OO design and analysis principles which are controlled by the designed user interfaces.

As with the manner of human-computer interaction, the concept of the user interface has been considered to extend beyond the simplistic representation of a tool or an interaction channel within the system. The graphic interface has been deemed as having shared responsibility for user-friendliness and cognitive ergonomics in the system implementation. Finally, to demonstrate the enhanced capability of data quality management, the prototype system has been demonstrated by performing different functions of data quality management such as data content browsing, querying and reporting data quality information.

REFERENCES

AUSLIG, 1992, (1992) Geodata Topo-250K Data User Guide, Australian Surveying and Land Information System Group, Canberra, Australia.

Deitel, H.M. and Deitel, P.J., 1994, *C++ How to Program*, (Prentice Hall: New Jersey), p. 485.

ESRI, 2000, *Using ArcCatalog*, (ESRI: Redlands, California).

Faulkner, C., 1998, *The Essence of Human-Computer Interaction*, (Prentice Hall: London, New York).

Fisher, P.F., 1993, Conveying Object-based Meta-information. In *Proceedings of the Auto-Carto 11 Conference*, Minneapolis, Minnesota, pp. 113–122.

Henderson-Sellers, B., 1997, *A Book of Object-oriented Knowledge*, (Prentice Hall: New Jersey).

Markham, R. and Rix, D., 1995, GIS Ergonomics and the Problems of Interface Design. In *Proceedings of First Joint European Conference on Geographical Information*, The Hague, pp. 539–546.

Medyckyj-Scott, D., 1991, User-oriented Enquiry Facilities. In *Metadata in the Geosciences*, edited by Medyckyj-Scott, D., Newman, I., Ruggles, C. and Walker, D., (Group D Publications: Leicestershire), pp. 85–112.

Nyerges, T.L., Mark, D.M., Laurini, T. and Egenhofer, M.J., 1995, *Cognitive Aspects of Human-Computer Interaction for Geographical Information System*, (Kluwer Academic: Amsterdam).

Qiu, J. and Hunter, G.J., 2000, Towards Dynamic Updating of Data Quality Information. In *Proceedings of the 4th International Symposium on Spatial Accuracy Assessment in Natural Resources and Environmental Sciences (Accuracy 2000)*, Amsterdam, edited by Heuvelink, G.B.M. and Lemmens, M. J. P. M., pp. 529–36.

Wong, D.W.S. and Wu, C.V., 1996, Spatial Metadata and GIS for Decision Support. In *Proceedings of the 29th Hawaii International Conference on System Sciences*, Wailea, Hawaii, pp. 557–566.

CHAPTER SEVENTEEN

Error Metadata Management System

Ernesto Gan and Wenzhong Shi

ABSTRACT

In many cases, users have difficulty extracting quality information about spatial objects graphically when coverage (or layer) level metadata is used. The intention of the Error Metadata Management System (EMMS) reported here is to provide detailed metadata information at feature level for spatial objects. The system gathers pieces of quality and temporal information of changes made during database maintenance. The EMMS also provides information on generalization for spatial objects, since the information is important for users wishing to know the accuracy of a dataset. This information facilitates data users in the data mining process, for instance. Users are thus able to make effective evaluation among features from different datasets. From the perspective of data producers, continuous improvements can be applied by incorporating more accurate and recent spatial objects into the database.

17.1 INTRODUCTION

It is rare for a data producer to revise a dataset from scratch. Partial revision of a dataset using various sources is usual in order to save time and cost. On the other hand, data accuracy and currency become more complex to report as multiple data capture methods are used. According to the Geospatial Positioning Accuracy Standards (FGDC, 1998a), accuracy of each method is separately recorded, but even if this guideline is applied on metadata, spatial variation of data quality across dataset coverage cannot be identified graphically. Further, currency and level of generalization are still revealed on a holistic basis. In this respect, metadata should be captured in feature level (Gan, 1999) to provide information on quality, temporality, data source, and processing step as well as generalization effects for identifying the suitability of a dataset in an area of interest. These pieces of information facilitate data users in data mining and knowledge discovery processes as described by Buttenfield et al. (2000).

The Error Metadata Management System (EMMS) reported here designates a set of metadata items per feature (listed in Table 17.1) to be captured automatically and promptly whenever a change is made. Some of the items can be found in the Feature Metadata Recommendations, issued by the Maine State Office of GIS (OGIS, 2000). The recommended feature metadata are: source, source originator, process, organization editing, date of edit, date of source, and a free text note. The organization editing and free text note are not captured under the EMMS, as the former is a single agent and the latter is difficult to standardize.

Feature metadata items in the EMMS can be categorized into quality information, temporal information, and generalization effects. Quality information consists of the five data quality elements: positional accuracy, attribute accuracy, logical consistency, completeness, and lineage held in the system. Temporal information (Guptill, 1995) is recorded to reflect selective updating of spatial objects. The system also considers the generalization effects on area and length of a feature during modification. Users can generate a graphic query and identify a subset of features from the dataset based on these pieces of information without obscuring them to evaluate the usability of a dataset (Rothenberg, 1996).

In this study, the EMMS has adopted the 1:20,000 digital topographic dataset from the Hong Kong Special Administrative Region (HKSAR) to illustrate the functions and benefits of quality metadata. The dataset is used because of its complexity of data sources and the underlying generalization processes for cartographic purposes. The homogeneous feature classes in the dataset can be gathered from different resolutions of information. This makes feature metadata more valuable and important in data mining. GIS users can obtain the most appropriate information objects (IOs) from a dataset which is enriched with feature metadata. Here, an IO refers to a spatial object and its attributes. The system is built on MicroStation/J and the database is implemented in Microsoft Access and Java MicroStation Development Language (JMDL). EMMS also caters for both the design and time factors in the descriptive study of usability of geospatial metadata (Gluck, 1997).

17.2 FEATURE METADATA

Metadata can be prepared at different levels at the discretion of data producers. It can be hierarchically categorized into (1) series level, (2) dataset level, (3) layer level, (4) feature class level, and (5) feature level. Metadata at the first four levels have limited information about symbolization, level of generalization, and other graphic design criteria. These pieces of information have not been itemized (Buttenfield and Weibel, 1999) in the Content Standard for Digital Geospatial Metadata (CSDGM; FGDC, 1998b). FGDC compliant metadata is based on a holistic view of a dataset which may consist of a number of files, layers, and feature classes. For example, the metadata of the B1000 dataset (a 1:1,000 basic topographic dataset) in the HKSAR is subdivided into a series level. To cover the whole region of Hong Kong, the series level contains 3,223 tiles. Currency of each tile is recorded in an external table.

Users cannot obtain precise information about a dataset from its metadata. For example, positional accuracy in the Water Mains Record Plans dataset is incorrectly prepared by the HKSAR's Water Supplies Department.

Horizontal_Positional_Accuracy_Report:
The 1:1000 topographic base was compiled according to the specifications of the Lands Department. Data for water work facilities are compiled according to survey data and site measurements and therefore the accuracy of the data varies and is unknown.

Positional accuracy shall be tested by an independent source of higher accuracy. If the test criteria are not met, the Spatial Data Transfer Standard describes three alternatives to determine the positional accuracy: (1) deductive estimate; (2) internal evidence; and (3) comparison to source. However, the accuracy report still relies on the decision of the data producer.

The CSDGM compliant metadata is sufficient to determine fitness-for-use. Data users can obtain information about data requirement, scale, file size, accuracy, format, and cost from it (SDVC, 2000a). If homogeneous feature class is available among heterogeneous datasets, it is unlikely that users can decide which dataset is appropriate. Meanwhile, metadata at feature level is more appropriate for evaluating currency and uncertainty of selected features across the coverage of dataset. The appropriateness-for-use can thus be justified from the level of generalization and the currency as compared with the costs of adoption.

17.3 DATA CONFLATION OF 1:20,000 TOPOGRAPHIC DATASET

The HKSAR's 1:20,000 topographic dataset was first digitized in 1996. Data conversion was accomplished by scanning working plates for head-up digitizing, with semi-automatic and automatic vectorization. The dataset at this stage was CAD-oriented spatial data without attribute information. It was originally used for cartographic map production. To effectively and efficiently record the changes of spatial objects, the 1:20,000 dataset has been continuously updated by reference to other data sources, including the 1:1,000 dataset, aerial photographs, engineering plans from the Highway Department, and gazettes. The process of data conflation or fusion in which data of different qualities are merged (Harding and Wilkinson, 1996) is necessary to improve overall data quality.

Recently, the 1:20,000 dataset has been further enriched by extracting attribute information from M10000 dataset (with 1:10,000 resolution) which includes rich and detail attribute information of spatial objects, especially the facility feature class. Facility features include various government facilities, hotels, cinemas, pagodas, foreign consulates, different types of educational institutes, and car parks, among other types of feature. This new version of the 1:20,000 dataset is composed of nearly 200 feature types in 12 layers in MGE format and 16 layers in Arc/Info format.

17.3.1 Attribute data

Since 1999, attributes have been incorporated into the 1:20,000 dataset by using Intergraph's MGE modules. The attributes are extracted from an internal M10000 dataset and are partially linked with external tabular data from the B1000 dataset. Linkage to the B1000 dataset, which includes street codes, building names, place names and geo-referencing, enriches the attribute information of the 1:20,000 dataset. The linkage is established via a unique geo-reference number.

Attribute data is updated from a variety of sources, such as partners of Mapping Information Services in an Internet project, publications from the private sector, telephone enquiries, and other government departments. The EMMS incorporates attribute accuracy information as tabular data with this dataset.

17.3.2 Positional data

The 1:20,000 dataset is not derived from other map series available in the HKSAR only. It is first converted from paper maps with continuous updating using a variety of sources such as ortho-photographs, satellite images, geodetic data, and photogrammetric data. The dataset is also derived from the 1:1,000, 1:5,000, and 1:10,000 digital datasets. These datasets may inherit generalization errors from their sources. With these characteristics, explicit storage of quality and temporal information is valuable to evaluate uncertainty of the dataset. The EMMS records errors associated with every manipulation on the spatial objects. These errors include such items as 3D positional accuracies, displacements, and aggregation errors.

Most generalization for the 1:20,000 dataset are accomplished by experienced operators. Automatic generalization is too complicated for the dataset because features have interlocking relationships. For example, a single displacement of an object may effect juxtaposed objects. Besides, automatic generalization software does not produce data of satisfactory quality. For example, CHANGE, developed by the University of Hannover, can complete only 50 percent of the generalization operations known to be present in the ATKIS data (João, 1998). Moreover, decisions to generalize are subjective; the outputs may be varied from time to time and person by person. Hence, it is necessary to capture the influences of quality information to IOs by means of error indicators.

17.4 ERROR INDICATORS

Accuracies of data sources vary from point to point, and from area to area. It is affected by data capture methods for the digital topographic dataset, but human errors are hard to measure. In the digitizing process, registration of control points can influence the accuracy of digitized point features. Especially, in manual digitizing, selection of points also effects the shapes and areas of digitized features. In order to achieve an acceptable accuracy (that is an accuracy ceiling), operator guidelines should be established. The accuracy ceiling is the lowest accuracy value for the method being used. In the EMMS, operators can input a higher accuracy value to reflect the working accuracy. Table 17.1 illustrates common data capture methods used in the Survey and Mapping Office for the Hong Kong 1:20,000 topographic dataset. From the table, a ceiling of positional accuracy of the methods are obtained with reference to the Basic Mapping System for the B1000 dataset (Chan, 1998). Other methods, such as semi-automatic vectorization of 1:20,000 working plates, and integration of proposed plans from the Planning Department, depend on the quality of sources and the methods used in digitization.

17.5 DATA LIFE CYCLE

Data life cycle of an IO, as with its geometrical part, can be categorized into three stages: (1) birth - first time input (initialization); (2) growth - further processing including displacement, line thinning, and other generalization processes (processing); and (3) death – a spatial object no longer exists (obsolescence). Figure 17.1 illustrates the life cycle of a spatial object in the EMMS. With advances in technology, it is now possible to dynamically capture feature level metadata throughout the data life cycle. In this way, the EMMS updates IOs automatically whenever a geometrical change to a spatial object is detected. Understanding the data life cycle provides preliminary information for system design and development of the EMMS.

Table 17.1 Positional accuracy for B20000 dataset

Data source		Data capture method	Positional accuracy ceiling in ground meters
1:1000	Title survey	Ground survey	0.0135 to 0.0178
	Topographic survey	Radiation survey	0.111
	Topographic survey	GPS	0.103
	Topographic survey	Tie measurement	0.173
	Engineering survey	Radiation survey	0.106
	Engineering survey	Tie measurement	0.166
	Photogrammetry	Photogrammetric plot	0.65
	Positive film	Digitizing	½ line gauge of the feature being digitized (e.g. coastline with line gauge 7, the worst case in digitizing with ½ line width is about 1.778 in ground meters)
1:1200	Positive film	Digitizing	½ line gauge of the feature being digitized
1:5000	Digital form	Direct integration	Depends on its source that B1000 and the generalization effects on accuracies
4000ft	Aerial photography	Head-up digitizing	0.65
8000ft	Aerial photography	Head-up digitizing	1.226

17.5.1 Birth stage

The birth of a spatial object is when the object is digitized into a GIS. The EMMS only caters for two pieces of specific temporal information: evidence time and capture time. When a feature is digitized, the system will automatically record the capture time - the time of digitization. A Java applet is used to explicitly input the evidence time and other pieces of information. The applet is executed after the geometric form has been created or digitized. The other pieces of information recorded consist of data source, data capture method, data capture parameters, positional accurcies, and attribute accuracy. Event time is logically possible to store in the EMMS database, but it is less likely that the occurrence time of a spatial object will be collected.

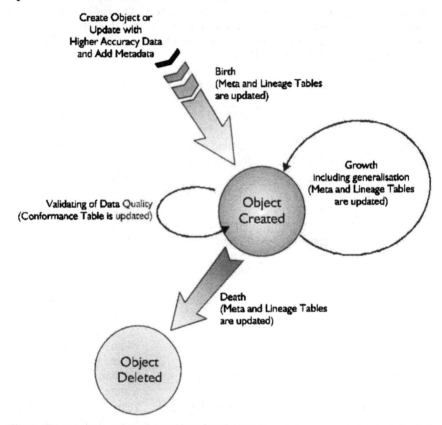

Figure 17.1 Automatic and parallel tracking of spatial data for updating error metadata in designated database tables.

17.5.2 Growth stage

Spatial objects will be further modified in order to retain clarity and aesthetic needs to comply with the 1:20,000 mapping specification. For instance, modifications may include various generalization techniques including collapse, displacement, enhancement, simplification, smoothing and aggregation. The EMMS contains three generalization tools (aggregation, collapse, and displacement modules) to capture changes effected by the generalization process. The system captures changes in area, length, and perimeter of influenced objects, as well as lineage information such as the method (or MicroStation command) applied in the process. This methodology avoids additional human interaction, which may introduce error into the system.

17.5.3 Death stage

The death stage occurs when a feature no longer exists in reality. The evidence time and lineage information are recorded automatically at this stage. Lineage information is significant in delivering a full historical picture in the metadata. During modification of a dataset, a spatial object may be deleted as a result of generalization processes. In this situation, the spatial object no longer exists in the dataset, but it can still be found in reality. Hence, it cannot be treated as dead in its data life cycle. To cope with this situation, the EMMS changes the deleted spatial object from MicroStation/J's primary class to the construction class.

17.6 SYSTEM DESIGN

17.6.1 Data flows

Figure 17.2 is an overview of the EMMS. The figure shows the system boundaries, external entities that interact with the system, and the major information flows between the entities and the system. From the data producers' perspective, spatial features can be added, updated, and deleted by operators. In the EMMS, the operators append related quality and temporal information to the IOs, unless they are generated by the system. Management concerns the accuracy of data, and quality assurance and control tools are available in the EMMS. These tools are capable of identifying any possibly obsolete data during revision, to query the data accuracy values when deciding whether to make a revision with higher accuracy data, and to assist with the CSDGM reporting.

From the data users' point of view, justification of the appropriateness-for-use of datasets is important. The users can use the quality assurance and control tools to select the geographic area of concern, and then, generate reports about currency and quality. These reports may be sufficient to make decisions on adoption.

17.6.2 Database design

The EMMS has been designed and developed to handle spatial data quality elements, temporal information, aggregation error information, generalization effects and information on data source, data capture methods, and data capture parameters. These are structured into five tables including two predefined supporting tables, as shown in Figure 17.3. Other user-defined attributes can be linked to a feature through the unique MSLINK number. A brief explanation of each field of the metadata database is shown in Table 17.2.

17.6.3 Capture of error metadata

Information about source and data capture method are inputted into the EMMS when a spatial feature is first digitized into the dataset. Figure 17.4 shows the interface for inputting these information.

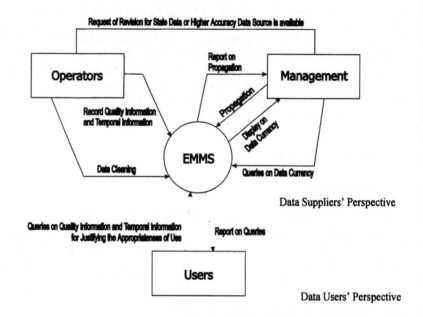

Figure 17.2 Context of the EMMS.

Much research effort has focused on positional accuracy (Shi, 1998; Shi and Liu, 2000), while the geometric changes in area, length, and perimeter are usually ignored. Typically, in generalization processes, an object changes from areal to point feature (collapse), and positional accuracy along a simplified coastline deteriorates (simplification). Figure 17.5 illustrates the process of amalgamation and Figure 17.6 shows details of the area changed for two buildings. The process of amalgamation will be recorded in the metadata records database for these objects. The geometric forms of the target buildings are still retained in the dataset, but they are changed to a construction class. The construction class features will be displayed only when the construction toggle in the view properties is set to off. The EMMS records every step in the process to ensure that no interim data quality information is omitted.

The EMMS has been developed to manage automatically generated or parallel tracked feature-based metadata following manipulation of any feature. A listener module is tactically programmed to detect the birth, growth, and death of a spatial object. For example, when a feature is digitized, the listener module will pop up a dialog box for inputting data source and quality information as shown in Figure 17.4. If a feature is deleted, the system will change the primary class feature to a construction class feature and then capture this modification in the relative tables. Before publicizing the dataset, data producers are required to check the dataset. Figure 17.7 shows an example of validating the quality information of a feature.

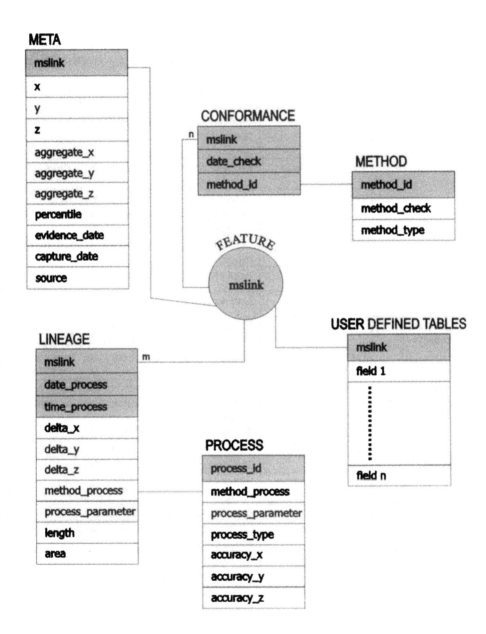

Figure 17.3 EMMS database structure. The shaded fields are the primary or combined primary keys.

Table 17.2 Description of the Quality and Temporal Information of IO described in EMMS.

Quality information and temporal information	Attributes	Description
Lineage	date_process	Processing date that is captured by the system
	time_process	Processing time that is captured by the system
	delta_x, delta_y, delta_z	Signed uncertainty in position on the process taken for calculation of the aggregation errors
	method_process	Process undertaken (e.g. head-up digitizing, generalization)
	process_parameter	Parameters of the method_process (for example, in scanning with MapSetter 4000, parameters are OFFSET, GAIN)
	length	Length or parameter of the IO before the process step
	area	Area of the IO before the process step
	source	Data source
Positional accuracy	x, y, z	Positional accuracy of individual features
	date_check	The date of positional accuracy checking
	method_check	Method used to check the positional accuracy
Attribute accuracy	percentile	The percentile ascertained to be a category of feature (e.g. feature categorized by remote sensing)
	date_check	The date of attribute accuracy checking
	method_check	Method used to check the attribute accuracy
Logical consistency	date_check	The date of consistency checking
	method_check	Statement of compliance among other features
Completeness	date_check	The date of completeness checking
	method_check	Statement of compliance with technical specification
Temporal	evidence_date	Date of evidence
	capture_date	Date of data capture
Others	aggregate_x, aggregate_y, aggregate_z	Aggregation errors from graphical manipulations on features
	accuracy_x, accuarcy_y, accuracy_z	Pre-defined accuracy ceilings for data capture methods

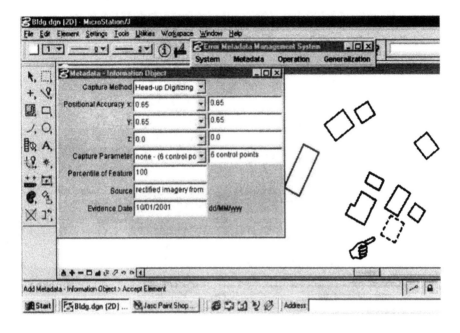

Figure 17.4 An example of digitizing a built-up object (indicated as in dashed-line). The user has to input capture method, positional accuracy, collaboration/configuration parameter, ascertained percentile of belonging category, data source and evidence date.

17.6.4 Other functions of the EMMS

Using functionality of MicroStation/J, metadata can be visualized through in-built commands. Lineage information, quality and temporal information can be retrieved with the Structured Query Language (SQL) via MicroStation/J's Visual Structured Query Language (VSQL) module as indicated in Figure 17.8.

Metadata at the object level facilitates calculation of aggregation errors of a single object by accumulating the differences of coordinates in each graphic modification. The EMMS has a customized module for this which updates the fields with the following equations:

$$\text{aggregate\_x} = \Sigma\,(\text{delta\_x})\,\text{n};$$ (17.1)

$$\text{aggregate\_y} = \Sigma\,(\text{delta\_x})\,\text{n; and}$$ (17.2)

$$\text{aggregate\_z} = \Sigma\,(\text{delta\_x})\,\text{n}.$$ (17.3)

Root Mean Square Error (RMSE), and maximum and minimum values of positional accuracy can be found through the EMMS. The EMMS performs these calculations by retrieving the META table (Figure 17.3) with a pre-defined selection set. The RMSE is useful for CSDGM reporting, as there is no higher independent source for the 1:20,000 digital topographic map series. Besides, comparison with the data sources is impossible, since varying subjective cartographic generalizations are applied to the different sources and spatial objects.

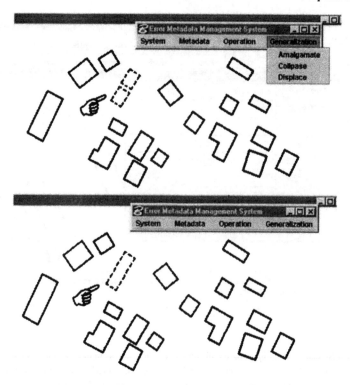

Figure 17.5 Two houses indicated as in dashed-line to the left before amalgamation (upper), and to the lower a new build-up area is formed after the process.

Figure 17.6 Lineage and temporal information as well as the area of the two build-ups are recorded.

17.7 EVALUATION OF THE EMMS

The database structure of the EMMS is simple, but vital to all parties who utilize a subset of cartographic objects from the digital dataset. Quality information and temporal information as well as aggregation errors can be retrieved. With the information, users are able to justify the appropriateness-for-use in adopting the dataset. This justification has a chain influence on the degree of accuracy in spatial analysis. Usefulness and functionality of the EMMS can be categorized into two areas: data producers, and data users, as well as the wider society.

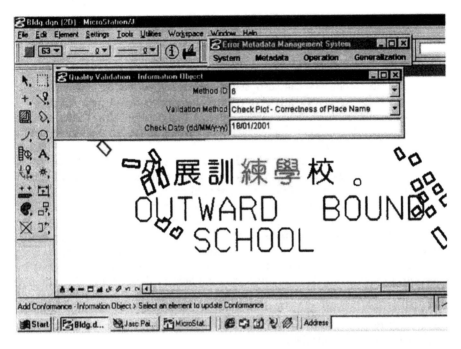

Figure 17.7 An example of adding quality checked information using Java applet.

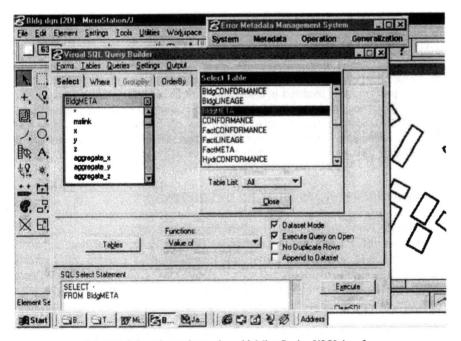

Figure 17.8 Querying and reporting with MicroStation VSQL interface.

17.7.1 Data producers - quality assurance

Currency of spatial data can be achieved with the tools provided by the EMMS. Data can be retrieved through queries on the temporal information of the metadata database. Data producers can also select features captured through a specific method. Revision with such spatial queries reduces the cost and time spent on re-collection of the required spatial data. Thus, a full revision on the map sheet is reduced to a partial revision. Consequently, it speeds up the revision cycle of the organization's products.

Spatial data may be integrated with other sources for upgrading the accuracy of the original dataset by replacement of spatial objects with those of higher accuracy. However, the substitute source should contain the temporal information of evidence time in order to make such comparison meaningful. With this pre-requisite, a more accurate feature can be adopted in the spatial database.

These controlling methods establish a base for quality assurance. Nevertheless, quality is dependent on the quality of the EMMS database. Data producers can achieve the advantages of Total Quality Management (TQM) by continuously improving their products with the help of the EMMS. The 1:20,000 dataset strategically captures every change in the spatial extent as soon as information is available from various sources. Meanwhile, spatial data can be captured from poor quality sources, and the representation replaced by the more accurate one when it is available.

17.7.2 Users – appropriateness-of-use

Metadata in the EMMS enhance data users' judgement on whether a dataset is appropriate to use in their analysis, and the level of generalization inaccuracies in any feature. Uncertainty of errors is estimated through an accumulative calculation from the process steps incurred in a feature. Attribute accuracy, completeness, and logical consistency can be retrieved for a further estimation on the appropriateness-for-use. This reduces the granularity problem of dataset generated from the CSDGM metadata. The more accurate the data used, the more accurate any estimation or forecasting formulated from it. Eventually, this is a benefit to society.

17.8 CONCLUSIONS AND FUTURE DEVELOPMENT

The EMMS is designed and developed for managing metadata of spatial objects at the highest level of metadata. It captures quality information and temporal information, data source, and processing step as well as generalization effects during manipulations of the dataset. These pieces of information relating to data source and data validation will be captured automatically with minimal input from operators. In the example of the Hong Kong 1:20,000 Digital Topographic Map Series, it is used to show the need of feature metadata especially when data are conflated. Other datasets like those based on 1:10,000 and 1:50,000 maps can also adopt the EMMS.

Feature metadata may not have instant benefits to data producers. However, in the long run, the error metadata can enrich a user's decision regarding the appropriateness-for-use of a dataset. It also facilitates the determination of positional error of GIS objects in the geometric form of point, line, and polygon. Quality control and quality assurance can also be implemented through the installation of the EMMS for data providers. Data users can ultimately select appropriate GIS data for a particular application based on the metadata in the EMMS.

Development of EMMS was motivated by the idea that feature level metadata is essential if data users are to make effective judgements when choosing an appropriate dataset. Further developments on how to use temporal GIS data may facilitate research on environmental analysis. Datasets will become larger, thus spatial data have to be maintained in a DBMS like Oracle Spatial and IBM's DB2 Spatial Extender. However, true GIS analysis functions are usually absent or limited in this full database format (SDVC, 2000b).

Feature metadata can be used in the automatic creation of CSDGM compliant metadata, both of which can be stored in XML format for web-based query on the degree of suitability for adoption (Mihaila *et al.*, 1999). Data users may also visualize the appearance of the database through the metadata (Zaslavsky, 2000).

17.9 REFERENCES

Buttenfield, B.P. and Weibel, R., 1999, *Visualizing the Uncertainty of GIS Data.* (http://www.colorado.edu/geography/courses/virtus/2pagers/babs.html)
Buttenfield, B.P., Gahegan, M, Miller, H. and, Yuan, M., 2000, *Geospatial Data Mining and Knowledge Discovery.* A UCGIS White Paper on Emergent Research Themes Submitted to UCGIS Research Committee, (http://www.ucgis.org/emerging/gkd.pdf)
Chan, K.K., 1998, Accuracy Report in the BMS. Unpublished material.
FGDC, 1998a, *Geospatial Positioning Accuracy Standards*, (July). Federal Geographic Data Committee, Washington, D.C., USA.
FGDC, 1998b, *Content Standard for Digital Geospatial Metadata*, Version 2.0, (January). Federal Geographic Data Committee, Washington, D.C., USA.
Gan, F., 1999, Error Metadata Management System with Application to Hong Kong 1:20,000 Digital Topographic Map Series, for the partial fulfilment of the MSc in Geo-information Systems, the Hong Kong Polytechnic University.
Gluck, M., 1997, *A Descriptive Study of the Usability of Geospatial Metadata.* (http://www.oclc.org/research/publications/arr/1997/gluck/gluck_frameset.htm)
Guptill, S.C., 1995, Temporal information. In Guptill, S.C. and Morrison, J.L, *Elements of Spatial Data Quality*, (New York: Elsevier), p153–165.
Harding, S.M. and Wilkinson, G.G., 1996, A strategic view of GIS research and technology development for Europe. In *Report of the Expert Panel Convened at the Joint Research Centre,* Ispra, 12–20 November 1996, Space Applications Institute. (http://www.ec-gis.org/exppanel.htm)
João, E.M., 1998, *Causes and Consequences of Map Generalization*, (London: Taylor & Francis).

Mihaila, G.A., Raschid, L., and Vidal M.E., 1999, Querying "quality of data " metadata. In *Third IEEE META-DATA Conference*, Bethesda.
(http://www.computer.org/proceedings/meta/1999/papers/65/gmihaila.html)

OGIS, 2000, Office of GIS, US, Feature metadata recommendation to the GIS Technical Group 10/10/2000.
(http://apollo.ogis.state.me.us/news/techn/flmeta/recommendation.htm)

Rothenberg, J., 1996, Metadata to support data quality and longevity. In *1st IEEE Metadata Conference*, Bethesda.
(http://www.computer.org/conferences/meta96/rothenberg_paper/ieee.data-quality.html)

SDVC, 2000a, *Data Sources*. Under the Metadata Education Project of the Spatial Data and Visualization Center, University of Wyoming.
(http://www.sdvc.uwyo.edu/metadata/source.html)

SDVC, 2000b, *Database Principles*. Under the Metadata Education Project of the Spatial Data and Visualization Center, University of Wyoming.
(http://www.sdvc.uwyo.edu/metadata/database.html)

Shi, W.Z., 1998, A generic statistical approach for modeling errors of geometric features in GIS. *International Journal of Geographical Information Science*, **12** (2), 1998, pp.131–143.

Shi, W.Z. and Liu, W.B., 2000, A stochastic process-based model for positional error of line segments in GIS. *International Journal of Geographical Information Science*, **14** (1), pp.51–66.

Zaslavsky, I., Marciano, R., Gupta, A., Baru, C., 2000, XML-based spatial mediation infrastructure for global interoperability. In *4th Global Spatial Data Infrastructure Conference*, Cape Town, South Africa.
(http://www.npaci.edu/DICE/Pubs/gsdi4-mar00/gsdi_iz.html)

CHAPTER EIGHTEEN

Maintaining Spatial Relativity of Utilities after DCDB Upgrade

Roger Merritt and Ewan Masters

ABSTRACT

When a digital cadastre database (DCDB) is moved to be more representative of the real world, spatially dependent data sets such as for utility assets like water, gas and electricity, are no longer mapped in their correct relative position. Utility companies and local government typically hold spatially dependent data sets that are affected by this problem. These organizations receive updated DCDB containing new subdivisions and compilations. They then have to identify areas in the new DCDB that have been repositioned and upgrade their dependent data's position to fit the new DCDB. This chapter sets the case for the use of the parametric least squares technique, and how this theory has been implemented to carry out a utility upgrade after the DCDB has been upgraded.

18.1 INTRODUCTION

Many utility and local government organizations in Australia have elected to locate their asset data for drainage, sewerage and cables, relative to the digital cadastral database (DCDB). Various placement rules have been used to derive coordinates for these assets. Some of these geometric constructions are described in more detail by Unkles (1992) and Lemon (1997). For example, it is very common for electricity, gas, water and sewerage to be placed at standard offsets from street frontages. Electricity may be at 2.0m from the boundary, whereas sewerage lines are set to 0.3m from the boundary. However, these rules are not usually stored with the data. As well, the utility data are not usually maintained by the same organization that maintains the DCDB.

The DCDB in Australia was itself originally digitized from various map sources. The accuracy of the DCDB is therefore dependent on the accuracy and scale of the original hardcopy mapping, and the skill of the digitizer operator. If 1:25,000 mapping was used, the accuracy of the DCDB is probably around 25 m. In urban areas, where 1:1000 scale mapping or larger was common, the accuracy will probably be at 1 m or better. However, in some cases through good digitizing practices, the DCDB actually looks visually much better than expected. For example, long straight lines appear straight, right angles are near square, and parcel dimensions appear close to being correct.

Errors in DCDB data tend to be highly correlated with neighbouring line work or topology. For example, a whole street frontage might be in error by 1 m, but the line will be straight. Therefore when quality measures are undertaken there is a need to compare relative accuracy of features as well as absolute positional accuracy. It is also possible to use this relative location information as part of an improvement process (Merritt *et al.*, 1998; Merritt and Masters, 1998).

In any case, it is becoming more common in Australia for the DCDB to be updated and upgraded at some point in time. Updates will generally add to, move or remove specific lines in the existing database, whereas an upgrade will move the coordinates of all graphics in the database. However, when a DCDB custodian upgrades the DCDB to make it more representative of reality, it is common for the correct relative location of the assets to be lost. This problem occurs because the utility asset themes are usually stored in layer-based geographical information systems (GIS), which do not store the relationships (Hebblethwaite, 1989; Wan and Williamson, 1994; Lemon,1997). Also, the DCDB custodians are usually different to the utility organizations, making the management of relationships between themes somewhat difficult.

This chapter explores a few of the geometrical relationships of this spatial problem, shows how the "lost" relationships can be restored in most cases, and explains how Parametric Least Squares can be used to re-align the correct relative locations of asset data to the newly upgraded DCDB .

18.2 THE UTILITY UPGRADE PROCESS

The utility upgrade scenario is built upon a network adjustment framework (Mikhail and Ackermann, 1976; Harvey, 1995). Merritt *et al.* (1998) describe an adjustment process to improve the accuracies of DCDB data. The technique develops parametric equations to represent the topological and spatial characteristics of a spatial data set, as well as allowing additional positional information to be integrated into the adjustment process. The most fundamental observation types are the distances and directions taken from the digitized lines in the spatial data. New data can be entered from various sources such as survey plans and field survey to enable the positional accuracy of the DCDB to be improved. The processes in this adjustment have been enhanced to adjust utility data as described below.

For the utility upgrade process the original "old" DCDB and utility data contain spatial objects (points, lines, polygons) that are assumed to be in their correct relative positional relationship. New data consist of the spatial objects from an upgraded DCDB that has moved with respect to the "old" themes. The problem is to move the graphics in the utility theme to the correct position with respect to the new DCDB.

Observations can be developed from the existing topological relationships in the "old" data sets. That is, lines in the "old" DCDB can be used to constrain the movement of objects in the adjustment. However, many points in this pseudo-network are common with points in the new cadastre. A map-matching process is required to identify these points. Once the matching process has been carried out, the beginnings of an adjustment can be built.

The positions of the points that match in the "old" and "new" DCDB's provide the base information to adjust the "old" to the "new". The new cadastre, which is treated as containing control points, is not allowed to move in the adjustment process. The directions and distances can be derived from the old cadastre and the asset themes, and the "matching" data in the old cadastre are substituted with their equivalent data in the new cadastre. The adjustment process using least squares estimation to minimize changes to the vector linework, will move any unmatched points in the "old" DCDB to an "adjusted" position in the "new" DCDB.

The one piece of information missing at this stage is any connection between the "old" DCDB and utility themes. Almost without exception, the utilities are not located on the boundary of any parcel in the DCDB. As there are no connecting directions and distances, that is "relationships" joining the DCDB and the utility theme, the old cadastre would move to the new cadastre in the adjustment process, but the asset theme would not move. The following section discusses how these relationships can be implemented in the adjustment process.

18.3 TOPOLOGICAL RELATIONSHIPS IN UTILITY UPGRADES

Upgrading utility themes is more complex than just applying shifts derived from nearby points in the cadastre. The following section will explore some of the more complicated geometrical problems that need to be addressed to deal with this problem more fully.

18.3.1 Inferred topology

Because the relative location of utility assets with respect to the DCDB is not explicitly stored, topological relationships must be derived from the existing database. We therefore refer to these derived relationships as *Inferred Topology*. For example, there is an "inference" that a utility asset located at the half angle of a bend in the cadastre, and at a certain offset, should be at that position after the shift to the new cadastre (Figure 18.1 in the colour plate section).

The inferred topology of a point can be expressed as a list of points that any point is topologically connected to. These lists of coordinates can be converted to directions (angles) and distances, which are the additional input data to the adjustment process, that will adjust these relationships to the matching points in the "old" and "new" DCDB's.

An *Inferred Topology Search Distance* constrains which spatial objects in the database have inferred topology with other objects. If this distance is too short then no inferred topology will be formed, which in turn will cause the asset data to be "left behind" in the adjustment. Setting this distance too large will mean the shifts on the asset will be partially based on points that are not nearby. For example, an extreme case would be to set this distance at 200 m in an urban situation. This would mean that the asset's new location would be affected by shifts in the back boundary of surrounding parcels, which had no relevance to the position of an asset when it was originally placed.

18.3.2 Bad topology

Bad topology refers to unexpectedly short lines occurring amongst lines of "normal" length for an area. Generally, there are many extra points in any spatial data set than may be necessary to represent a real-world feature. The reasons why these extra points exist vary, and some points may be necessary whilst others are not. For example, short lines are caused by the GIS technique of storing arcs as a series of short chords and also by the use of tiling to store contiguous themes into smaller, more manageable, rectangular tiles. Some examples are given in Figure 18.2.

Whatever the reason for these extra points occurring, there is no easy automatic way of knowing if these points should be there or not. Although it is possible to find them and display them, putting an operator onto the job of removing them is usually not an option due to the costs involved with an operator doing the work.

There is little concern with bad topology where these extra points form lines with the same magnitude as generally occurs in the theme. However, every unnecessary point provides the potential for slight deviations from say a straight line. Where these points form lines significantly shorter than those generally in the theme, unstable shifts can be caused in the adjustment process. In some cases, where the adjustment shifts are far greater than the length of the line, they can cause divergence (the point will not converge to its correct location). There is no exact rule in this regard, as it is case dependent. However, it is generally wise to adjust from a situation where shifts are smaller than the lengths of lines, but this may not always be possible.

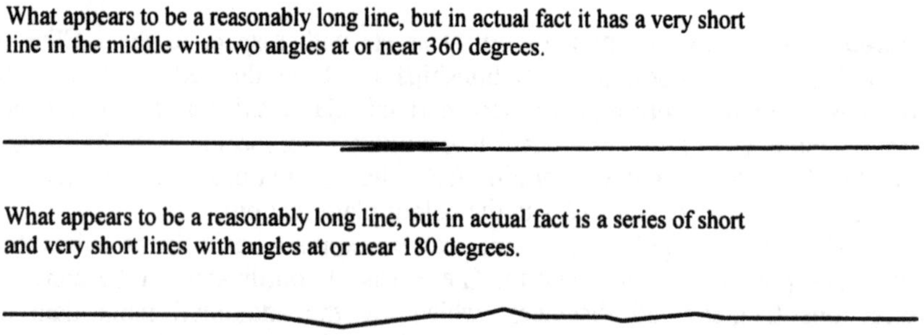

What appears to be a reasonably long line, but in actual fact it has a very short line in the middle with two angles at or near 360 degrees.

What appears to be a reasonably long line, but in actual fact is a series of short and very short lines with angles at or near 180 degrees.

Figure 18.2 Examples of bad topology.

In utility data, short lines commonly connect closely located equipment and represent feeders going into property. Hence, it is important that these points have their relative placement preserved in an adjustment situation. This problem has been solved using the following process.

Whenever a short line is found, as defined by a user tolerance, additional directions and distances are stored to all points connected to the first point (start node) on the short line, as well as to all points connected to the other point (end node). This has the effect of bracing the short line's points, much like a surveyor's braced quadrilateral surveying technique. It also has the effect of adding longer lines to the observation set.

18.3.3 Vertical topology

Using coordinates in one theme to define features in other themes creates *Vertical Topology*. Vertical topology creates problems for layer-based GIS technology. An example of vertical topology would be using the DCDB to define administrative boundaries. Sets of coordinates in the DCDB are duplicated into the administrative theme, causing the administrative theme to plot over the top of the DCDB. The difficulty is that these points are copied, and an update in the DCDB has to be somehow duplicated into the administrative theme to maintain this verticality between themes. Due to the lack of tools and the complexity of this work, this job is often done manually. More details of the problem are described by Lemon (1997).

This vertical topology problem is handled by changing the model and structure of the spatial data. The typical GIS model stores polylines as a series of sequential coordinates through which the line is drawn. The *Spatial Adjustment Engine* used in this work stores each coordinate once only and assigns an identification number to each point. Each spatial theme is held in a separate table and each record stores two point numbers through which a line is drawn. The standard GIS model is definitely three or so times faster at redraws, but lacks the data modelling needed for the type of adjustment used here or to manage vertical topology.

18.3.4 Implied vertical topology

Implied Vertical Topology occurs when spatial objects should share common points in the database but actually do not. Implied vertical topology, on the surface, is the same thing as vertical topology, however it has to be handled differently. Vertical topology occurs when two features share common locations. Implied vertical topology occurs, for example, when a line (line1) is drawn from a location on another line (line2), but the other line (line2) does not contain a vertex/point with the starting coordinates of first line (line1). Such an example occurs in water and sewerage pipes data as shown in Figure 18.3.

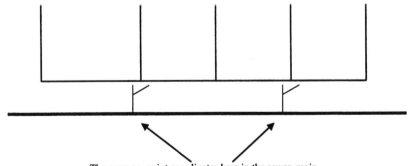

There are no point coordinates here in the sewer main.

Figure 18.3 Example of implied vertical topology.

For database modelling purposes, a main pipe is stored as one item in the database, and as one item in the GIS. It does not matter how many side-pipes are connected to it, the pipe has the one set of dimensions, the one grade, coefficient of friction, etc. In the GIS, two points are used to represent the pipe, one at the beginning and one at the end.

Each side-pipe is 'snapped' to the main so that it draws correctly, but there is no topological connection. The 'implication' is that there is vertical topology that must be maintained between the side pipes and the main pipe, even though there is no topological relationship in the database.

This problem has been dealt with as follows. The first step happens in the data preparation stage. A procedure is run which goes through each line in each theme and finds all points which are within a very small tolerance of being on the line. Directions are stored which act in the adjustment to keep the point on the line. However, because the shifts at the two ends on the main are rarely the same as those part of the way along the main, it is likely that the start point for the feeder pipes will move a little from the main pipe. The second step in the process occurs after the adjustment. This process simply finds the same points and snaps them back onto the main pipe.

18.3.5 Preserving long lines

Long straight lines occur in various features, and even occur across features. For example, overhead power lines are made up of long straight lines between anchor points, broken by pole locations. Each pole can be safely assumed to be on a perfectly straight line due to the very large tensions in the cable. Each straight line in an overhead power line is independently anchored at both ends due to the tension used to reduce sag. The poles between these anchor points are placed on a straight line to reduce lateral tensions, which would cause the pole to lean.

Sewerage and water pipes are manufactured and laid in long straight lines with constant grade, but when these are modelled as having only the start and end points they do not need to be treated to preserve their straightness. If they do not contain vertices between the two ends there can be no deflection (Figure 18.4).

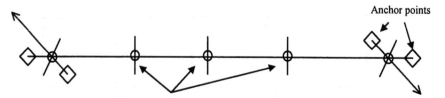

There are extra coordinates in the overhead power line.

Figure 18.4 Example of long straight lines.

This situation is dealt with in two steps. The first step happens in the data preparation stage. A procedure is run which goes through each line in each theme and finds all lines that contain a minimum number of points that form a straight line within a small tolerance. Additional observations are stored which act in the adjustment to keep all the points on the line.

However, because the shifts on individual points along the line are rarely the same as those for the total line, it is likely that some points will move a little from the line. For example, these varying shifts might occur due to the front boundaries in the DCDB being straightened up, or because the anchor is located at the half angle in the cadastre and the angle in the cadastre has been changed. The second step in the process occurs after the adjustment. This process simply finds the same lines, derives the line of best fit, and snaps these points back onto the line. Extra care has to be taken with points that form the intersection of two or more long lines.

18.3.6 Preserving associativity in rural areas

It is possible for utility services to run for several kilometres down a perfectly straight rural road, with one property on each side. In these situations there simply are no nearby points in the cadastre with which to form inferred topology. However, it is expected that the point from which the utility service enters the property will remain the same distance from the road frontage(s).

This problem is overcome by checking that sufficient observations (directions and distances) are formed for each point. If the service is at the end of a utility line, the only observations that can be formed are the direction and distance back to the previous point on the service. There are no nearby points in the old cadastre, so observations cannot be formed to build observations from inferred topology. However, there are one or two nearby lines, being the road boundaries. The method adopted finds these lines, builds directions and distances to their end points, and adds these observations to the ones from the point at the other end of the utility service. This provides proportioning both along and across the road, depending on variations in the road length and width, thus the relative position is preserved correctly.

18.3.7 Preserving the location of (annotation) points

Isolated points do not normally occur in utility themes, as they typically represent the flow of water, electricity, gas, sewerage, and hence make up a network of connected lines. However, some assets such as street furniture and street signs could be represented as points.

The most common unconnected point in GIS is the annotation point. There is usually such a point for every polygon. Often an algorithm in the GIS generates these locations, but there are many instances where they have been manually placed for cartographic reasons. The potential to lose the cartographer's placement of annotation makes it important to treat annotation points properly in an upgrade situation.

If the shift from an adjustment moves the polygon perimeter such that the centroid or the annotation point is outside the polygon, then subsequent plots will mislabel the polygons and various spatial analysis techniques will be disabled. For example, moving attributes from one polygon theme to another using the centroids in one theme to select the matching centroid in the other will not work. Depending on the future usage of the theme, this may be a considerable problem.

Isolated points, such as annotation, can be treated initially as points on the end of a service, and hence searches are possible for nearby lines with which to form implied topology. The search distance for neighbouring lines can be iteratively expanded in stages up to 100m. If one or more nearby lines are found, directions and distances are formed to two points on each of those lines. The search ends at 100m, as any point more than 100m away from any line can be considered spatially independent of those lines, that is the point was probably not carefully located against nearby assets or parcel boundaries.

18.4 THE RESULTS

Figure 18.5 in the colour plate section shows an overlay of the "old" and "new" DCDB for a sample region used to prototype the adjustment technique. It is noticeable that shifts are up to 10 m (half road width) and in various directions. The road in the middle near the top of the diagram shows a complex curve in the "old" with a simpler larger radius curve in the "new".

Figure 18.6 in the colour plate section shows the "new" DCDB with the "old" unadjusted utility data. It is noticeable that the utility data do not match the parcel boundaries very well, with connections encroaching into the parcels or missing them. Figure 18.7 in the colour plate section shows the "New" DCDB with the "Adjusted" Utility data.

18.5 CONCLUDING REMARKS

The parametric least squares adjustment technique can be used to upgrade DCDB's and with appropriate modifications upgrade associated utility themes. The various mechanisms for building observations that this technique needs has been explained for the various geometrical combinations of linework found in various databases investigated at this stage.

The technique has been successfully implemented to operate in a personal computer / Windows environment. The approach taken assumes that all changes in the database are upgrades to the coordinates and not actual changes in parcel boundaries. These types of update would need to be detected by some other process implemented in the maintenance and quality assurance procedures of the utility organization. However, we would expect the types of updates that would affect the utility upgrade process to be a very small proportion of the total number of updates that occur. For example, most updates would be subdivisions of existing parcels. Such changes would not affect this upgrade process unless the "old" parcel boundaries are moved in the subdivision.

The parametric adjustment technique used to upgrade DCDB data has been extended to successfully model utility upgrades. Recent tests on large data sets indicate that the technique as implemented can successfully implement *inferred vertical topology* and automatically move more than 99% of the utility data to match the DCDB in a spatial database. The small percentage of unsuccessfully moved spatial objects in the database requires interpretation from an operator as in a conventional manual update process.

The development of an application that allows for further experimentation in "real-world" environments is available for download, for evaluation purposes only, from the Internet Location http://www.SpatialWeb.com.au. The download comes with sample data from South Australia, courtesy of Department of Environment and Heritage and South Australian Water.

REFERENCES

Harvey, B.R., 1995, *Practical Least Squares and Statistics for Surveyors*, Monograph 13, (Sydney: School of Geomatic Engineering, The University of New South Wales).

Hebblethwaite, D.H., 1989, Concepts for coping with a shifting cadastral model. *The Australian Surveyor*, **32** (5), pp. 486–493.

Lemon, D., 1997. *The Nature and Management of Positional Relationships within a Local Government Geographic Information System*, Unisurv S53, (Sydney: School of Geomatic Engineering, The University of New South Wales).

Merritt, R. and Masters, E.G., 1998, DCDB Upgrade: The Sutherland Council Project, 1998. In *Proceedings of AURISA98*, (Canberra: Australasian Urban and Regional Information Systems Association), cdrom.

Merritt, R., Keats, G. and Masters, E.G., 1998, Upgrading Cadastre using survey accurate measurements. In *Proceedings of OZRI12*, (Perth: ESRI Australia), pp. 1–15.

Mikhail, E. and Ackermann, F., 1976, *Observations and Least Squares*, (New York: University Press of America).

Unkles, H., 1992, Owning the solution provides the power. In *Proceedings of AURISA 92*, (Canberra: Australasian Urban and Regional Information Systems Association), pp. 158–163.

Wan, W. Y. and Williamson, I. P., 1994, Problems in maintaining associativity in LIS with particular reference to the needs of the utility industry. *The Australian Surveyor*, **39** (3) pp. 187–193.

CHAPTER NINETEEN

Spatial-thematic Accuracy and Domestic Burglary in the Safer Cities Programme in England[1]

Ho Chung Law and Peter F. Fisher

ABSTRACT

The Safer Cities Programme in England as a whole implemented over 3,600 crime prevention schemes in 20 cities between 1988-1995 (total costing £30 million). The large-scale evaluation of the Programme's impact on domestic burglary has estimated that, overall, schemes of the Safer Cities Action reduced burglaries by 56,000 and were cost-effective (a saving of about £31 million). Using two cities, Bristol and Coventry, within the Safer Cities Programme as a case study, this research aims to explore some of the accuracy issues in the GIS processing involved in the evaluation. This chapter (1) describes how spatio-thematic accuracy can be estimated using Monte Carlo and dasymetric methods within the context of the Safer Cities Programme Evaluation; (2) thereby provides a precise quantitative statement on the errors involved in the geographical data processing; and (3) examines how spatial errors may affect the conclusion of the Evaluation using multi-level modelling. On average, the results show that the overlay method used in the Evaluation has over-estimated the household counts by 3.6% and 5% for Bristol and Coventry respectively. Subsequently, the Safer Cities Programme Evaluation has underestimated the action intensity by −0.8 and −9% and the burglary risk by −7% and −5% (for Bristol and Coventry respectively). Multi-level modelling shows that the mean errors due to the spatial interpolation estimated by the Monte Carlo dasymetric method are −1.5%, 2.3% and 0.7% for Coventry, Bristol, and the two cities combined respectively. In all cases, these are well within the standard errors generated by the overlay method. It is concluded that spatial and thematic errors have no significant impact upon the conclusions of the Safer Cities Programme Evaluation. However, spatial analyses show that potential burglary hot spots might have been missed as a result of such errors in crime pattern analysis. The analysis of the error distribution shows that a geographical area would have a higher error rate if it has: dense population; is near the city centre; or has an irregular geographical boundary. The implications in GIS applications, and crime prevention for decision and policy makers are discussed.

---

[1] The views expressed are those of the authors and do not necessarily represent those of the Home Office or any other government department.

19.1 Introduction

Spatial data quality has been recognised as an important issue (Goodchild and Gopal, 1989; Guptill and Morrison, 1995), and implementation of research methods to a large GIS application has been relatively unexplored. In order to address this issue, a six-year research programme was formulated between the Home Office Research Development and Statistics Directorate (RDS) and the Midland Regional Research Laboratory, University of Leicester between 1993-1999 (Law, 1999). The summary of the results was reported at the first *International Symposium on Spatial Data Quality* (Law and Fisher, 1999). This chapter aims to clarify the methodology issues and expand further on the discussion on the implication of the results. For the benefit of the readers, the context of the research is reiterated in Section 19.2. The research questions were formulated within the scope of the study (Section 19.3). Section 19.4 describes the methodology used for spatial error modelling. The results of the error estimation and its effects on the Safer Cities Programme evaluation are described in Section 19.5, and finally Section 19.6 provides the conclusion and discussion of the case study.

19.2 Context

The Safer Cities Programme in England was inaugurated in 1988 and wound up in Autumn 1995. Altogether, it implemented over 3,600 highly diverse and geographically scattered crime preventive schemes in 20 cities and cost about £30 million. Safer Cities was set up as part of Action for Cities, the Government's wider program to deal with the multiple social, physical and economic problems of some of our larger urban areas. The objectives of Safer Cities were to reduce crime, lessen fear of crime, and create safer cities within which economic enterprise and community life could flourish (Home Office, 1993). The Programme impact evaluation by the Home Office RDS aimed to evaluate the extent of change, if any, in crime pattern and their perception, which could be attributed to the effects of Safer Cities action. The evaluation required a GIS to combine diverse sources of data sets: crime types per police beats; Enumeration Districts (EDs); and their 'neighbourhoods'. The analyses involved modelling changes in burglary risk and linking the changes to the Safer Cities action intensity (*action scores* for short). The research findings on domestic burglary schemes showed that overall, the Safer Cities Action reduced burglary and was cost-effective. Simply implementing action in a police beat reduced local risks by nearly 10%. In total, it has been estimated that £31 million was saved from the prevention of 56,000 burglaries (for details of evaluation design, see Ekblom *et al.*, 1994, 1996a, 1996b; Law and Ekblom, 1994; Law, 1997, Law, 1998; and for the nature of action, see Law and Ekblom, 1996).

Although the results of the Evaluation show that the Safer Cities Programme has an effect upon crime reduction such as burglary risk, the complexity of the data manipulation involving GIS operations in the evaluation process raised several data quality issues. First, the data collection of such large-scale research was problematic. There were missing data in some of the Safer Cities and data input was not an error-free process for such a large amount of data sets.

Second, in order to produce beat-level data for the analysis from the Census and the Index of Local Conditions, the beat boundaries were digitised from beat maps, then overlaid with the smaller EDs using ARC/INFO. However, police beats usually bore no relationship to other administrative territories such as wards or EDs. The boundaries of beats and their constituent EDs do not always match to form a one-to-many relationship (Figure 19.1).

Figure 19.1 An example of Beat-ED overlay operation in one of the Safer Cities
(Coventry, West of the city centre; Beats indicated by thicker lines).

Third (and similar to the second), the actions and survey interviews were linked relationally via ED-ID using GIS. However, the survey originally used 1981 EDs as a spatial unit (because the Before Survey was carried out prior to the 1991 Census), while the rest of the data used the 1991 EDs. It is a well known phenomenon that Census boundaries change. This required transformation of the 1981 ED-IDs to 1991 ED-IDs. Initially this was achieved by a look-up table between the two supplied by OPCS, but we found the cut-off points between their overlays insufficiently accurate for our purposes, and the fact that the relationship between 1981 and 1991 EDs was often not 'one-to-one' but 'many-to-many' raised the question of spatial accuracy of these data transformations.

Fourth, the police beat boundaries changed over time (for example, Bristol). Fortunately the change tended to be the amalgamation of smaller beats into a larger beat (the so-called super-beat for the purpose of spatial database management). The beats varied widely in size and population, with averages of 230 hectares and over 2,200 households. 'True' beats averaged 180 hectares and 1,700 households; superbeats 600 hectares and 6,300 households. The fact that beat boundaries change might seem a simple administrative task and to have required no more than keeping the spatial database up-to-date. However, this did raise the question of the basic unit of analyses for crime statistics. It is related to a well-know problem - called the Modifiable Areal Unit Problem, MAUP (Openshaw and Taylor 1980; Openshaw, 1984). The ecological fallacy due to the MAUP known as 'aggregation bias' is also well reported in the field of criminology (Langbein and Lichtman, 1978).

A further problem is that of areal interpolation since the ability to define alternative aggregations of the same area means that statistical measures need to be interpolated from one aggregation to another (Openshaw and Taylor, 1980; Goodchild and Lam, 1980; Lam, 1983; Openshaw, 1984; Fotheringham, 1989; Flowerdew and Openshaw, 1987; Goodchild *et al*, 1993; Law and Fisher, 1995).

The MAUP also highlights a basic spatial representational issue. It raises the question as to how many zones a given region should represent in order to present a meaningful map. For example, Robinson *et al.* (1995) show how different maps could be presented from the same sets of data by adjusting different selections of choropleth class intervals. While there may be no GIS error involved, there is the question of interpretation errors.

Thus if users were not aware of all of the above issues, the problems of the spatial accuracy might remain hidden. These might cast doubts upon the validity and reliability of the conclusion of the Evaluation. These issues will be further explored in a more specific way and form the scope of the research presented in this chapter.

19.3 Scope of the study

Judging from the theoretical importance of the spatial accuracy issues and their practical implications upon policy decision making as exemplified by the Safer Cities Programme Evaluation, the primary aim of this research is to *explore the data accuracy issues in GIS processing.*

The objectives of this case study are to:

- assess the impact of the spatial uncertainty upon the Safer Cities Programme Evaluation;
- provide a comprehensive quantitative account for the possible errors in the GIS operation for the evaluation;
- determine whether the errors were significant or not.

In achieving the above aim and objectives, it is necessary to explore the validity of the geographical assumptions made; and to simulate the process of analysis through the error propagation so that more precise error statements can be made about the effect of the Safer Cities Programme. The research will have important implications both to the conclusion of the evaluation as knowledge for policy making within the specific context of the Safer Cities Programme, and for the use of GIS in general.

In order to achieve the above objectives, two cities (Bristol with total population = 371,020 and Coventry, population = 293,141) were selected for further investigation on error propagation within the GIS processing. The selection was based on their characteristics, representativeness and data availability at the start of the study.

19.4 Methodology

Based on a broad literature review (Law, 1999, Law and Fisher, 1995), and in particular, using Lanter and Veregin's (1992) paradigm and Fisher and Langford's (1995) assessment of areal interpolation methods, an innovative methodology has been developed to handle the spatio-thematic accuracy issues in the evaluation of the Safer Cities Programme. The following procedures have been formulated to implement the methodology and to assess the spatio-thematic accuracy of the spatial interpolation.

- Identify the error index.
- Perform a quick evaluation of the range of errors using simple methods (such as area weighting) to check whether further assessment is required (if so, proceed to Step 3; else stop).
- Develop error propagation functions within the data transformation processes using Monte Carlo dasymetric method.
  - Get the satellite imagery.
  - Calibrate the satellite image using the source zones to estimate the attribute values of interests.
  - Estimate the attribute values in the target zones using the calibrated satellite image as a dasymetric map and Monte Carlo simulation.
  - Test the utility of Step 2 by assessing spatio-thematic accuracy in the GIS application.

The Monte Carlo technique was implemented in Lisp-Stat object programming (Tierney, 1990). The changes in burglary risk were measured at the police beat level; the action scores represent the amount of money in £ spent on burglary preventive action per household. Since data of the household density at ED level were derived from the 1991 Census data, the household density at beat level had to be derived by means of the spatial interpolation using EDs as the source zone (S) and beats as the target zone (T). Unlike Fisher and Langford's method, both the source zones and the target zones in the Safer Cities Programme were pre-defined, and thus could not be simulated freely. So for ED-beat areal interpolation in our case study, the elementary zones E can be defined as EDs which can be aggregated into any Zone A, and the target zones T as beats. The source zones S are the aggregated EDs, i.e. S=A. Using land-use maps of classified Landsat TM imagery as the control zones C, C can be superimposed with T to form a binary representation of populated and unpopulated pixels. Since the population of E is known, the populations of S and T are also known (Figure 19.1).

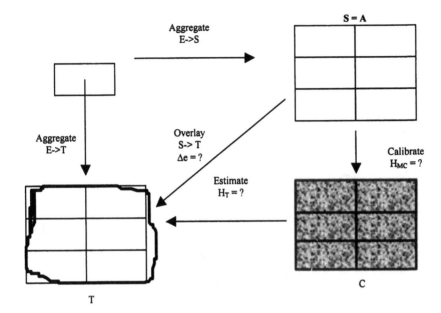

Figure 19.2 Monte Carlo simulation and dasymetric mapping with satellite imagery to estimate population (based on 1991 census data). E: elementary zones (ED); A: aggregated zones; T: target zones (beats); and C: control zones (satellite imagery).

The triangle EAT (or SET) effectively defines the spatial interpolation problem of this research; and the triangle CAT (or SCT) offers the solution to that problem. The latter (SCT) can be alternatively viewed in terms of the common map overlay (layer-based representation, Lanter and Veregin, 1992; Shi, 1994) as shown in Figure 19.2.

The variable of interest Δe is the error of the spatial interpolation using the overlay method. This can be computed as follows:

$$\Delta e = H_{MC} \sim H_{OV} \tag{19.1}$$

where Δe is the error of spatial interpolation from S to T
             H_{MC} household count by means of Monte Carlo estimation
             H_{OV} household count by means of the overlay method

Assuming a non-biased Monte Carlo estimator, the RMSE error due to the simulation after the calibration with ED references would be approximately equal to the standard deviation σ of the simulation. The true household count H_T can be estimated by means of the household counts obtained from the Monte Carlo simulation in terms of the household counts by means of the overlay method (H_{OV}), and its error estimation (Δe) and standard deviation by means of the Monte Carlo method (σ):

$$H_T = H_{OV} \pm \Delta e \pm \sigma \tag{19.2}$$

The above equation is consistent with the error model associated with the spatial variation proposed by Journel and Huijbregts (1978), Burgess and Webster (1980a, 1980b), Webster (1985), Burrough (1986), Davis (1986), Heuvelink *et al.* (1989), and Lodwick *et al.* (1990). Comparing to their regionalised variable theory, the first term in the right hand side of the equation is equivalent to a function describing the structural component of the spatial (thematic in this case) variable, the second term is a spatial-correlated random component denoting the stochastic, locally-varying spatially-dependent residuals from the first, and the third term is a residual error (white noise) that is spatially uncorrelated having a mean zero.

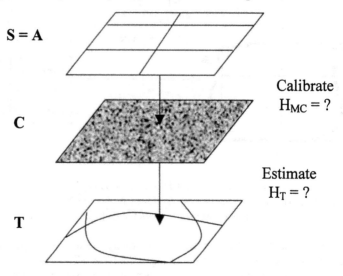

Figure 19.3 Layer-based representation of the Monte Carlo dasymetric method.

By defining different zonal areas using Monte Carlo techniques, a precise statement about areal interpolation error can be made. Thus a whole range of error distribution can be worked out in terms of RMSE, not only by comparing the population of A and C, but also C and T.

To test whether the above results have any significant implication or not to the conclusion of the evaluation, the error figures can be re-run through the muti-level model. The simplest version of the variance components model is used (Bryk and Raudenbush, 1992; Jones, 1992; Woodhouse *et al.*, 1992; Goldstein, 1995) where beat-years (Level 1) are nested into j beats (Level 2):

$$y_{ij} = \beta_{0j} + \beta_1 x_{ij} + u_j + e_{ij} \qquad\qquad (19.3)$$

where y_{ij} represents the burglary risk [$\text{asin}\,((B/H)^{1/4})$]
                     x_{ij} action score
                     β_{0j} intercept
                     β_1 coefficients of the action score
              u_j and e_{ij} are the random variables at level 2 and 1 respectively.

19.5 Results and analyses

The results of the simulation for Coventry are summarised in Table 19.1, which shows the household population estimated by ARC/INFO method and the Monte Carlo dasymetric method (MC), the RMSE of the Monte Carlo dasymetric method, the error rate (in percentage) of ARC/INFO method (error%) as estimated by the Monte Carlo dasymetric method.

Table 19.1 Estimated household population and error rate in Coventry and Bristol

Beat	ARC/INFO	MC (h)	RMSE	Error%
Total (Coventry)	116,593	109,084	9,622	6.88
Total (Bristol)	155,422	159,853	5,560	-2.77
Mean (Coventry)	2,486	2,321	214	6.88
Mean (Bristol)	2,467	2,537	88	-2.77

For Coventry, out of a total of 47 beats, the total number of households estimated by the ARC/INFO method was 116,593. Using the Monte Carlo dasymetric method, the more accurate number of households was 109,084 ± 9,622. This represents error of 7% ± 1% as a whole. The error ranged from a minimum 1% error (excluding zero population) to the maximum error of 96%!

For Bristol, out of the total 63 beats, the total number of households estimated by the ARC/INFO method was 155,422. Using the Monte Carlo dasymetric method, the more accurate number of households was 159,853 ± 5,560. As a whole, this represents the error of -3%. The pattern of the error variation is similar to Coventry, but with a wider range: from a minimum 0% error (excluding zero population) to the maximum error of 233% (beat 33)! This may be due to the larger population of households in Bristol than Coventry. The extremely large error rate seems to occur at the city centre beat (Beat 33). This does not really matter very much as the actual population in the city centre is very small.

The action intensity scores (S) in each beat-year were re-calibrated taking into account the error (e) for each beat using the following formulas.

$$S' = S \pm eS \tag{19.4}$$

where S' is the 'calibrated' action scores

Since the action scores represent the amount of money spent per household, the over-estimation of the household density would actually under estimate the amount of the action scores. The results show that this is indeed the case. The over-estimation of household density (5% for Coventry and 3.6% for Bristol) has resulted in the under-estimation of the cost of the Safer Cities action by -9% and -0.8% for Coventry and Bristol respectively. The amount of the resultant error of each individual beat depends not only on the initial errors from the GIS spatial interpolation, but also on whether the beat is involved in the further processing (scoping in this case) and the value of the attribute (action score).

19.5.1 Geographical analysis of burglary risk - crime pattern analysis

Similar analyses applied to the outcome measure – burglary risk, suggest a mean under-estimation of the burglary risk by -7% for Bristol and -5% for Coventry. Here the "burglary risk" is defined as *the number of domestic burglaries per 100 households in each beat, in each year.* The spatial pattern of the burglary risk for each year can be visualised on a series of thematic maps similar to the error analyses described in the last two chapters. The process of such inspection is also known as crime pattern analysis - an increasingly important task in crime prevention as a result of the impact of using GIS. Emerging crime patterns can be identified by inspecting these maps over the six year period for the two cities.

For Bristol, Beats 19, 20, and 63 consistently had the highest burglary risk over the six year period. Beats 19 and 20 were densely populated areas just outside the city centre (North East). Beats 32, 34, and 35 had no domestic burglary at all because they were city centre beats and had zero household counts. Beats 6, 8, 18, 32, 43, 46, 57, 58, 60, 61 and 62 consistently enjoyed relatively low burglary risk. Most of these beats were countryside and park areas and had low population density. Beats 3 and 17 changed from no burglary to relatively high burglary risk over time. This was probably due to a new development of residential housing in these areas.

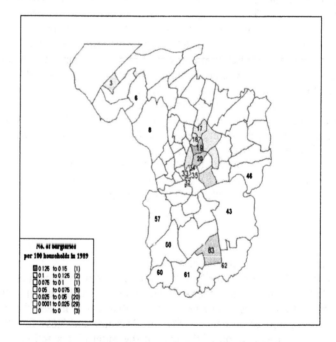

Figure 19.4 Burglary risk in Bristol 1989 (overlay method).
Numbered beats are mentioned in the text.

The thematic maps from the Monte Carlo dasymetric method follow a similar pattern. An exception is beat 33, which is shown up as a hot spot on the Monte Carlo thematic map but not on the overlay map. This is a small beat just next to the city centre (West of Beat 35) and has the highest estimated error (see Chapter 8). An important implication of such an observation is that a less accurate map may miss a potential crime hot spot. This may have a further implication in resource allocation for crime prevention. For easy reference the maps showing the burglary risk in 1989 as estimated by both methods are shown in Figures 19.4 and 19.5.

For Coventry, the areas with the highest burglary risk were Beats 29 (just below Beat 30 at the city centre), 33, 43, and 101 between 1987-92, and their burglary risk fell in 1993. With a few exceptions, most of the rest of the beats (in particular West of the city) had relatively low burglary risk. Beats 31, 37 and 38 changed into high-risk areas in 1990-1992 (during the period of the Safer Cities Programme). Beats 37 and 38 received no Safer Cities action. There is no evidence of geographical displacement for these two beats as they are not the neighbourhood of the action beats. Beat 31 received high action intensity and was relatively near to the high-risk area (Beat 33). It is next to the city centre beat and is one of the densely populated areas.

Again, the Monte Carlo dasymetric maps show a similar pattern except Beats 21 and 22. These two beats constitute one of the burglary hot spots and yet it has completely been missed out on the overlay maps throughout the period. As an example, see Figures 19.6 and 19.7. The implication is clear as discussed earlier in Bristol. The mis-identification of the hot spots might cause inappropriate allocation of resources such as the Safer Cities Programme fund. The beats that are not perceived as high crime areas would attract less preventive action than they actually deserve.

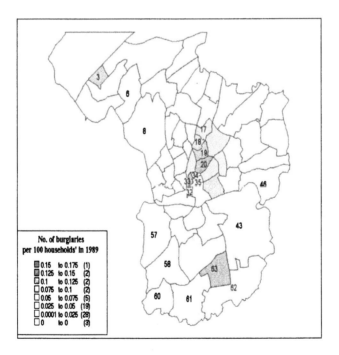

Figure 19.5 Burglary risk in Bristol 1989 (Monte Carlo dasymetric method).
Numbered beats are mentioned in the text.

In general, the patterns of burglary risk in the two cities are consistent with the finding by Shaw and McKay (1972), that distance from the city centre tends to have a small negative relationship with the burglary risk. High population density is one of the obvious important factors (Braithwaite, 1979). Most high-risk areas are relatively near to the city centre which, together with their accessibility, form the 'awareness space' of the potential motivated offenders (Brantingham and Brantingham, 1984; Beavon *et al.*, 1994). This can also be interpreted in terms of the routine activity theory (Cohen and Felson, 1979) and proximal circumstances (Ekblom, 1994). However the spatial correlation and the nature of disadvantaged areas are complex issues, most likely multi-dimensional (Hirschfield, 1994), and are outside the scope of this study.

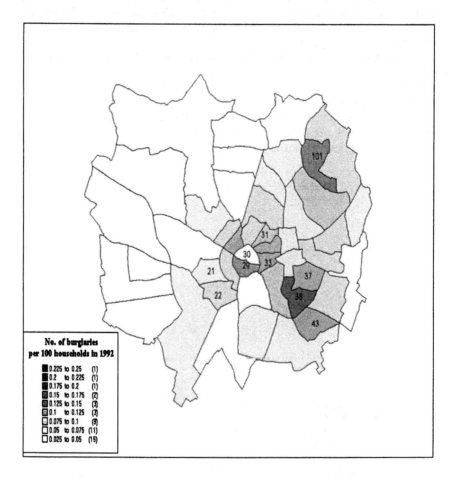

Figure 19.6 Burglary risk in Coventry 1992 (overlay method)
Numbered beats are mentioned in the text.

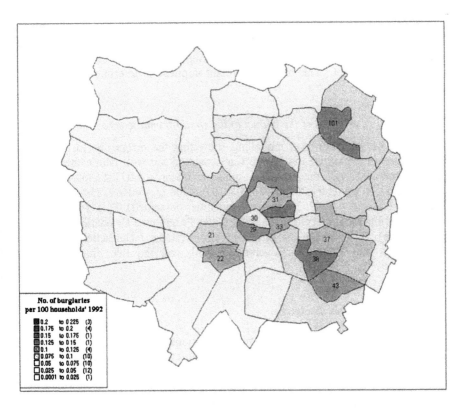

Figure 19.7 Burglary risk in Coventry 1992 (Monte Carlo dasymetric method).
Numbered beats are mentioned in the text

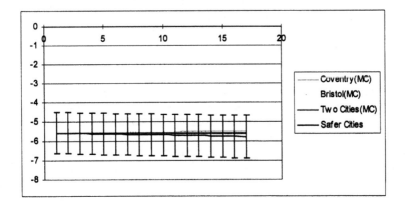

Figure 19.8 Comparison between the overall Safer Cities effect with the Two Cities effect.

The results of the Monte Carlo dasymetric method suggest a mean under-estimation of the burglary risk by -7% for Bristol and -5% for Coventry. Crime pattern analysis shows that potential hot spots might have been missed as a result of such under-estimation. This has an important implication in crime prevention for decision and policy makers.

19.5.2 Statistical analysis of the impact of the spatio-thematic errors

The results of the multi-level modelling show that the errors due to the spatial interpolation estimated by the Monte Carlo method are well within the standard errors generated by the overlay method (mean: 0.02136536>-0.00738>-0.02136536 for Coventry; 0.04698854 > 0.009906 > -0.04698854 for Bristol; 0.02629 > 0.00338 > -0.02629 when the two cities combined). This is also the case when the results are projected to the Safer Cities Programme Evaluation as a whole (Figure 19.8). On average the estimated errors (at the mean action scores) are -1.5%, 2.3% and 0.7% for Coventry, Bristol, and the two cities combined respectively.

19.6 Conclusions and discussion

The results have a number of important implications for (1) the evaluation of the Safer Cities Programme (Section 19.6.1); (2) the crime pattern analyses and the evaluation of future crime preventive action (Section 19.6.2); and (3) spatial data quality assessment (Section 19.6.3). Recommendations for future research and development are summarised in Section 19.6.4.

19.6.1 Implications for the evaluation of the Safer Cities Programme

Taking the spatial and thematic errors into account (i.e. using Monte Carlo dasymetric method) the difference between the two methods is too small to have a significant impact upon the conclusion of the Safer Cities Programme evaluation (p = 0.2 for Coventry; 0.08 for Bristol; and 0.76 when the two cities are combined; c.f. p = 0.108 for all Safer Cities with d.f. = 1 in all cases). *It is concluded that no significant impact due to spatial and thematic errors upon the conclusion of the Safer Cities Programme evaluation has been found.*

This conclusion is encouraging to both the evaluation team and the policy makers. The research finding suggests that the methodology used by the Safer Cities Programme evaluation is robust enough to cope with the spatio-thematic error. From now on, we can use the evaluation strategy developed from the Safer Cities Programme with confidence, and apply it to the future evaluation of crime prevention initiatives and GIS application. If necessary, the methodology developed in this research can be used to test the validity of the future application as an extra-quality assurance.

There are at least three reasons why no significant impact can be detected in the evaluation as a whole (at the 'global' aggregated scale) while the estimated error varies widely across different beats (at the 'local' disaggregated scale). First, as the number of data sets used in a GIS analysis increases, the accuracy of the result decreases due to the aggregation effects (see Veregin, 1989). This finding confirms this general conjecture. Second, the multi-level modelling shows that the risk-action relationship between the overlay and Monte Carlo dasymetric methods tends to interact in the multi-level modelling and that the overall effect of the underestimation of the action intensity and burglary risk seem to cancel each other out when the two cities are combined.

Third, because this research has only focused on the spatio-thematic error in GIS processing, any other types of errors would have been excluded from the scope of this case study. For example, in this case study, it is *assumed* that there are no gross error blunders in all the input data. However, such an assumption may be unrealistic for this application with a very large amount of data sets. Some errors might have occurred in the early stages of data capture such as data collection and input. For instance, on a closer re-examination of Figure 19.1 (top left hand corner of the map), it seems likely that there might be some gross error such as mis-registration in the digitising stage. This may account for the large error margin within the evaluation of the Safer Cities Programme. So we must not be too optimistic about the scale of error within the evaluation just because this case study shows that the aspects of GIS processing (spatial interpolation in particular) have not affected the conclusion significantly.

19.6.2 Implications for crime pattern analyses and the evaluation of future crime preventive action

The crime pattern analysis in this case study shows that potential hot spots might have been missed as a result of such under-estimation. This has an important implication for decision and policy makers in terms of crime prevention and resource allocation. If more accurate high-risk areas were identified, decision-makers would be able to target those areas with more appropriate resources and crime preventive strategies. The observation also implies that a spatial unit as large as a beat with aggregated data is susceptible to the spatio-thematic error and therefore is not appropriate for crime pattern analysis. Alternative approaches with smaller spatial units such as postcodes and enumeration districts would give a more accurate spatial pattern (for example, Brunsdon 1989; Ratcliffe and McCullagh, 1998).

The analyses also show that the Safer Cities effect varies across different cities as exemplified by Bristol and Coventry. Coventry appears to have a greater action effect upon the burglary risk reduction than Bristol. It may be generalised that the Safer Cities schemes of some Safer Cities may be more "successful" than the others. Combining the data sets from all the Safer Cities in the analysis tend to dilute the effect of the more 'successful' Safer Cities schemes. With such hindsight, it would have been better to analyse each city individually as exemplified by this case study. This would have enabled the evaluation team to 'comb' the more 'successful' Safer Cities schemes and investigate further the mechanism behind the success (as well as the failure) of Safer Cities schemes (rather than the Programme as a whole). While it is justified for the evaluation to combine all the cities in the multi-level model to assess the impact of the Safer Cities Programme as a whole (as required by the Treasury), analysing the effect city by city would have enabled the policy makers to assess the effectiveness of the Programme in certain cities. This would also help researchers to unpack the mechanism of the preventive processes and raise further research questions such as: why is it that some cities are more successful in implementing crime prevention initiatives than others?

19.6.3 Implications for the data quality assessment in GIS processing

Spatial analyses of the error distribution (Law, 1999; Law & Fisher 1999) show that a geographical area would have a higher error when it has:
- dense population;
- nearness to the city centre; or
- an irregular geographical boundary.

Further techniques need to be developed to assess the spatial structure of errors in practical applications.

Since there are no formal agreed methods for individual applications as it depends on the context, there is a need to *develop a standard of best practice* as exemplified by this research.

19.6.4 Summary of recommendations

While many spatial interpolation methods and their accuracy assessment have been developed, more work is needed on the integration of statistical methods for spatial data quality measurement. To summarise, the recommendations for future research and development are listed as follows (in no particular order).
- Develop system functions to help users to assess data accuracy as part of the GIS standard.
- Develop meta-database to include data quality as part of its attribute.
- Develop spatial sampling strategy for other disciplines such as social research.
- Integrate GIS with other systems such as statistical analytical system.
- Use GIS as a tool for testing criminological theories standardised error reporting.
- Provide a range of evaluation methods as part of the GIS functionality.
- Develop further techniques to assess the spatial structure of errors in practical applications.
- Investigate the optimal unit of analyses.
- Investigate alternative approaches for spatial representation such as probability surfaces.
- Implement a standard of best practice.

REFERENCES

Beavon, D., Brantingham, P. and Brantingham, P., 1994, The influences of street network on the patterning of property offences. In R. V. Clark (Ed.) *Crime Prevention Studies.* Vol. II. Monsey, (New York: Criminal Justice Press).

Braithwaite, J., 1979, *Inequality, Crime and Public Policy,* (London: Routlege & Kegan Paul: London)

Brantingham, P. and Brantingham, P., 1984, *Patterns in Crime*, (New York: MacMillian).

Brunsdon, C., 1989, Spatial Analysis Techniques Applied to the Local Crime Pattern Analysis, PhD thesis. Newcastle University.

Bryk, A. S. and Raudenbush, S., 1992, *Hierarchical Linear Models - Applications and Data Analysis*, (Newbury Park, CA: Sage).

Burgess, T. M. and Webster, R., 1980a, Optimal interpolation and isarithmic mapping I. The semivariogram and punctual kriging. *Journal of Soil Science,* **31**, pp. 315–330.

Burgess, T. M. and Webster, R., 1980b, Optimal interpolation and isarithmic mapping II. Block kriging. *Journal of Soil Science,* **31**, pp. 333–344.

Burrough, P. A., 1986, *Principles of Geographical Information Systems for Land Resources Assessment*, (Oxford: Clarendon Press).

Cohen, L. E. and Felson, M., 1979, Social change and crime rate trends: a routine activity approach. *American Sociological Review,* **44**, pp. 588–608.

Davis, J. C., 1986, *Statistics and Data Analysis in Geology.* Second Edition, (New York: John Wiley & Son).

Ekblom, P., 1994, Proximal Circumstances: A Theory-Based Classification of Crime Prevention. Crime Prevention Studies, 2., (London: Home Office).

Ekblom, P., Howes, D. and Law, H.C., 1994, Scoping, scoring and modeling: linking measures of crime preventive action to measures of outcome in a large, multi-site evaluation using GIS and multilevel modeling. In *Proceedings of the GISRUK*, pp. 123–132.

Ekblom, P., Law, H.C. and Sutton, M., 1996a, *Domestic Burglary Schemes in the Safer Cities Programme.* Home Office RSD Research Findings No. 42, (London: Home Office).

Ekblom, P., Law, H.C. and Sutton, M., 1996b, *Safer Cities and Residential Burglary*, Home Office Research Study 164, (London: HMSO).

Fisher, P. F. and Langford, M., 1995, Modelling the errors in areal interpolation between zonal systems by Monte Carlo simulation. *Environmental & Planning A,* **27**, pp. 211–224.

Flowerdew, R. and Openshaw, S., 1987, A review of the problems of transferring data from one set of areal units to another incompatible set. *Research Report. NRRL*, Lancaster & Newcastle, England.

Fotheringham, A. S., 1989, Scale-independent spatial analysis. in the accuracy of spatial databases. In *The Accuracy of Spatial Databases,* edited by M. F. Goodchild and S. Gopal, (London: Taylor & Francis), pp. 221–228.

Goldstein, H., 1995, *Multilevel Statistical Models*, (London: Edward Arnold).

Goodchild, M. F. and Gopal, S., 1989, *Accuracy of Spatial Databases*, (London: Taylor & Francis).

Goodchild, M. F. and Lam, N. S-N., 1980, Areal interpolation: a variant of the tradition spatial problem. *Geoprocessing,* **1**, pp. 297–312.

Goodchild, M. F., Anselin, L. and Deichmann, U., 1993, A framework for the areal interpolation of socioeconomic data. *Environment & Planning A,* **25**, pp. 383–397.

Guptill, S. C. and Morrison, J. L., (Eds), 1995, *Elements of Spatial Data Quality.* (Oxford: Elsevier Science).

Heuvelink, G. B. M., Burrough, P., and Stein, A., 1989, Propagation of error in spatial modelling with GIS. *International Journal of Geographical Information Systems,* **3** (4), pp. 303–322.

Hirschfield, A., 1994, Using the 1991 population Census to study deprivation. *Planning Practice and Research,* **9** (1), pp. 43–54.

Home Office, 1993, *Safer Cities Progress Report, 1991–1992,* (London: Home Office).

Jones, K., 1992, Multi-level modeling. In *Survey and Statistical Computing,* edited by Westlake, A., Banks, R., Payne, C. and Orchard, T., (New York: North-Holland).

Journel, A. G. and Huijbregts, Ch. J., 1978, *Mining Geostatistics,* (New York: Academic Press).

Lam, N. S-N, 1983, Spatial interpolation interpolation methods: a review. *American Cartographer,* **10**, pp. 129–149.

Langbein, L. I. and Lichtman, A. J., 1978, *Ecological Inference,* (Beverly Hills: Sage).

Lanter, D. P. and Veregin, H., 1992, A research paradigm for propagating error in layer-based GIS. *Photogrammetric Engineering & Remote Sensing,* **58**, pp. 825–833.

Law, H. C., 1997, From action to outcome of the Safer Cities Programme. In *Proceedings of the AGI 97*, Birmingham.

Law, H. C., 1998, Error: the neglected area, but does it matter? In *Proceedings of the AGI 98,* Paper: 8.1, AGI & Miller Freeman.

Law, H. C., 1999, *Spatial-thematic Accuracy in the Evaluation of English Safer Cities Programme.* PhD thesis, Leicester University.

Law, H. C. and Ekblom, P., 1994, Application of GIS: Evaluation of Safer Program. In *Proceedings of the AGI 94.*

Law, H. C. and Ekblom, P., 1996, Describing the nature of Safer Cities Action using GIS. In *Geographical Information - From Research to Application,* edited by M. Rumor, R. McMillam, and H. F. L. Ottens, *Proceedings of the Second Joint European and Exhibition on Geographical Information,* Vol. 2, IOS Press, pp. 1007–1016.

Law, H. C. and Fisher, P. F., 1995, Error modeling in areal interpolation: a case study on the data precision of the Safer Cities Evaluation. In *Proceedings of the GIS Research UK*, GISRUK'95.

Law, H. C. and Fisher, P. F., 1999, Spatial accuracy in the Evaluation of the Safer Cities Programme in England. In *Proceedings of the International Symposium on Spatial Data Quality.* Hong Kong Polytechnics University.

Lodwick, W. A., Monson, W. and Svoboda, L., 1990, Attribute error and sensitivity analysis of map operations in geographical informations systems: suitability analysis. *International Journal of Geographical Information Systems,* **4** (4), pp. 413–428.

Openshaw, S. and Taylor, P. J., 1980, Statistical Application in the Spatial Science. In *A million or So Correction Coefficients: Three Experiments on the Modifiable Area Unit Problem,* edited by Wrigley, N. (London: Pion), pp. 127-144.

Ratcliffe, J. and McCullagh, M., 1998, Aoristic crime analysis. *International Journal of Geographical Information Science,* **12,** (7), pp. 751–764.

Robinson, A.H., Morrison, J.L., Muehrcke, P.C., Kimerling, A.J., and Guptill, S.C., 1995, *Elements of Cartography*, 6th Edition, (New York: Wiley and Sons).

Shaw, C. and McKay, H., 1972, *Juvenile Delinquency and Urban Areas*, (Chicago: University of Chicago Press).

Shi, W., 1994, *Modelling Positional and Thematic Uncertainties in Integration of Remote Sensing and Geographical Information Systems*. Publication 22. Enschede, (Netherlands: ITC).

Tierney, L., 1990, *Lisp-Stat an Object-oriented Environment for Statistical Computing and Dynamic Graphics*, (NewYork: John Wiley & Sons).

Veregin, H., 1989, Error modelling for the map overlay operation. In *The Accuracy of Spatial Databases*, edited by M. F. Goodchild and S. Gopal, (London: Taylor & Francis).

Webster, R. , 1985, Quantitative spatial analysis of soil in the field. In *Advances in Soil Science,* Vol. 3, (New York: Spring-Verlag).

Woodhouse, G., Rasbash, J., Goldstein, H. and Yang, M., 1992, *A Guide to ML3 for New Users,* (London: Institute of Education, University of London).

CHAPTER TWENTY

Monte Carlo Simulation of Long-term Spatial Error Propagation in Forestry Databases

Michael Weir

ABSTRACT

Once established, spatial databases to support forest management need to be maintained to support current operations, to provide a record of past activities and to evaluate the possible outcomes of management decisions. This chapter describes the simulation of long-term spatial error propagation in a database covering a forest area in Austria. Various scenarios of error propagation were produced using Monte Carlo simulation. Arcs in each of the five states of the database were perturbed with different levels of error, depending on both the type of feature and the date of data acquisition. In this way, the simulation modelling not only reflected the degree of uncertainty associated with different linear features but also accounted for the likelihood that surveys on separate occasions may be of different quality. Results of this simulation modelling indicate that an initially high quality database can become significantly degraded if it is amended with low quality data. The amendment of an initially low quality database with high quality data, however, provides a comparatively small improvement in the overall quality of the database. The chapter concludes by demonstrating how the benefits obtained from improving data quality may be assessed in relation to the cost of database maintenance.

20.1 INTRODUCTION

The need for reliable and diverse spatial information to support planning and decision making has been instrumental in the evolution of a tradition of surveying and mapping by forest management agencies and, more recently, by their widespread implementation of GIS technology (Weir, 2000). As spatial databases increasingly are used in complex spatial modelling and decision support tasks, the uncertainty inherently associated with forest resource data may have potentially detrimental implications, varying from direct financial impacts on business decisions to more intangible effects on sustainable resource management. Given the size of their investment decisions, forest agencies clearly must consider the financial impact of any reduction in the quality of the spatial database used to support planning and management (Keefer, 1994; Prisley, 1994).

All forest areas are subject to gradual and abrupt changes as a result of both natural causes and human activity. It follows that forest databases have to be regularly updated. At the same time, the dynamic nature of forest resources requires that non-current data be retained to monitor changes and to support the development of predictive models to assess the outcomes of different management options (Lowell, 1994). Because different categories of spatial data, from a variety of sources, are treated as separate, temporally inconsistent layers in a conventional GIS, forestry agencies will benefit from efforts to implement temporal GIS (TGIS) in which spatio-temporal data are handled in a more structured manner (Langran, 1992a). To be effective over the long periods of time involved in forest management, however, the quality of a TGIS database should not deteriorate significantly with each successive update. This chapter describes investigations into the long-term propagation of spatial error in forestry databases.

20.2 APPROACH

Traditionally, forest management plans are prepared every ten years. Because the widespread introduction of GIS has only taken place during the last two decades, it follows that forest management agencies have yet to acquire experience of long-term database maintenance. Consequently, it is not yet possible to assess the long-term effects of spatial error propagation under operational conditions. The research described in this chapter therefore followed an approach based on the use of historical map data covering part of the Kobernausserwald, a forest area in Austria. Although the Kobernausserwald has been surveyed and mapped on various occasions since 1872 (Weir, 1997), this has been carried out on a regular basis only during the past 50 years. The maps used in this study therefore comprised an original survey, undertaken in 1947, and four subsequent revisions. All maps were available in the form of 1:10,000 scale sheets on stable film. These were used to trace changes in the forest over a 40-year period. Using the original survey as a base state, the database was amended by adding new objects digitized from revised versions of the management map (Figure 20.1). All digitizing employed the same set of control points throughout the 40-year period. During digitizing, every arc was coded with a feature label and a time stamp to indicate its temporal validity. With each amendment, non-current data were retired to archive files, thus emulating the maintenance of a spatio-temporal database comprising a base state and four amendments (Bulger and Hunt, 1991; Langran, 1992b).

In the absence of the original survey observations and detailed information about the methods employed in the production and revision of the three map series, it was not possible to use analytical methods (Drummond, 1995) to evaluate the long-term effects of spatial error propagation. Instead, Monte Carlo simulation was applied to model the effects of error propagation from the base state through the four amendments. Various authors (for example, Fisher, 1991; Heuvelink, 1998) have shown the usefulness of Monte Carlo simulation as a means of predicting spatial data quality. Circular Map Accuracy Standard (CMAS) values were defined to represent high, medium and low levels of spatial error as typically applied by forest management agencies in defining quality standards (Weir, 2000). Three categories of information were considered: (1) cadastral boundaries, (2) forest roads and compartment boundaries, and (3) forest stand boundaries (Table 20.1). The original, "error-free", vector data were perturbed using a random number generator and a look-up table relating CMAS values to each combination of attribute code and time stamp. In perturbing the individual points on an arc, three different cases of error behaviour may be considered (Griffith, 1989). The results presented here refer to the case in which individual points have zero covariance and are not autocorrelated.

Each realization of the vector data was rasterized using a 1m gridcell. Cumulative overlay of 1000 raster realizations resulted in raster models of the error bands (Caspary and Scheuring, 1992; Shi, 1998) associated with the type and date of every linear feature in each of the five temporal states of the database (Figure 20.2).

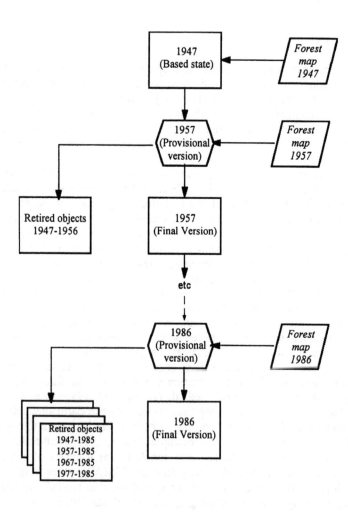

Figure 20.1 Emulation of a spatio-temporal database.

Table 20.1 CMAS values (m) for different categories of linear features in typical forest databases.

Accuracy level	Cadastral boundaries	Roads and compartments	Stand boundaries
high	1.25	2.5	5
medium	2.5	5	10
low	5	10	20

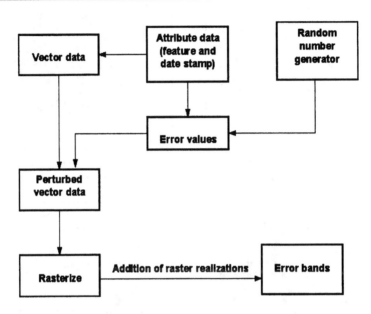

Figure 20.2 Monte Carlo simulation of error bands.

The results were presented as a series of maps visualizing the error bands associated with five scenarios for long-term error propagation:

- low, medium and high levels of error quality throughout the life of the database;
- a high quality base state, amended with low quality data (Figure 20.3);
- a low quality base state, amended with high quality data (Figure 20.4).

Although the Monte Carlo simulation approach associates the probability distribution of the errors with each arc, the examples shown in Figures 20.3 and 20.4 are simplified for clarity, and show the 90% confidence regions.

20.3 RESULTS

Although the areas of the error bands indicate the amount of error present in each state of the database, it is the areas *unaffected* by the error bands that indicate the improvements or deterioration in data quality resulting from error propagation. From the original unperturbed coordinates, the total "true" area shown in Figures 20.3 and 20.4 is 77.9ha. Table 20.2 shows the "certain" area that is not covered by the error bands for each scenario. These areas were obtained by pixel counts of the result of the Boolean overlaying of the error bands on raster representations of the original, unperturbed polygon data.

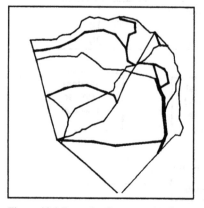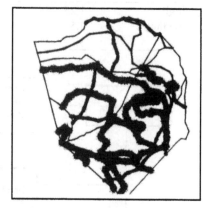

Figure 20.3 Example of error propagation over 20 years from an initially high-quality base state (left), subsequently twice amended with low-quality data (right).

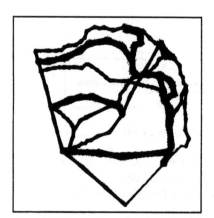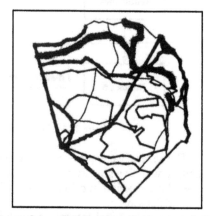

Figure 20.4 Example of error propagation over 20 years from an initially low-quality base state (left), subsequently twice amended with high-quality data (right).

Table 20.2 Area (ha) unaffected by error bands

Base state quality	Amendment quality	1947	1957	1967	1977	1986
Low	Low	55.2	46.6	41.0	44.0	42.1
Medium	Medium	64.6	59.2	56.0	57.8	57.0
High	High	70.2	67.1	65.5	66.4	66.0
High	Low	70.2	58.0	49.0	51.8	49.7
Low	High	55.2	54.2	56.2	57.1	57.4

Analysis of the data in Table 20.2 indicated that, in the case of consistent levels of error, the overall quality of the database deteriorated over time and that this deterioration was greatest in the case of the low quality data. This decline can be attributed to the fact that the total length of linear features is relatively low in the early states of the database. The greater fragmentation of the forest stands recorded by the time of the later states, together with the construction of new forest roads, results in an increase in the length of the stand boundaries, and thus in the total area occupied by the error bands.

In the case of a spatial database initially compiled to a relatively high standard of accuracy but subsequently updated with low quality data, overall quality declines rapidly from around 90% to about 65% with the first two amendments (Figure 20.5). In the earlier states, high quality data are retired and replaced by low quality data. Thereafter, only permanent features, such as the cadastral boundary, remain to contribute high quality data. After 1967, therefore, the overall quality remains relatively stable, but at a level only slightly higher than that of the low quality database. The data associated with a low quality base state updated with high quality data show only a very slight increase in overall quality, from 70.9% in 1947 to 73.7% in 1986. As in the case of the "high-low" scenario, permanent features carried over from the base state continue to exert an influence, thus preventing the overall quality from reaching the level associated with that of consistently high quality data. Nevertheless, after the first amendment, the quality of the "low-high" scenario is better than that of the "high-low" scenario.

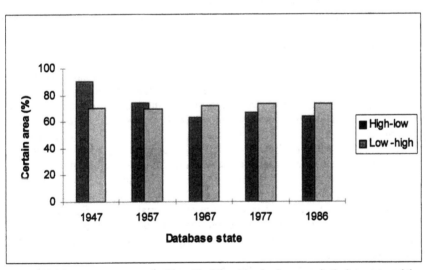

Figure 20.5 Long-term error propagation, with different levels of accuracy in the base state and the subsequent amendments.

Although the simulation of the error bands was found to be a useful means of visualizing spatial data quality, it provided only global estimates of database quality. To overcome this limitation, the coefficients of variation (CV) in the areas obtained from 150 realizations of the individual polygons were used to test for statistically significant differences between the error propagation scenarios. Pair-wise comparison of the area CV values for the consistently low quality and the "low-high" scenarios indicated that the first amendment of an initially low quality database with high quality data leads to a statistically significant improvement in overall quality. Likewise, comparing the consistently high quality data with a low quality amendment indicated a statistically significant deterioration in quality. The comparison of the high-low and low-high scenarios after the first amendment yielded an interesting result, namely that there is no significant difference in the quality of the two approaches to database establishment and updating. This result agrees well with that obtained from the analysis of error bands (Figure 20.5), which shows that the trend lines of the two scenarios intersect between the first (1957) and second (1967) amendments. These results must, however, be treated with caution. Although the study area is broadly typical of forest areas in many parts of Europe, the limited extent and specific local conditions of the study area mean that reported quantitative results, such as map accuracy values and error percentages, clearly are not generally applicable. In other forest areas, and selecting other error values, the approach may yield different results.

20.4. MANAGEMENT IMPLICATIONS

Although information, including GIS databases, may be regarded as an economic good, the assessment of its economic value can be difficult (De Man, 1989). In the case of database establishment and maintenance, financial analysis may be used to determine the net benefit of improved spatial data quality through a "with" and "without" comparison (Gittinger, 1982). In this approach, the benefits and costs that arise from the implementation of a project (in this case, the introduction of methods to improve spatial data quality) are compared with those that arise without the project. Because most forest management agencies are directly involved in primary data acquisition, they are in the fortunate position of being able to exert a considerable degree of control over the quality of their spatial databases. Indeed, many forest management agencies already work to, or at least aim at, "near professional" standards of surveying and mapping (Weir, 2000). In contemplating the use of very precise methods for spatial data acquisition, however, "the main challenge related to value-added information is to collect more information as long as the marginal benefit of it is higher than the marginal costs of getting the information" (Päivinen and Solberg, 1995, p.59). A cost benefit analysis (CBA) of improving spatial data quality therefore requires two main inputs: first, a model that adequately relates the costs of different methods of data acquisition to their achievable spatial accuracy and, second, a means of expressing the benefits of improved quality in monetary terms.

Although published data on the cost and accuracy of spatial data acquisition by forest management agencies are sparse, some limited cost data and reasonable assumptions concerning data quality have been used to develop a plausible, non-linear model to illustrate the CBA approach. In this model, approximated by a power series, increasingly stringent accuracy tolerances result in sharply rising data acquisition costs per km^2. A typical example is when a resource management agency opts to commission an aerial survey rather than purchase satellite imagery. The assessment of benefits is, however, more difficult. Nevertheless, in the case of forest management, there are some situations in which improvements in spatial data quality can be directly translated into financial benefits (Keefer, 1994). In commercial forestry, the value of land is principally considered in terms of its ability to produce timber. Prisley and Smith (1991) have demonstrated how the quality of the spatial data used in conjunction with forest inventories has a direct impact upon the certainty associated with assessments of timber volume and thus resource value. In general, however, the benefits of improved spatial data quality are less tangible and must be considered in relation to their impact upon the wide range of forest management activities that are supported by maps or spatial databases.

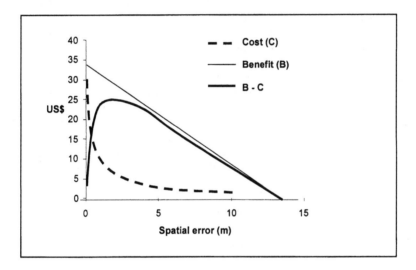

Figure 20.6 Example of financial analysis.

For the purpose of this research, financial benefit was estimated in terms of the annual operational cost associated with the forest area. Because the area that is unaffected by the error bands increases linearly with a reduction in locational error, the benefits per km^2 also show a linear gain. In the financial analysis, the costs (C) and benefits (B) associated with different quality levels are compared. The optimum quality level is then determined where B-C is a maximum (Figure 20.6). When, as is this research, the interval between successive amendments is several years, benefits measured in terms of annual operating cost must be calculated over several years and discounting applied to account for the time value of money.

Although the CBA approach offers promise as a means of selecting appropriate levels of data quality to support forest management, further work is required to develop sound models of both benefits and costs. As the range of GIS applications in forestry expands, it becomes more difficult to evaluate in monetary terms the tangible and intangible benefits of the technology in general and of spatial data quality in particular. Likewise, environmental conditions, technical procedures and local salary levels exert a strong influence on production costs. For this reason, not only must the development of cost-quality models to support financial analysis consider the particular circumstances of individual forest management agencies, but also, they will need to be modified from time to time as new products and techniques for spatial data acquisition become available.

20.5 CONCLUDING REMARKS

Geographical information systems are now generally accepted to be indispensable for the effective management and conservation of forest resources, and spatio-temporal databases show particular promise as instruments for monitoring forest change. The power and functionality of GIS technology has improved dramatically in recent years and forest management agencies may use their spatial databases to support an increasingly wide range of applications, often with different data and quality requirements. Quality, as "fitness-for-use", thus becomes an increasingly broad concept that depends on a range of time-dependent factors such as levels of technological development and economic circumstances. In this respect, there is no guarantee that the quality of current spatial data acquisition efforts will meet future user requirements. Investigations into the question of long-term spatial error propagation in GIS may contribute to efforts to ensure and maintain the quality of spatial data to support the long-term management of forest land. In many tropical countries, however, this land is under increasing threat and forest resources are declining at alarming rates. Although improvements in spatial data quality generally increase the value of a spatial database, the increased cost of such improvements may be beyond the present means of forest management agencies in some developing countries. In the face of serious deforestation and forest degradation, these agencies must therefore consider relaxing relatively costly and time-demanding quality specifications rather than delay the implementation of an effective tool to support resource management.

REFERENCES

Bulger, D. and Hunt, H., 1991, The forest management decision support system project. *The Forestry Chronicle*, **67** (6), pp. 622–628.

Caspary, W. and Scheuring, R., 1992, Error-bands as measures of geometric accuracy. In *Proceedings EGIS'92*, (Utrecht: EGIS Foundation), pp. 226–233.

De Man, E.W.H., 1989, Value added of a geographical information system. In *Proceedings of the International Conference on Computers in Urban Planning and Urban Management*, (Hong Kong: Centre of Urban Studies and Urban Planning, University of Hong Kong), pp. 315–320.

Drummond, J.E., 1995, Positional accuracy. In *Elements of Spatial Data Quality*, edited by Guptill, S.C. and Morrison, J.L., (Amsterdam: Elsevier), pp. 31–58.

Fisher, P.F., 1991, Modelling soil map-unit inclusions by Monte Carlo simulation. *International Journal of Geographical Information Systems*, **5** (2), pp. 193–208.

Gittinger, J.P., 1982, *Economic Analysis of Agricultural Projects*, (Baltimore: Johns Hopkins University Press).

Griffith, D., 1989, Distance calculations and errors in geographic databases. In *The Accuracy of Spatial Databases*, edited by Goodchild, M.F. and Gopal, S., (London: Taylor & Francis), pp. 81–90.

Heuvelink, G.B.M., 1998, *Error Propagation in Environmental Modelling with GIS*, (London: Taylor & Francis).
Keefer, B.J., 1994, Impact of spatial accuracy on business decisions: a forest management perspective. In *Proceedings of the International Symposium on the Spatial Accuracy of Natural Resource Data Bases*, (Bethesda: American Society for Photogrammetry and Remote Sensing), pp. 18–23.
Langran, G., 1992a, Forest monitoring using temporal GIS techniques. In *Proceedings GIS'92*, (Ottawa: Ministry of Supply and Services) Section B1, pp. 1–7.
Langran, G., 1992b, *Time in Geographic Information Systems*, (London: Taylor & Francis).
Lowell, K., 1994, Probabilistic temporal GIS modelling involving more than two map classes. *International Journal of Geographical Information Systems*, 8 (1), pp. 73–93.
Päivinen, R. and Solberg, B., 1995, From raw data to value-added forestry information. In *Proceedings of the XX IUFRO World Congress*, Subject Group S6.3-00, edited by Kempf, A. and Brown V., (Birmensdorf: Swiss Federal Institute for Forest, Snow and Landscape Research), pp. 59–64.
Prisley, S. P., 1994, Why natural resource information managers must be concerned about spatial accuracy. In *Proceedings of the International Symposium on the Spatial Accuracy of Natural Resource Data Bases*, (Bethesda: American Society for Photogrammetry and Remote Sensing), pp. 24–34.
Prisley, S.P and Smith, J.L., 1991, The effect of spatial data variability on decisions reached in a GIS environment. In *GIS Applications in Natural Resources*, edited by Heit, M. and Shortreid, A., (Fort Collins: GIS World) pp. 167–170.
Shi, W., 1998, A generic statistical approach for modelling error of geometric features in GIS. *International Journal of Geographical Information Science*, 12 (2), pp. 131–143.
Weir, M.J.C., 1997, A century of forest management mapping, *Cartographic Journal*, 34 (1), pp. 5–12.
Weir, M.J.C., 2000, Spatial data acquisition by forest management agencies. In *Quantifying Spatial Uncertainty in Natural Resources: Theory and Applications for GIS and Remote Sensing*, edited by Mowrer, H.T. and Congalton, R.G., (Chelsea: Ann Arbor Press), pp. 103–110.

Epilog:
A Prospective on Spatial Data Quality

In this book, three areas of spatial data quality were addressed, namely theoretical models for uncertain geographic information systems (GIS); methods for handling spatial data quality; and applications studies concerning spatial data quality. Each chapter reported a recent advance for one of the areas. There are still many open problems in spatial data quality, and thus much need for further research. In this epilog, we intend to discuss some of these potential research issues.

It is risky to speculate about the future of a discipline that is in the early stages of development. However, to a certain extent we understand what is required but not well developed at this moment for the field of spatial data quality, and what is therefore the likely basis of a research agenda. For example, we need a fundamental theory to support the development of methods for handling uncertainties in spatial data and analysis; we need various methods to measure, evaluate, and analyze the quality of spatial data in GIS; the grand aim would be GIS software that can handle uncertain data and information automatically within the system - from capture, editing, storage, transmission, query, and analysis to final decision support; and environments need to be developed for GIS data providers and users in which awareness and tracking of quality is considered to be essential for the application and use of spatial data in GIS.

THEORETICAL STUDIES

GISci is inter-disciplinary and therefore is associated with a number of theoretical frameworks. For example, GISci is related to computer science, geography, surveying, cartography, mathematics, statistics, environmental science, social science, and others. Is this a complete set of theoretical frameworks? How do they relate to each other? How can GISci be developed further based on these frameworks? It seems we cannot fully answer these questions at this stage.

Within GISci, a theory of uncertainties in spatial data and analysis should be one of the most fundamental components of the discipline. This is because uncertainties and errors are associated with almost all aspects of spatial data and analysis in GIS. For example, the original GIS data may possess different levels of accuracy due to the errors associated with them; errors and uncertainties may propagate and accumulate through analytical operations within a GIS such as overlay or buffer analysis; the results of spatial queries may be subject to uncertainties in both spatial data and linguistic interpretation; and spatial decision-making on the basis of geographic information can be affected by uncertainties and errors. Therefore, theoretical understanding of uncertainties in spatial data and analysis should be one of the most fundamental components of GISci.

How should a more comprehensive theory of uncertainties in spatial data and analysis be developed? Perhaps we can consider the following two alternative approaches. One is to adopt and integrate relevant theories from other disciplines.

Several existing theories, such as probability theory, the mathematical theory of evidence, fuzzy set theory, geostatistics, and spatial statistics are clearly relevant.

Among these, probability theory can be used to handle uncertainty caused by random errors. Evidence theory is an extension of probability theory, and it has the potential to solve problems in spatial decision-making. The primary strength of fuzzy set theory is its ability to handle inexactness associated with vagueness. Thus, fuzzy set theory can be very relevant in natural-language-related spatial queries, and natural-language-based classifications. Spatial statistics is arguably a more general theory than non-spatial statistics, since the latter can be seen as imposing unrealistic assumptions regarding homogeneity and independence. These existing theoretical frameworks are complementary to each other, to a certain extent. Since each of them can deal only with some aspects of uncertainty in spatial data and analysis, they can be considered as frameworks to be adopted and applied to the handling of uncertainties in spatial data and analysis.

The second approach is to develop a totally new theory of uncertainties in spatial data and analysis. In this case, theory must be developed from limited foundations - by defining the problem, formulating hypotheses, and gathering empirical support. This second alternative of developing new theory is clearly much more complex than the first alternative. Which of the two methodologies we should choose in developing theory on uncertainty in spatial data and analysis is still an interesting question. The first option seems to be easier and more realistic; but the second may lead to a theory that focuses specifically on uncertainties in spatial data and analysis. This will be a more focused and systematic theory, if it can be developed.

METHODS DEVELOPMENT

In a GIS, we have spatial and non-spatial data, spatial query and analysis on the basis of that data, and decision support based on existing or derived information. In the past few years, the research focus was mainly on the quality of spatial data themselves, such as positional and attribute errors. The research also extended to the quality of temporal data and temporal resolution. However, uncertainties among the logical relationships of spatial objects, particularly topological relationships, were less studied. Furthermore, uncertainties in spatial analysis, such as buffer and overlay operations, also attracted less attention. These are clearly directions to be studied further in the future for spatial data quality and analysis. Among these directions, the following research issues may have higher priority and can be tackled in the near future.

Error indicators for spatial data

The quality of spatial data in GIS can be reported in various ways, such as by graphical visualization; by giving the values of the error metrics that address a particular aspect of uncertainty; and by simulating specific uncertainties as input to a spatial analysis or data processing procedure. Among these, evaluating error metrics is one of the most straightforward methods for representing the quality of spatial data. The quality of the entire data set can be represented by one or several values of the metrics. In previous research, error metrics for positional and attribute error were addressed. Error metrics for positional error were developed based on the statistical approaches used in surveying. On the other hand, error metrics for attribute error in GIS data were mainly adopted from the methods for estimating errors of classification of remotely sensed images, and attribute error metrics for vector-GIS data were relatively less addressed. Furthermore, error metrics for other aspects of spatial data in GIS, such as temporal uncertainty and topological relationships, also did not attract much attention, and should be the focus of further research. Finally, all this past research ignored the spatial arrangement (spatial autocorrelation) which is now widely recognized by those in the research community as being present in the errors, and no attempt has been made to define how this should be recorded and reported.

Error propagation and accumulation in spatial analysis

Error models for spatial data in GIS have been studied quite intensively in the past few years. However, the nature of error propagation and accumulation in spatial analysis, such as overlay and buffer operations, have attracted less attention. Although there have been several reported studies of error propagation for raster-based spatial analysis, which is relatively easier to deal with, error propagation in vector-based GIS, which is a frequently used technique in GIS, was less addressed and needs to be further investigated. Furthermore, error analysis for buffering operations in a vector-based GIS also needs to be further studied.

Uncertain spatial relationships

Many types of relationship may exist among spatial objects. Among these, topological relationships are widely used in GIS. In the past few years, for vector-based GIS, our focus was mainly on the quality of individual spatial objects themselves, such as points, lines, and polygons, rather than on uncertain relationships among them. Error models were developed based on certain assumptions about nodal distributions, such as random error distributions. Several models were proposed, such as probability models, error distribution models, and confidence models for vector objects in GIS. On the other hand, error models of spatial relationships between objects were less addressed. Although there has been some research on this issue - for example, that described by Winter and Bittner in Chapter 3 and by Shi and Guo in Chapter 4 - at this stage, many uncertain relationships among spatial objects cannot be modeled and therefore need to be further studied.

Besides these, there are many other issues that need further investigation, such as the visualization of spatial data quality, uncertainties in spatial query, and spatial decision-making based on uncertain information.

Towards an uncertain GIS

The final value of the methods developed for handling uncertainties in spatial data and analysis should be realized in a GIS environment. This is because GIS is the computerized information system where theoretical and technical ideas about GISci are materialized and implemented. In a normal GIS we have functions for data import, digitization, editing, storage, spatial query, spatial analysis, visualization, and decision support. These functions were designed for handling spatial data without much concern about the effects of uncertainties, either because it was assumed that no problems of uncertainty existed, or because there were no adequate methods developed for handling uncertainties in spatial data at that time. It might be an appropriate time for us to consider developing an uncertain GIS, if the methods for handling uncertainties in spatial data and analysis have reached a sufficient stage of maturity. Such a GIS should be able to model individual uncertain objects and uncertain relationships among objects, to support uncertain data editing, to support uncertain queries, to conduct uncertain spatial analysis (e.g., uncertain overlay or buffer operations), to visualize uncertainties in the spatial data and results from a spatial analysis, to provide metadata that include metrics of uncertainty, and to support decision-making with uncertain information.

APPLICATION ISSUES

In the application of GIS technology, a number of issues need to be considered to cope with uncertainties in spatial data and analysis. First, users of spatial data should be aware that errors exist within their data, of different types and magnitudes. These may include, for example, random errors, systematic errors, and errors that are gross in nature. The quality of spatial data should be taken into consideration in determining whether the data sets should be selected or not – in other words, their fitness for a particular use in a GIS application. Demands for higher accuracy spatial data may lead to higher costs in purchasing or capturing the data. On the other hand, a lower quality of spatial data may lead to unacceptable inaccuracy in derived information, or the risk of a wrong decision being made being too high due to the poor quality of the data. Second, high costs may be incurred in measuring the quality of spatial data. For example, to provide quality information for positional accuracy of a GIS data set, we need to have reference measurements of higher quality, which leads to extra costs that can be very high. Also if higher quality information exists then among users and data providers, there must the temptation to simply use it to replace the original lower quality information, rather than regarding the uncertainty information alone as increasing the value of all the information.

Quality information is an important component of spatial data infrastructure and as such should be included in GIS, although there are many difficulties in implementing quality measurement in many real-world GIS. Metadata provide a technical solution for storing and managing quality information about spatial data. Both technological and non-technological problems exist in including data quality information in GIS. Technologically, we need to define and standardize error metrics, and to devise ways of capturing their values. Non-technologically, we need to educate users to be aware of and use quality information, and also to find extra funding to support the measurement of quality.

Finally, we need more study of the implications of poor data quality in different situations. There are too few studies which take basic geographical information being used in a decision context together with data quality information and propagate the quality information through to the decision. Using Monte Carlo simulation it is possible to generate many different outcomes of the spatial analysis and so to produce many different potential decisions or recommendations. Do typical levels of data quality in geographical information lead to significantly different outcomes when the data is being applied such that the decision becomes unstable? Are the risks if the initial decision is wrong too great for the political consequences? These questions are barely addressed but are fundamental to the use of any the data quality information.

TRENDS

Basic research on spatial data quality will lead to improved theory regarding uncertainties in spatial data and analysis, as a fundamental part of the theories of GISci. This theoretical development may be one of the major foci for GIScientists in the future. In developing methods for measuring and evaluating spatial data quality, the focus must move from the modeling of *individual* spatial objects to the modeling of uncertain *relationships* among spatial objects, and furthermore from the handling of individual *static* data to *dynamic* processes. In the implementation and application of data quality information, non-technological issues, such as the cost of measuring data quality and awareness of errors in spatial data, are even more crucial than the technological issues.

While theory is likely to advance in many interesting directions, several severe impediments remain that will have to be overcome if the vision of an uncertainty-aware GIS is to be realized. First, it is often argued by skeptics that uncertainty is present in all data, and that attention to uncertainty specifically in the case of GIS may be unwarranted, if little attention is paid to uncertainty in other types of data, information systems, and analysis toolboxes. Clearly the case must be made for attention to uncertainty in the specific case of spatial data. This case has to rest on the importance of spatial data quality in decision-making; on the problems caused by varying levels of uncertainty in data sets that are merged in GIS; and on the relationships between costs and uncertainties discussed in the previous section. It is also potentially important to evaluate the uncertainties in those other types of data, information systems and toolboxes.

Second, one of the unstated but nevertheless real attractions of maps and spatial databases is their ability to present a clarified, simplified view of a world that is actually complex and confusing. People clearly *prefer* such a view, which is undermined by explicit attention to uncertainty. People, and particularly decision-makers, may not want to know about uncertainty, and may see in GIS an attractive simplicity. This is certainly reflected in the views expressed by the major GIS vendors – customers do not ask for uncertainty-aware technology. Against this viewpoint lies the argument that uncertainty constitutes an Achilles Heel of GIS – a problem that, if not addressed, may undermine the whole basis of the technology. A court is likely to hold that a GIS user should make reasonable efforts to deal with uncertainty, and is likely to take a dim view of regulations or decisions that are based on GIS data for which issues of uncertainty have not been addressed.

Finally, any theory of uncertainty is undoubtedly complex, and difficult for many users to comprehend. Users are unlikely to adopt a technology based on theories of uncertainty unless the theoretical and conceptual basis is relevant to them and easily comprehended. Much progress has been made in developing methods of visualization and simulation that have comparatively simple foundations, but much more remains to be done.

The chapters of this book present a cross-section of current research on spatial data quality. In the future, we look forward to:

- continued interest in the topic in the GISci research community, and continued progress on methods and theory;
- greater attention to quality through improved methods for capturing, presenting, and updating metadata;
- clarification of the complex relationship between data quality and data acquisition cost; and
- greater willingness on the part of data producers to measure and describe essential elements of data quality.

Most fundamentally, however, progress in understanding spatial data quality will require a concerted attack on two basic beliefs: that information shown on maps and captured in digital spatial databases is *correct*, and that numerical information from computers is somehow endowed with inherent authority.

Wenzhong Shi
Michael F. Goodchild
Peter F. Fisher

Index